Ravikumar Chinthaginjala · Pawel Sitek ·
Nasro Min-Allah · Kenji Matsui ·
Sascha Ossowski · Sara Rodríguez
Editors

Distributed Computing and Artificial Intelligence, 21st International Conference

Springer

Editors
Ravikumar Chinthaginjala
Vellore Institute of Technology University
Vellore, Tamil Nadu, India

Nasro Min-Allah
Department of Computer Science
Imam Abdulrahman Bin Faisal University
(KSA)
Dammam, Saudi Arabia

Sascha Ossowski
University Rey Juan Carlos
Madrid, Spain

Pawel Sitek
Kielce University of Technology
Kielce, Poland

Kenji Matsui
Osaka Institute of Technology
Osaka, Japan

Sara Rodríguez
Biotechnology, Intelligent Systems
and Educational Technology (BISITE)
Research Group
University of Salamanca
Salamanca, Spain

ISSN 2367-3370 ISSN 2367-3389 (electronic)
Lecture Notes in Networks and Systems
ISBN 978-3-031-82072-4 ISBN 978-3-031-82073-1 (eBook)
https://doi.org/10.1007/978-3-031-82073-1

This Springer imprint is published by the registered company Springer Nature Switzerland AG
The registered company address is: Gewerbestrasse 11, 6330 Cham, Switzerland

Lecture Notes in Networks and Systems 1259

The series "Lecture Notes in Networks and Systems" publishes the latest developments in Networks and Systems—quickly, informally and with high quality. Original research reported in proceedings and post-proceedings represents the core of LNNS.

Volumes published in LNNS embrace all aspects and subfields of, as well as new challenges in, Networks and Systems.

The series contains proceedings and edited volumes in systems and networks, spanning the areas of Cyber-Physical Systems, Autonomous Systems, Sensor Networks, Control Systems, Energy Systems, Automotive Systems, Biological Systems, Vehicular Networking and Connected Vehicles, Aerospace Systems, Automation, Manufacturing, Smart Grids, Nonlinear Systems, Power Systems, Robotics, Social Systems, Economic Systems and other. Of particular value to both the contributors and the readership are the short publication timeframe and the world-wide distribution and exposure which enable both a wide and rapid dissemination of research output.

The series covers the theory, applications, and perspectives on the state of the art and future developments relevant to systems and networks, decision making, control, complex processes and related areas, as embedded in the fields of interdisciplinary and applied sciences, engineering, computer science, physics, economics, social, and life sciences, as well as the paradigms and methodologies behind them.

Indexed by SCOPUS, EI Compendex, INSPEC, WTI Frankfurt eG, zbMATH, SCImago.

All books published in the series are submitted for consideration in Web of Science.

For proposals from Asia please contact Aninda Bose (aninda.bose@springer.com).

Preface

The transfer of technology in areas like distributed computing and artificial intelligence remains a significant challenge, which is why such contributions will be given special attention at this symposium. Distributed computing is playing an increasingly crucial role in modern signal and data processing, information fusion, and electronics engineering, including fields like e-commerce, mobile communications, and wireless devices. The application of artificial intelligence in distributed environments, in particular, is becoming a highly valuable and economically promising element.

The 21st International Symposium on Distributed Computing and Artificial Intelligence 2024 (DCAI 2024) serves as a platform for exchanging ideas between scientists and professionals from academia and industry, which is essential for developing systems that meet contemporary societal needs. Over the past decade, research in intelligent distributed systems has progressed significantly, leading to the deployment of many effective applications. Currently, technologies such as the Internet of Things (IoT), Industrial Internet of Things (IIoT), big data, blockchain, and distributed computing in general, are continuously evolving due to substantial research and technical efforts from both universities and businesses. With computing systems ranging from personal laptops to edge, fog, and cloud computing systems available for parallel and distributed computing, this conference provides the ideal forum to present innovative techniques applied to complex problems across these domains.

This year's technical program will present both high quality and diversity, with contributions in well-established and evolving areas of research. Specifically, 138 papers were submitted, by authors from 43 different countries (Algeria, Angola, Argentina, Austria, Belgium, Brazil, Bulgaria, Chile, Colombia, Croatia, Cyprus, Czechia, Denmark, Ecuador, France, Germany, Greece, Hungary, India, Iran, Ireland, Italy, Japan, Jordan, Luxembourg, Mexico, Morocco, Netherlands, Norway, Pakistan, Poland, Portugal, Qatar, Romania, Russia, South Korea, Spain, Sweden, Switzerland, Tunisia, United Arab Emirates, UK, USA), representing a truly "wide area network" of research activity. The DCAI'24 technical program has selected 36 full papers in the main track and, as in past editions, it will be special issues in ranked journals such as Sensors, IJIMAI, Smart Cities, Electronics, Future Internet Systems and Advances in Distributed Computing and Artificial Intelligence Journal. These special issues will cover extended versions of the most highly regarded works. Moreover, DCAI'24 Special Sessions have been a very useful tool to complement the regular program with new or emerging topics of particular interest to the participating community.

This symposium was organized by the University of Salamanca and was held from 26 to 28 June 2024 in the city of Salamanca (Spain). We would like to thank all the contributing authors, National Associations (AEPIA, APPIA, LASI), the sponsors (AIR Institute), the funding supporting of the project "COordinated intelligent Services for Adaptive Smart areaS (COSASS), Reference: PID2021-123673OB-C33, financed by

MCIN/AEI/10.13039/501100011033/FEDER, UE., and finally, the organization members and the program committee members for their hard work, which was essential for the success of DCAI'24.

June 2024

Ravikumar Chinthaginjala
Pawel Sitek
Nasro Min-Allah
Kenji Matsui
Sascha Ossowski
Sara Rodríguez

Organization

Advisory Board

Kenji Matsui	Osaka Institute of Technology, Japan
Yuncheng Dong	Sichuan University, China
Francisco Herrera	University of Granada, Spain
Tan Yigitcanlar	Queensland University of Technology, Australia
Tiancheng Li	Northwestern Polytechnical University, China

Program Committee Chairs

Ravikumar CV	Vellore Institute of Technology University, India
Pawel Sitek	Kielce University of Technology, Poland
Nasro Min Allah	Imam Abdulrahman Bin Faisal University (KSA), Saudi Arabia

Special Sessions Chairs

Rashid Mehmood	King Abdulaziz University, Saudi Arabia
Goreti Marreiros	ISEP/GECAD, Portugal

Organizing Committee Chairs

Sascha Ossowski	King Juan Carlos University, Spain
Sara Rodríguez	University of Salamanca, Spain
Guillermo Hernández González	University of Salamanca, Spain
Laura Grande Pérez	University of Salamanca, Spain

Program Committee

Adel Boukhadra	National high School of Computer Science, Algeria
Alberto Fernandez	University Rey Juan Carlos, Spain
Alfonso González-Briones	University of Salamanca, Spain

Amel Borgi	ISI/LIPAH, Université de Tunis El Manar, Tunisia
Amin M. Khan	UiT The Arctic University of Norway, Norway
Ana Belén Gil González	University of Salamanca, Spain
Ana Faria	ISEP, Portugal
Ander García	Visual Communication Technologies VICOMTech
André Zúquete	University of Aveiro, Portugal
Angel Martin Del Rey	Department of Applied Mathematics, Universidad de Salamanca, Spain
Angélica González Arrieta	Universidad de Salamanca, Spain
António J. R. Neves	University of Aveiro, Portugal
António Jorge Do Nascimento Morais	Universidade Aberta, Portugal
Araceli Queiruga-Dios	Department of Applied Mathematics, Universidad de Salamanca, Spain
Arkadiusz Gola	Lublin University of Technology, Poland
Aurélie Hurault	IRITENSEEIHT, France
Benedita Malheiro	Instituto Superior de Engenharia do Porto and INESC TEC, Portugal
Bozena Wozna-Szczesniak	Institute of Mathematics and Computer Science, Jan Dlugosz University in Czestochowa, Poland
Carlos Alejandro De Luna-Ortega	Universidad Politecnica de Aguascalientes, Mexico
Carlos Carrascosa	GTI-IA DSIC Universidad Politecnica de Valencia, Spain
Carmen Benavides	University of León, Spain
Cesar Analide	University of Minho, Portugal
Dalila Durães	University of Minho, Portugal
Edgardo Bucciarelli	University of Chieti-Pescara, Italy
Eleni Mangina	UCD, Ireland
Elisa Huzita	State University of Maringa, Brazil
Emilio Serrano	Universidad Politécnica de Madrid, Spain
Ester Martinez	Universidad de Alicante, Spain
Fábio Marques	University of Aveiro, Portugal
Felix Freitag	Universitat Politècnica de Catalunya, Spain
Fernando De La Prieta	University of Salamanca, Spain
Fidel Aznar Gregori	Universidad de Alicante, Spain
Florentino Fdez-Riverola	University of Vigo, Spain
Francisco A. Pujol	Specialized Processor Architectures Lab, DTIC, EPS, University of Alicante, Spain
Francisco García-Sánchez	University of Murcia, Spain

Francisco Javier Calle	Departamento de Informática. Universidad Carlos III de Madrid, Spain
Friederike Wall	Alpen-Adria-Universitaet Klagenfurt, Austria
Giner Alor Hernandez	Instituto Tecnologico de Orizaba, Mexico
Giovanni De Gasperis	DISIM, Università degli Studi dell'Aquila, Italy
Guillermo Hernández	University of Salamanca, Spain
Gustavo Isaza	University of Caldas, Colombia
Gustavo Santos-Garcia	Universidad de Salamanca, Spain
Holger Billhardt	Universidad Rey Juan Carlos, Spain
Ichiro Satoh	National Institute of Informatics, Japan
Irina Georgescu	Academy of Economic Studies, Romania
Isabel Praça	GECAD/ISEP, Portugal
Jesús García Herrero	University Carlos III de Madrid, Spain
Jona Beysens	CSEM
Jose M. Molina	Universidad Carlos III de Madrid, Spain
José Ramón Méndez Reboredo	University of Vigo, Spain
José Ramón Villar	University of Oviedo, Spain
Jose-Luis Poza-Luján	Universitat Politècnica de València, Spain
Juan Gomez Romero	University of Granada, Spain
Juan Pavón	Universidad Complutense de Madrid, Spain
Juan-Luis Posadas-Yague	Universitat Politècnica de València, Spain
Kenji Matsui	Osaka Institute of Technology, Japan
Li Weigang	University of Brasilia, Brazil
Lourdes Borrajo	University of Vigo, Spain
Luis Fernando Castillo Ossa	University of Caldas, Colombia
Luiz Romao	Univille, Brazil
Mar Pujol	Universidad de Alicante, Spain
Mariano Raboso Mateos	Facultad de Informática, Universidad Pontificia de Salamanca, Spain
Masaru Teranishi	Hiroshima Institute of Technology, Japan
Mauricio Orozco-Alzate	Universidad Nacional de Colombia, Colombia
Michal Wozniak	Wroclaw University of Technology, Poland
Michifumi Yoshioka	Osaka Pref. Univ., Japan
Miguel A. Vega-Rodríguez	University of Extremadura, Spain
Miguel Angel Patricio	Universidad Carlos III de Madrid, Spain
Naoufel Khayati	COSMOS Laboratory - ENSI, Tunisia
Nuno Silva	DEI & GECAD - ISEP - IPP, Portugal
Pablo Negre	University of Salamanca, Spain
Paulo Mourao	University of Minho, Portugal
Paulo Novais	University of Minho, Portugal
Paulo Vieira	Instituto Politécnico da Guarda, Portugal
Pawel Pawlewski	Poznan University of Technology, Poland

Pedro Sousa	University of Minho, Portugal
Philippe Mathieu	University of Lille, France
Radu-Emil Precup	Politehnica University of Timisoara, Romania
Rafael Corchuelo	University of Seville, Spain
Raffaele Dell'Aversana	Università "D'Annunzio" di Chieti-Pescara, Italy
Reyes Pavón	University of Vigo, Spain
Rui Camacho	University of Porto, Portugal
Sara Rodríguez	University of Salamanca, Spain
Sascha Ossowski	University Rey Juan Carlos, Spain
Sergi Robles	Universitat Autònoma de Barcelona, Spain
Shimpei Matsumoto	Hiroshima Institute of Technology, Japan
Stefan-Gheorghe Pentiuc	University Stefan cel Mare Suceava, Romania
Tayuka Yoshihiro	Faculty of Systems Engineering, Wakayama University, Japan
Tiago Pinto	IPP, Portugal
Vicente Julian	Universitat Politècnica de València, Spain
Worawan Diaz Carballo	Thammasat University, Thailand
Yeray Mezquita	University of Salamanca, Spain
Yun Zhu	Shaanxi Normal University, China
Zbigniew Banaszak	Warsaw University of Technology, Faculty of Management, Dept. of Business Informatics, Poland
Zhen Zhang	Dalian University of Technology, China
Zhu Wang	XINGTANG Telecommunications Technology Co., Ltd., China
Zita Vale	GECAD/ISEP/IPP, Portugal

Organizing Committee

Fernando De la Prieta	University of Salamanca, Spain
Javier Prieto Tejedor	University of Salamanca and AIR Institute, Spain
Ricardo S. Alonso Rincón	AIR Institute, Spain
Alfonso González Briones	University of Salamanca, Spain
Pablo Chamoso Santos	University of Salamanca, Spain
Javier Parra	University of Salamanca, Spain
Belén Pérez Lancho	University of Salamanca, Spain
Ana Belén Gil González	University of Salamanca, Spain
Ana De Luis Reboredo	University of Salamanca, Spain
Angélica González Arrieta	University of Salamanca, Spain
Angel Luis Sánchez Lázaro	University of Salamanca, Spain
Emilio S. Corchado Rodríguez	University of Salamanca, Spain

Guillermo Hernández	University of Salamanca, Spain
Liliana Durón	University of Salamanca, Spain
Laura Grande	University of Salamanca, Spain
Carmen Losada	AIR Institute, Spain
Albano Carrera González	AIR Institute, Spain
Marta Plaza Hernández	University of Salamanca, Spain
Beatriz Bellido	University of Salamanca, Spain
María Alonso	University of Salamanca, Spain
Diego Valdeolmillos	AIR Institute, Spain
Carlos Álvarez	AIR Institute, Spain

DCAI 2024 Sponsors

Sponsors

Organizers

Support from Research Associations, Institutions & Labs

Contents

Adaptive Traffic Light Control Through Reinforcement Learning Based on Sensor Integration

Pedro Uribe-Chavert[1(✉)], Luis López-Cuerva[1,2,3], Juan-Luis Posadas-Yagüe[3], and Jose-Luis Poza-Lujan[3]

[1] Doctoral School, Universitat Politècnica de València, Camino de vera, sn, 46022 Valencia, Spain
pedurcha@doctor.upv.es, luilocu2@alumni.upv.es
[2] Informatics School, Universitat Politècnica de València, Camino de Vera, sn, 46022 Valencia, Spain
[3] Research Institute for Industrial Computing and Automatics, Universitat Politècnica de València, Camino de vera, sn, 46022 Valencia, Spain
{jposadas,jopolu}@upv.es

Abstract. This paper explores intelligent traffic light management advancements, focusing on controlling intersection traffic opening times. The decision-making process is influenced by factors such as traffic density. The information for these decisions is gathered from sensors placed on the streets, whose accuracy can vary. Data collected are processed to aid control agents in decision-making. The paper proposes an intersection control algorithm that operates under the assumption of lacking sensorisation. To balance raw sensor data, control nodes implement a reinforced learning algorithm to select the most suitable combination of sensors to improve traffic parameters. The paper also introduces a method for calculating traffic density by combining sensors with imprecise data. This research contributes to intelligent traffic management by providing a novel approach to intersection control and traffic density calculation.

Keywords: Intelligent Traffic Light Management · Reinforced Learning Algorithm · Traffic Density Calculation

1 Introduction

In the field of traffic control in cities, a great deal of progress has been made in intelligent traffic light management. Intelligent traffic light control focuses primarily on controlling the traffic opening time (green) at street intersections. The decision about the appropriate opening time at each intersection depends on several factors [9]. Some of these factors are traffic density, attention to personal vehicles, such as bicycles or electric scooters, and emergencies, such as accidents or priority vehicles. To regulate city traffic, decisions must be made based

R. Chinthaginjala et al. (Eds.): DCAI 2024, LNNS 1259, pp. 1–10, 2025.
https://doi.org/10.1007/978-3-031-82073-1_1

on the information available from the streets adjacent to the intersections to be regulated. The information comes from sensors placed in the street, and the accuracy of these sensors can vary.

The data collected by the sensors must be processed to obtain information for control agents to make good decisions. The sensors' type, number, and accuracy are the keys to having accurate and valuable data for processing and making intelligent decisions in traffic management. Processing involves using control nodes that can be organised based on different paradigms [4].

Consequently, intersection management involves decision-making based on several traffic parameters, such as vehicle queue length or waiting time. To know these parameters, sensors are necessary to obtain the behaviour of vehicles. In a situation where Vehicle to Infrastructure (V2I) communication is not implemented, the infrastructures (I2I), through sensors, must provide the traffic parameters to Infrastructure [12].

This article proposes an intersection control algorithm that works with the usual parameters but with the assumption of lacking sensorisation. The control nodes implement a reinforced learning (RL) algorithm to balance the raw sensor data. The node selects which combination of sensors on the street is the most suitable to improve the traffic parameters. Another contribution of the article is calculating traffic density based on combining sensors with imprecise data.

This paper is organised as follows: the next section reviews the distributed traffic control environment to define the control area used; it then describes how traffic parameters are obtained from basic sensors, followed by details of the reinforcement learning employed; SUMO experiments simulating sensors and an intersection environment are shown; conclusions and the way forward are presented.

2 Distributed Computation

Having distributed control nodes with computing capabilities means these nodes can be connected to other nodes (or agents) to coordinate their data and control actions. Many physical and logical architectures of distributed intelligent computing are characterised by the connections between the agents at each control node.

In a centralised computing system, only one agent or entity manages all the intersections to coordinate them in real time to deal with changing traffic patterns. This control type can effectively manage traffic flow across a city or large area. However, it may need to be more responsive to local changes in traffic patterns at individual intersections [5]. Agents can be distributed in a sub-area of distributed computing consisting of autonomous agents linked in a shared network. These sub-areas are the basis of cluster computing, which involves using a group of linked computers that work together as if they were a single entity. However, the closest form of computation to the control node is when these nodes have agents with sufficient intelligence capacity to make local decisions; this is the basis of Fog/Edge computing systems.

In Fog/Edge computing, the control nodes are equipped with sensors and actuators to manage the traffic light at each intersection. This type of control is often referred to as "smart" or "adaptive" traffic light control. It uses Internet of Things (IoT) devices, sensors, and agents to compute and adapt traffic light controls based on real-time traffic data [19]. Currently, cooperation between paradigms implies that the computing agents of the same system can be distributed in centralised nodes coordinated with Edge control nodes. The first approach is Cloud Computing, which involves using a network of remote servers hosted on the Internet to store, manage, and process data. It has been used in intelligent traffic control frameworks using local traffic smart servers and remote cloud servers to improve traffic signal processing time and in sensor-cloud systems for intelligent traffic control [13].

For instance, different computing paradigms could be used for other aspects of traffic data processing, such as data collection, feature extraction, and traffic prediction. Jungle computing is a relatively new concept that uses a wide variety of different computing paradigms simultaneously [11].

To support systems in which different computation paradigms must collaborate, [15] presents an architecture in which heterogeneous control nodes share "control areas" to compute decisions to manage the intersections of streets. Control areas can exchange information among themselves At a basic level, values of traffic parameters such as density or waiting times can be exchanged. Additionally, agents can interchange decisions made in control, such as an increase or decrease in traffic light time. At the highest level, even instructions can be exchanged, such as ordering from one control area to another to increase or decrease the passage time. In this last case, when one control area must give instructions to another control area, this is when we enter entirely a Jungle Computing paradigm. Suppose two control areas wish to give instructions to each other. In that case, there are two main paradigms: follow a hierarchy or establish cooperation. The hierarchical Paradigm approach organises distributed processing in a vertical delegation format, allowing management functionality across different system levels. The Cooperative Paradigm emphasises semantically rich delegation, empowering both agents to control specific elements with limited instructions. This paradigm leverages high-level goals and contextual data, allowing agents to operate autonomously and cooperatively towards a common goal.

In conclusion, studying different distributed computation paradigms and their application to traffic light control reveals a complex interplay of technologies and strategies. From centralised systems to Fog/Edge computing and Cloud Computing, each approach offers unique advantages and challenges. Developing more effective and responsive traffic light control systems is possible by understanding these paradigms and how they can work together. The following section explains how these different paradigms have been adapted to be simulated and validated.

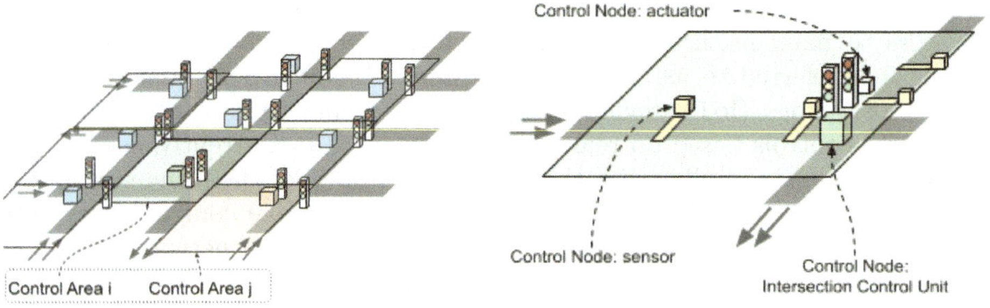

(a) (b)

Fig. 1. (a) represents a Manhattan-style neighbourhood where each intersection is a control area managed by an Intersection Control Unit (ICU). Each control area can send or receive data (parameters or decisions) or even instructions from other control areas. (b) Represents a control area of a street intersection with the control nodes (sensors and actuators) and one of the control nodes acting as the ICU.

3 Environment

The core of the architecture is the control area. In the case of traffic adaptation, there is a control node in each control area whose agent acts as an Intersection Control Unit (ICU) [3]. The control nodes of each control area are responsible for reading data from the sensors and transferring said data to the actuators. The role of the ICU is to manage these data and instructions and communicate them to the ICUs, who must be informed when making decisions by consensus or sending orders in the case of working in a hierarchical paradigm where an ICU has priority over another. In any case, one of the keys is the information obtained through sensor data. This data can provide the required information, such as traffic density, vehicle type, and speed [10].

Figure 1a shows the environment in which the control areas act. Figure 1b illustrates the configuration of the initial scenario to be deployed in the SUMO traffic simulator. This scenario, serving as the primary test-bed for a subsequent, larger-scale simulation, comprises an intersection with two ingress streets of two lanes each and two egress streets featuring two lanes each. The scenario, sourced from [1], underwent a modification with the integration of sensors. Specifically, each lane is equipped with two Induction Loop E1 sensors (as per SUMO [7]). These sensors were chosen due to the unavailability of the preferred ultrasonic sensors in SUMO. The placement of these sensors is strategically at a certain distance from the intersection and exclusively on the ingress lanes.

The deployment of sensors in this experiment aims to eschew reliance on SUMO's provision of potentially speculative and challenging-to-acquire data, such as the Average Waiting Time per vehicle and Queue Length per street. However, such parameters will be utilised for verification purposes. Through data harvested from the sensors, this study seeks to ascertain the three fundamental traffic variables: mean speed, traffic flow, and density [8]. This paper endeavours to estimate these fundamental traffic variables, proposing a method to derive them directly from sensor data.

The Synchronisation Level facilitates communication across different ICUs [14], enhancing inter-junction coordination by providing preceding and succeeding intersections with supplementary control information. This synchronisation mechanism becomes increasingly crucial in multi-lane environments, enabling the interconnection of junctions for optimised traffic management.

Within the fundamental traffic variables, the average speed v_t adapted to the sensors is defined as the average speed of the vehicles calculated in a space determined by a set of sensors, some placed near the traffic light and others far away. Instead, the weighted harmonic average of the speeds will be used to determine the average speed of the street, varying the weighting factor (w) depending on the distance to the intersection sensor as is represented in Eq. 1

$$HSpeed = \frac{\sum_{i=1}^{n} w_i}{\sum_{i=1}^{n} \frac{w_i}{v_i}} \tag{1}$$

Where HSpeed is the Harmonic speed, w is the weighting factor for each vehicle, and v is the speed for each vehicle.

Occupancy (*occ*) another fundamental traffic variable, adapted to the sensors, is defined as the percentage of time an induction loop is occupied by a vehicle, offering an estimate of road occupancy and, by extension, traffic density. Urban settings pose challenges for occupancy measurement due to partial vehicle detection over sensors at red lights, resulting in "unrealistic" values. In an innovative method, density will be calculated differently based on the number of vehicles between the sensors closest to and farthest from the intersection over the area encompassed by this segment, shown in Eq. 2. Therefore, density can be derived directly from occupancy by knowing the number of vehicles that have passed through each sensor.

$$D = \frac{\sum_{i=1}^{far} N_{veh_i} - \sum_{j=1}^{near} N_{veh_j}}{(D_{far} - D_{near}) \cdot width} \tag{2}$$

Where N_{veh_i} is the number of vehicles that have passed the farthest sensors, N_{veh_j} is the number of vehicles that have passed the nearest sensors, $D_{far} - D_{near}$ is the distance between the farthest sensor and the nearest sensor and width is the width of the street.

Traffic flow (I) defined as the number of vehicles passing a reference point per unit of time is directly ascertained by the induction loops So the equation is represented as 3

$$I = \frac{\sum_{i=1}^{far} N_x}{\Delta t} \tag{3}$$

Where I is the traffic flow, N_x is the number of vehicles that have passed a sensor and t is the total time that have passed.

4 Reinforcement Learning

Classical approaches to traffic light control rely on robust assumptions about the traffic dynamics present on the roadway [17]. However, the reality usually deviates from these assumptions due to non-deterministic factors of the road participants, such as the way drivers interact with pedestrians and bicycles. The main advantage of reinforcement learning is handling complex environments without making a priori assumptions about their topology and dynamics. This is achieved by learning the optimal traffic light control policies from a trial-and-error process. In addition, reinforcement learning is highly adaptable and generalizable to multiple scenarios [20] along with the ability to deal with uncertain environments (e.g., failure of sensors that report the state of the environment). However, this technique has two clear disadvantages: it requires the ability to have simulation scenarios and large data sets and it has a high computational time.

Reinforcement learning problems are usually formulated as Markovian Decision Processes (MDPs) [18]. MDPs, an environment, is modelled as a set of states and actions, and the system's goal is to control the system so that some performance criteria are maximised. Formally, this framework can be defined by a set of terms $\langle \mathcal{S}, \mathcal{P}, \mathcal{A}, \mathcal{R}, \gamma \rangle$ [2]. $\mathcal{S}$ is the state space, a finite set of Markov states of the environment. The agents use this state to decide what action to take. $\mathcal{A}$ is a set of legal actions the agent can take. $\mathcal{P}$ is the probability of transitionate from an state s_t to an state s_{t+1} using an action a_t. $\mathcal{R}$ is the reward obtained by the agent by realising an action a_t in a state s_t. γ is the discount factor applied to factor the importance of the expected future and immediate rewards. The goal of reinforcement learning is for the agent to find a policy of actions π that maximizes the final reward. To learn this policy, we have employed the dynamic programming algorithm Q-learning [16].

Four variables derived from traffic have been employed to represent the state space $\mathcal{S}$. The selected encodings to represent the variables are one-hot encoders and vectors have been used. The state of the traffic lights (green or red) and whether the next timestep the traffic light will exceed the maximum time have been encoded as one-hot encoders. Meanwhile the number of observed vehicles waiting in a traffic light and the observed average speed have been encoded as vectors. For both observed average speed and observed number of waiting vehicles, we take into account two measures: a first measure that indicates the state of the variables in a position near the traffic light and a second measure that indicates the state of the variables at a more distant position. Equation 4 shows how we have calculated the speed and Eq. 5 how we have calculated the observed number of waiting vehicles. Equation 6 defines the reward function used throughout the reinforcement learning process.

$$Obs_{speed} = [\sum_{c \in C} s_{cn}, \sum_{c \in C} s_{cf}] \tag{4}$$

$$Obs_{n\_veh} = [\sum_{c \in C} n\_veh_{cn}, \sum_{c \in C} n\_veh_{cf}] \tag{5}$$

Where C is the set of sensors associated with a traffic light, n states the sensor is part of the near sensors set, f states the sensor belongs to the furthest sensor set, s is the mean time average speed of vehicles measured by an induction loop in SUMO [7] during the current interval and n_veh is the number of vehicles that passed the detector during the current interval. Where D is defined in Eq. 2

$$Obs_{density} = -\overline{D} \tag{6}$$

To test the environment as described in Sect. 2, the learning outlined in this section will be applied. To assess how this approach enhances current methods, the next section will detail the experiments and results obtained.

5 Experiments and Results

To verify the system, a simulation of a control area described in Fig. 1b has been simulated using SUMO. In the first test, SUMO was provided with a vehicle entrance following a Weibull distribution [6] that follows the formula 7.

$$f(x) = (-ln(U))^{1/a} \tag{7}$$

where a corresponds to 0.5 and U being a uniform distribution (0,1]. The function follows the graph shown in the Fig. 2

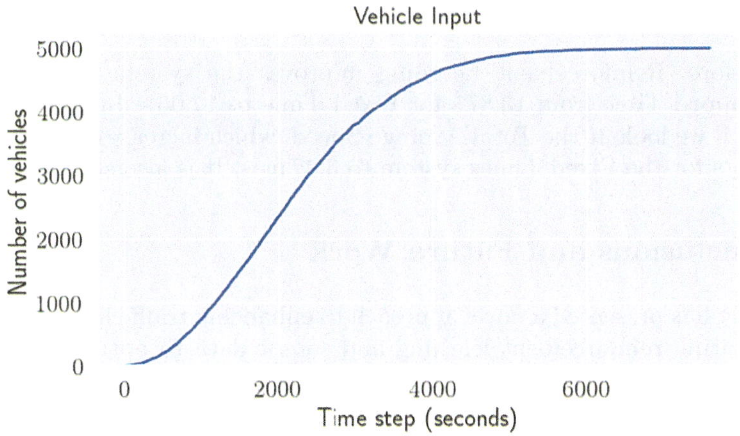

Fig. 2. Number of vehicles that have entered in the simulation per time step

This experiment will compare two systems: a fixed-time system and a Reinforcement Learning system. The Fixed Time system is the system that currently exists in most systems. Fixed Time means that the traffic signal plans of the network are split into time intervals according to the expected volume flow along the day. In this scenario, the north-south and east-west traffic lights have equal service times for the crossing, which are 15 s for green and red. Reinforcement learning is proposed in this crossing to reduce congestion, as explained in the previous section.

The Total Stopped Time parameter and the Total average Speed of each vehicle will be taken to compare these results. The variables used in this comparison, Total Stopped Time and Total Average Speed, are directly obtained from Sumo variables, ensuring the credibility and transparency of the data. Total Stopped Time is when the vehicle spends stationary or at a deficient speed in the scenario, in seconds. Total Average Speed is the average time of each vehicle in the simulation, in m/s.

The observation functions used for the reinforcement learning simulation were 4, 5 and 6. After careful analysis using both observation functions, it was found that the function of 6 yielded the best results, significantly improving the system's performance.

Table 1. Total Stopped Time and Total Average Speed for Fixed Time and RL simulations

	Total Stopped Time (s)			Total Average Speed(m/s)		
	Mean	Max	Min	Mean	Max	Min
Fixed Time	13.87	36.00	0.00	3.16	8.29	0.00
RL	12.06	36.00	0.00	3.47	8.25	0.00

Therefore, Reinforcement Learning improves the system by reducing the Total Stopped Time from 13.87 s for Fixed Time to 12.06 s. In addition, it also improves if we look at the Total Average Speed, which increases from an average of 3.16 m/s for the Fixed Times system to 3.47 m/s. It is also shown in Table 1.

6 Conclusions and Future Work

This study has presented a novel approach to enhancing traffic light management by integrating reinforcement learning and sensor data to optimise traffic flow. This approach is useful when the number of sensors or the accuracy of the sensors is limited. Using a combination of sparse sensor inputs and a reinforced learning algorithm, we developed an intersection control system that adapts to real-time traffic conditions, significantly improving traffic parameters such as density and flow.

Sensors' type, number, and location are vital aspects to consider. Working on simulations with sensors and not with simulation data allows an approximation to the reality that facilitates the design of the control system.

Our research contributes to the broader field of intelligent traffic management systems by introducing a method for estimating traffic density using low accuracy sensor data, instead of simulated high-precision data that SUMO provides. This method enhances the accuracy of traffic control decisions. It provides a scalable solution that can be adaptable to varying levels of sensor availability in urban environments. The experimental results, obtained through simulations in SUMO, underline the potential of our approach to reduce congestion and improve overall traffic efficiency. Specifically, the reinforced learning algorithm experimented, provides more accurate traffic density calculations and, consequently, more informed traffic light control decisions.

However, our work has limitations. The current model relies on specific sensor distribution and traffic pattern assumptions, which may sometimes only apply. Furthermore, the scalability of our solution across different urban settings with varied traffic dynamics remains to be thoroughly tested.

Future work will address these limitations by exploring more sophisticated reinforcement learning models that dynamically adjust to diverse traffic conditions and sensor configurations. Additionally, we aim to implement our algorithm in a real-world environment to evaluate its performance in practical scenarios. By further refining our approach, we hope to contribute to developing more intelligent, more resilient urban traffic management systems capable of responding to the challenges of modern cities. Recognising the evolution of distributed computing and artificial intelligence in traffic management, our research marks a step towards fully adaptive, intelligent traffic control systems. As we advance, the prospects for significantly improving urban mobility and reducing congestion through AI-driven solutions appear increasingly promising.

Acknowledgements. Work supported by the Spanish Ministry of Science and Innovation MICINN Project: CICYT PRECON-I4: "Predictable and reliable information systems for Industry 4.0" TIN2017-86520-C3-1-R.

References

1. Alegre, L.N.: SUMO-RL (2019)
2. Bouktif, S., Cheniki, A., Ouni, A., El-Sayed, H.: Deep reinforcement learning for traffic signal control with consistent state and reward design approach. Knowl.-Based Syst. **267**, 110440 (2023). https://doi.org/10.1016/j.knosys.2023.110440
3. Chen, L., Englund, C.: Cooperative intersection management: a survey. IEEE Trans. Intell. Transp. Syst. **17**(2), 570–586 (2015)
4. Fan, X.: Deep learning for intelligent traffic sensing and prediction: recent advances and future challenges. CCF Trans. Pervasive Comput. Interact. **2**(4), 240–260 (2020). https://doi.org/10.1007/s42486-020-00039-x
5. Gershenson, C.: Self-organizing traffic lights download pdf. Complex Systems **16**(1) (2005)
6. Lai, C.D., Murthy, D., Xie, M.: Weibull distributions and their applications. In: Springer Handbooks, pp. 63–78. Springer (2006)
7. Lopez, P.A., et al.: Microscopic traffic simulation using sumo. In: The 21st IEEE International Conference on Intelligent Transportation Systems. IEEE (2018). https://elib.dlr.de/124092/

8. Mena-Oreja, J., Gozalvez, J.: On the impact of floating car data and data fusion on the prediction of the traffic density, flow and speed using an error recurrent convolutional neural network. IEEE Access **9**, 133710–133724 (2021)
9. Modi, Y., Teli, R., Mehta, A., Shah, K., Shah, M.: A comprehensive review on intelligent traffic management using machine learning algorithms. Innovative Infrastruct. Solutions **7**(1), 128 (2022)
10. Poza-Lujan, J.L., Uribe-Chavert, P., Posadas-Yagüe, J.L.: Low-cost modular devices for on-road vehicle detection and characterisation. Des. Autom. Embed. Syst. **27**(1), 85–102 (2023)
11. Seinstra, F.J., et al.: Jungle Computing: Distributed Supercomputing Beyond Clusters, Grids, and Clouds, pp. 167–197. Grids, Clouds and Virtualization pp (2011)
12. Tahir, M.N., Leviäkangas, P., Katz, M.: Connected vehicles: V2V and V2I road weather and traffic communication using cellular technologies. Sensors **22**(3), 1142 (2022)
13. Tang, C., Wei, X., Liu, J.: Application of sensor-cloud systems: smart traffic control. In: Security, Privacy, and Anonymity in Computation, Communication, and Storage: 11th International Conference and Satellite Workshops, SpaCCS 2018, Melbourne, NSW, Australia, December 11-13, 2018, Proceedings 11, pp. 192–202. Springer (2018)
14. Tomar, I., Sreedevi, I., Pandey, N.: State-of-art review of traffic light synchronization for intelligent vehicles: current status, challenges, and emerging trends. Electronics **11**(3), 465 (2022)
15. Uribe-Chavert, P., Posadas-Yagüe, J.L., Poza-Lujan, J.L.: Proposal for a distributed intelligent control architecture based on heterogeneous modular devices. In: Distributed Computing and Artificial Intelligence, Volume 2: Special Sessions 18th International Conference 18, pp. 198–201. Springer (2022)
16. Watkins, C.J.C.H., Dayan, P.: Q-learning. Mach. Learn. **8**(3), 279–292 (1992). https://doi.org/10.1007/BF00992698
17. Wei, H., Zheng, G., Gayah, V., Li, Z.: Recent advances in reinforcement learning for traffic signal control: a survey of models and evaluation. SIGKDD Explor. Newsl. **22**(2), 12–18 (2021). https://doi.org/10.1145/3447556.3447565
18. Wiering, M.A., Van Otterlo, M.: Reinforcement learning. Adapt. Learn. Optim. **12**(3), 729 (2012)
19. Wu, Q., Wu, J., Shen, J., Yong, B., Zhou, Q.: An edge based multi-agent auto communication method for traffic light control. Sensors **20**(15), 4291 (2020)
20. Xing, H., Chen, A., Zhang, X.: RL-GCN: traffic flow prediction based on graph convolution and reinforcement learning for smart cities. Displays **80**, 102513 (2023). https://doi.org/10.1016/j.displa.2023.102513, https://www.sciencedirect.com/science/article/pii/S0141938223001464

Improvement of Anomaly Detection Through Enhanced Feature Extraction in TimesNet

Hiroyoshi Nagahama[1(✉)], Michifumi Yoshioka[1], Katsufumi Inoue[1], Masayoshi Todorokihara[2], and Keishi Omori[2]

[1] Graduate School of Informatics, Osaka Metropolitan University, Osaka, Japan
sd24025e@st.omu.ac.jp, {s21386n,inoue}@omu.ac.jp

[2] Microdevices Operations Division, Seiko Epson Corp., Suwa, Japan
{todorokihara.masayoshi,omori.keishi}@exc.epson.co.jp

Abstract. Accurate detection and classification of machine anomalies are essential tasks for performing machine maintenance. Recently, deep learning methods have been widely utilized for anomaly detection to improve performance. For this study, we utilized TimesNet as a feature extractor. TimesNet had achieved state-of-the-art performance in time series analysis tasks, including anomaly detection, in 2023.

TimesNet enables efficient feature extraction by segmenting and stacking the vibrations at critical periods during the preprocessing. However, we found that TimesNet lacks exactness in the period length used for the segmentation. Thus, we additionally proposed two methods to address this issue and compared the accuracy of our proposed models to that of TimesNet. As a result, the proposed methods outperformed the original TimesNet.

Keywords: Anomaly Detection · Deep Learning · TimesNet

1 Introduction

Anomaly detection is essential for monitoring machine conditions and maintenance. Due to the wide range of adaptable equipment and the multitude of detectable anomaly types, vibration data is commonly utilized for anomaly detection [1].

In recent years, deep learning models have been employed to realize accurate anomaly detection. In previous studies, deep learning models, such as Transformer-based models, CNN-based models etc., have been applied for anomaly detection [2–4]. In 2019, He et al. [3] applied Temporal Neural Network (TCN [5]: A CNN-based network for time series analysis) for anomaly detection and achieved better performance on three real-world datasets. In 2022,

Supported by Graduate School of Informatics of Osaka Metropolitan University & Microdevices Operations Division of Seiko Epson Corporation.

R. Chinthaginjala et al. (Eds.): DCAI 2024, LNNS 1259, pp. 11–20, 2025.
https://doi.org/10.1007/978-3-031-82073-1_2

Xu et al. proposed the Anomaly Transformer with Anomaly-Attention mechanism. They achieved state-of-the-art(SoTA) results on six unsupervised time series anomaly detection benchmarks [4]. In 2023, a novel time series analysis model called TimesNet achieved SoTA in five tasks including anomaly detection [6]. TimesNet enables efficient feature extraction by segmenting and stacking the vibrations at critical periods during preprocessing.

At the start of this study, we chose to adopt TimesNet model because it was considered to be the SoTA method. In implementing TimesNet, we found that TimesNet lacks accuracy in the period length used for feature extraction, preventing TimesNet from properly extracting the features of vibration data. To address this issue, we propose a method that alleviates the problem of the TimesNet encoder and compare its accuracy in both anomaly detection and anomaly classification.

2 Related Studies

In recent years, many anomaly detection methods using deep learning models, such as CNN-based, Transformer-based, etc., have been proposed. Many of them are unsupervised or semi-supervised models.

For example, TCN (Temporal Convolutional Network) is applied for anomaly detection by He et al. [3]. They utilized prediction errors as indicators for anomalies, resulting in a better performance than the original TCN in three open real-world datasets. Regarding Transformer-based models, Anomaly Transformer, which introduces Anomaly Attention to Transformer, achieved SoTA on six unsupervised time series anomaly detection benchmarks in 2022 [4]. In 2023, a novel time series analysis model called TimesNet achieved SoTA in five tasks including Anomaly detection. TimesNet encoder, focusing on the multi-periodicity of real-time series data, extracts short-term and long-term dependency of multi-important periods. Since TimesNet was the SoTA model when this study was launched, we applied it to anomaly detection on another dataset.

The Structure of TimesNet. The structure of TimesNet encoder is shown in Fig. 1. The encoder of TimesNet is composed of several modules called "TimesBlock". TimesBlocks are stacked with residual connection, which enables models to learn only the difference from the original input variations [7].

TimesBlock. An overview of TimesBlock is also shown in Fig. 1.

Initially, the vibration data input into TimesBlock is transformed to representations of frequency domain by fast Fourier transformation. These frequency domain representations are then averaged along the channel direction. After that, top-k frequencies with the largest amplitudes are selected. Let $f_1, f_2, ..., f_k$ represent top-k frequencies and T represent the input series length. The extraction periods $p_1, p_2, ..., p_k$ are calculated using Equation (1) for each respective frequency. It is important to note that this process disregards the decimal values of the extraction periods.

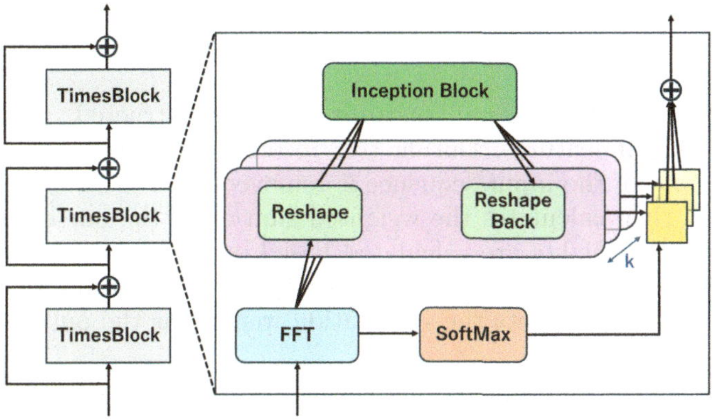

Fig. 1. The Structure of TimesNet Encoder [6]

$$p_1, p_2, ..., p_k = \lfloor \frac{T}{f_1} \rfloor, \lfloor \frac{T}{f_2} \rfloor, ..., \lfloor \frac{T}{f_k} \rfloor \quad (1)$$

The following feature extraction processes are shown in Fig. 2.

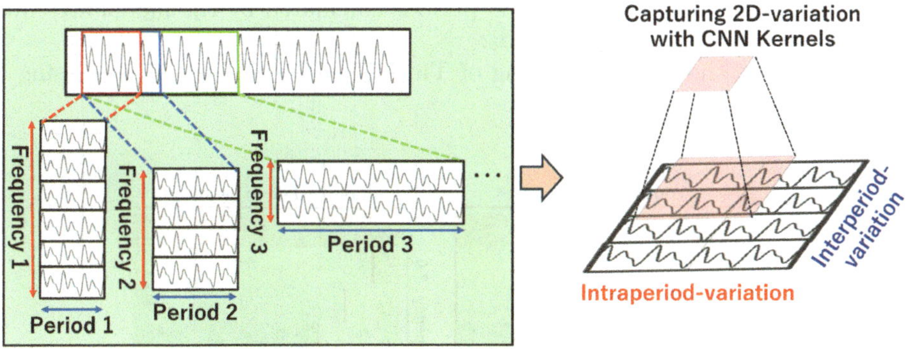

Fig. 2. Feature Extraction with CNN [6]

Next, the vibration data is padded with zeros to ensure it is divisible by each extraction period. Subsequently, the data is divided into segments using extraction periods $p_1, p_2, ..., p_k$ and then stacked. This transformation enables the arrangement of short-term dependencies of vibrations in the row direction and long-term dependencies in the column direction. Through this process, k sets of two-dimensional vibration data are generated.

Two-dimensional vibration calculated in the above processing is input into a convolutional neural network(CNN). For this CNN, TimesBlock uses the Parameter-efficient Inception Block, which was originally part of GoogleNet.

GoogleNet achieved SoTA performance in the ILSVRC 2014 competition [8]. The output of the CNN is two-dimensional features that have the same shape as its input.

Afterward, one-dimensional features are reconstructed from the two-dimensional output features. During this reconstruction process, any padding added to the end of the input sequence is removed.

The model then calculates the weighted sum of all the one-dimensional features. Summation weights are calculated based on the intensity information of frequency spectra. Intensity of each frequency spectrum is input into SoftMax function, and the model gets the summation weights as the output of the SoftMax function.

If we denote each one-dimensional feature as $\boldsymbol{x}'_1, \boldsymbol{x}'_2, ..., \boldsymbol{x}'_k$, and the weights corresponding to each frequency as $w_1, w_2, ..., w_k$, then the aggregated summation of all the features $\boldsymbol{x}_{all}$ is calculated by Eq. (2). This sum serves as the output of TimesBlock.

$$\boldsymbol{x}_{all} = \sum_{i=1}^{k} \boldsymbol{x}'_i \times w_i \tag{2}$$

Problems of Timesblock. TimesBlock, serving as the feature extractor, calculates the period corresponding to each frequency based on Equation (1). However, due to the truncation, the extraction periods calculated by the model are equal to or shorter than the actual periods.

We compare the ideal processing of TimesBlock to the actual processing in Fig. 3.

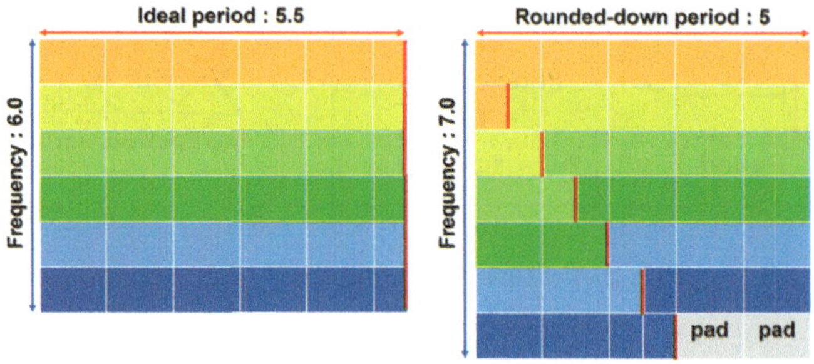

Fig. 3. Comparison of ideal and real processing (Ideal processing on the left, real processing on the right)

Assuming an input sequence length of 33 and a frequency of 6, Eq. (1) yields an extraction period $p = \lfloor 33/6 \rfloor = 5$, while the ideal extraction period is 5.5. Each color corresponds to an ideal period, and red vertical lines indicate the ideal end of each segment, which is ideally aligned vertically. However, in the figure on the right side of Fig. 3, the red line is not aligned vertically because signals are divided using a rounded-down period. This misalignment causes TimesBlock to potentially overlook some information when extracting features at the target frequencies.

3 Proposed Method

To address the issue that rounded-down periods, employed through feature extraction in TimesNet, may affect the performance of the model, we have made two improvements.

3.1 Extraction Period with Decimal Values (Proposed Method 1)

In this method, we apply extraction periods with decimal values to create two-dimensional vibrations.

Let T denote the input length of the data, f_i denote the i-th extraction frequency, and X_{1D} denote the input data. The extraction period is calculated by Eq. (3). Note that decimal values are not truncated in this calculation.

$$p_i = \frac{T}{f_i} \tag{3}$$

Then, we calculate the rounded-up period p_{ceil} using Eq. (4).

$$p_{ceil} = \lceil \frac{T}{f_i} \rceil \tag{4}$$

To begin with, we utilize Eqs. (5) and (6) to compute the starting point ($Head$) and ending point (End) of the j-th segment respectively.

$$Head = Round((j-1) \times p_i) \tag{5}$$

$$End = Round(j \times p_i) \tag{6}$$

We subsequently extract a portion of the input data, commencing from the $Head$ and concluding at the End using Eq. (7). Here, the function $Round()$ returns the rounded value of the input number.

$$Segment \leftarrow X_{1D}[Head : End] \tag{7}$$

Equation (7) represents the process of transmitting data from the $Head$-th to the End-th of the input data X_{1D} to the Segment. The extraction length is denoted by $End - Head$.

Since the length of the segments needs to be the same, we aligned the length of all the segments by filling out the shorter segments with a zero. If the extracted segment is shorter than p_{ceil}, it is filled with zero using Eq. (8). Here, $Concat()$ is a function that returns the result of concatenating two input vibration data sequences.

$$Segment \leftarrow Concat(Segment, 0) \tag{8}$$

In the end, overlay the data extracted from the j-th row in Eq. (9) onto the output data.

$$Output \leftarrow Stack(Output, Segment) \tag{9}$$

By repeating the process from Eq. (3) to Eq. (9) for $j = 1, 2, ..., f_i$, we obtain a two-dimensional matrix data X_{2D}.

Padding with Identical Variation (Proposed Method 2). In proposed method 1, we handle the problem of TimesNet by filling up missing parts of the variation with zeros. However, there was a concern that filling the variation with zeros would disrupt the continuity of variations and impede proper feature extraction. To address the aforementioned concerns, an additional refinement has been introduced. Specifically, in the processing described by Eq. (8) of proposed method 1, we propose a method to fill in missing parts with identical signal value. The process described above is represented in Eq. (10).

$$Segment \leftarrow Concat(Segment, X_{1D}[End]) \tag{10}$$

3.2 The Expected Effect of the Proposed Methods

We compare the 2D input tensors to the convolutional layer under specific condition in original TimesNet and proposed methods 1 and 2. Figure 4 illustrates the input tensors to the convolutional layer for each method with an input length T of 33 and a frequency f_i of 6. For the purpose of explanation, Fig. 4 also visualizes the region where CNN can process.

First, the same points of extraction periods are not aligned vertically in the original TimesNet, while the same points of the periods are roughly aligned in both of the proposed methods. It is thought that this refinement helps CNN extract both intraperiod-variation and interperiod-variation.

In addition, CNNs typically utilize padding to compensate for processing at the edges of images. As a result, efficient feature extraction cannot be achieved at the edges of the images. In TimesNet, the regions heavily affected by lack of exactness in the extraction periods are within the processable region of CNN. In contrast, in the proposed method, extraction errors tend to accumulate at the edges, where CNNs are inherently poor at processing. Consequently, it is presumed that the impact of errors on extracted features is reduced in the proposed method.

Based on the two reasons mentioned above, we expect the proposed method to extract features more efficiently than the original TimesNet method.

4 Experiment

We conducted experiments using vibration data obtained from the dry pump dataset to verify the effectiveness of the proposed method.

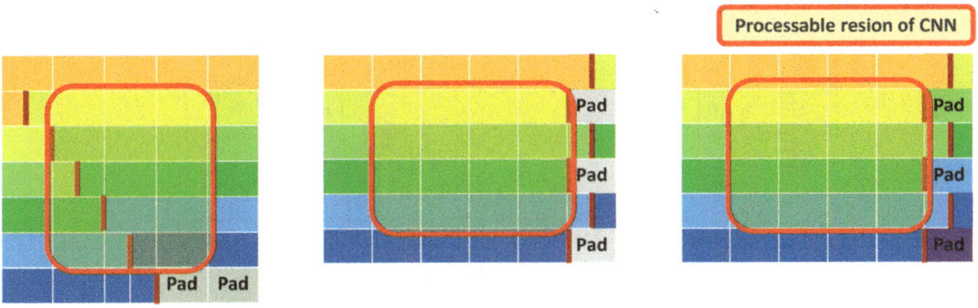

Fig. 4. Comparison of input 2D tensor (baseline model on the left, proposed method 1 in the middle, proposed method 2 on the right)

4.1 The Detail of Dry Pump Dataset

The dry pump to be measured is equipped with a main pump and a mechanical booster pump, each of which has its own bearing and gears. To measure the vibration during operation, a 6-axis digital output IMU sensor (Seiko Epson M-G354, acceleration sensor bandwidth 200 Hz) was fixed on the top surface of the main pump, with the sensor's X-axis aligned with the axis direction, the Y-axis aligned with the lateral direction, and the Z-axis aligned with the vertical direction.

To secure the fixture, we made use of a double-sided adhesive tape (strong adhesive thin type) for fixation, ensuring it did not affect the measurements. We applied UART as the communication protocol for data acquisition and gained 3-axis acceleration data, angular velocity data, and temperature data for 5.5 s at 2000 samples per second using dedicated logger software installed on a PC. Only the acceleration data was used in conducting this experiment.

Based on the dry pump data in normal conditions, we artificially created the failure data by shifting the phase of the X-axis acceleration by sixty degrees. Figure 5 depicts the amplitude spectrum of the X-axis vibration measured from the dry pump. As shown in Fig. 5, the data have peaks at 85 Hz, 314 Hz, and 398 Hz, and by shifting the phase of each frequency, three types of failure data were generated. The data was input into the model by slicing it into segments with a width of 96 points and shifting one point at a time. The data was divided into training, validation, and test sets in a ratio of 7:1:2.

4.2 Experiment Workflow

Figure 6 illustrates an overview diagram of the experimental procedure. Initially, the deep learning model is trained using only normal data, with the task of predicting future values of the time series. Subsequently, the trained encoder is utilized to extract features from the test data. During feature extraction, the normal data which was treated as test data during training, as well as all anomaly data are utilized as test data. Finally, the extracted features are classified by a classifier such as SVM.

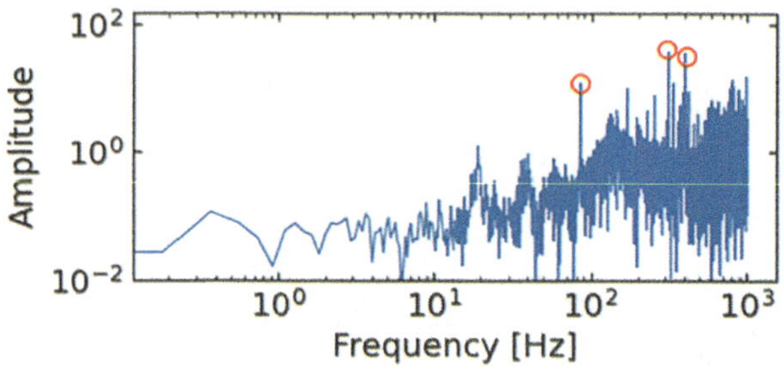

Fig. 5. Amplitude Spectrum of Drypump Dataset

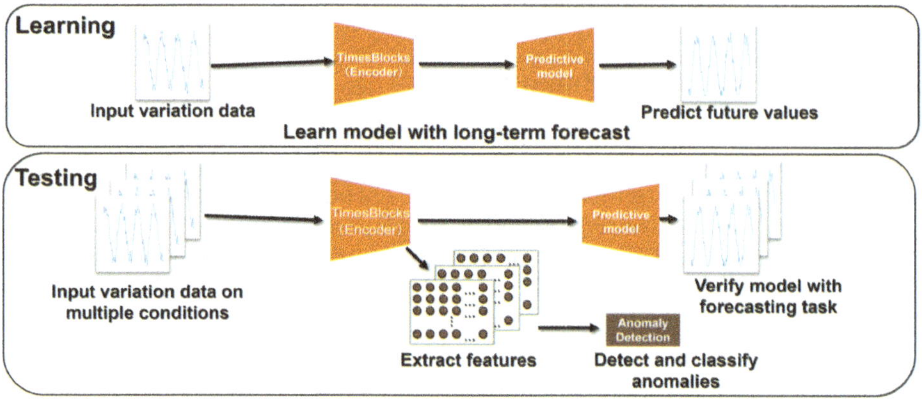

Fig. 6. Experiment workflow

4.3 Evaluation of Anomaly Detection Performance with One-Class SVM

In this section, we input the features into the One-Class SVM and evaluated the detection performance of anomaly data for each anomaly mode. We trained the TimesNet with normal data for a maximum of 15 epochs. When inputting anomaly data into the trained model, we input the features extracted from the encoder into One-Class SVM and determined whether it is anomalous.

In the training of One-Class SVM, we utilized extracted features of normal data. In testing phase, we employed extracted features of anomalous data. In terms of evaluation metrics, AUC (Area Under The Curve) in case that the distance from the discriminative boundary is treated as anomaly severity was applied.

Time-series learning ends with a maximum of 15 epochs of training (training stops if the validation loss does not decrease for five consecutive epochs), with both the input length and prediction length set to 96.

Table 1 shows the results of anomaly detection performance using both the original model and the proposed model as the feature extractor. For all three anomaly modes, the proposed method exhibited higher AUC compared to the baseline. Based on this experiment, it is evident that the alleviation of the cutout errors has led to a more organized input to the model, enabling efficient extraction of vibration features and facilitating detection.

Table 1. Comparison of anomaly detection performance between baseline and proposed methods (Best Accuracy: **Bold**, 2nd Accuracy: Underlined)

Model	85 Hz	314 Hz	398 Hz
Baseline	0.870 ± 0.008	0.881 ± 0.010	0.934 ± 0.003
Proposed Method 1	$\mathbf{0.874 \pm 0.006}$	$\underline{0.904 \pm 0.006}$	$\mathbf{0.953 \pm 0.005}$
Proposed Method 2	$\mathbf{0.874 \pm 0.006}$	$\mathbf{0.905 \pm 0.007}$	$\underline{0.950 \pm 0.006}$

4.4 Evaluation of Anomaly Classification Performance with SVM

In this section, SVM was employed to evaluate the classification performance among different anomaly modes based on the extracted features. Similar to the evaluation using One-Class SVM in Sect. 4.3, the encoder was trained on normal data for the time-series prediction task, and then the test data was input to the encoder of the trained model to obtain features. These features were subsequently fed into the SVM for classification among the anomaly modes.

Table 2 shows the results of classification performance among anomaly modes. In the classification between 85 Hz/314 Hz and 398 Hz/85 Hz, proposed method 2 exhibited the best classification performance, followed by proposed method 1 with the second-highest accuracy. In the classification between 314 Hz/85 Hz, the baseline model yielded slightly better results than the proposed methods. On average, the classification accuracy of proposed method 2 was the highest, followed by the classification accuracy of proposed method 1. Based on these results, it appears that the feature extractor of the proposed method extracts features more precisely, resulting in easier classification.

The reason for the unsatisfactory classification accuracy of the proposed method in specific anomaly modes remains unexplored.

Table 2. Comparison of anomaly classification performance between baseline and the proposed method (Best Accuracy: **Bold**, 2nd Accuracy: Underlined)

	Anomaly Modes			
Model	85/314	314/398	398/85	Average
Baseline	73.7 ± 5.1	$\mathbf{97.9 \pm 1.2}$	95.6 ± 3.4	89.1
Proposed Method 1	$\underline{76.3 \pm 3.9}$	96.9 ± 0.9	$\underline{97.3 \pm 0.9}$	$\underline{90.2}$
Proposed Method 2	$\mathbf{77.3 \pm 4.1}$	$\underline{97.3 \pm 0.8}$	$\mathbf{97.4 \pm 0.6}$	**90.7**

5 Conclusion

In this study, we focused on addressing an issue of TimesNet. We proposed methods to utilize decimal length of periods and then verified the effectiveness of our approaches through anomaly detection experiments.

For the dry pump dataset, our proposed methods outperformed the baseline in two key tasks:

1. Anomaly detection using One-Class SVM
2. Anomaly mode classification with SVM

Our future work involves visualizing the decision-making process of the model and explore what in fact makes the difference. Besides, we will evaluate the proposed model with other dataset and other time series analysis tasks.

References

1. ISO 13380. Condition monitoring and diagnostics of machines - General guidelines on using performance parameters (2002)
2. Wen, L., Li, X., Gao, L., Zhang, Y.: A new convolutional neural network-based data-driven fault diagnosis method. IEEE Trans. Industr. Electron. **65**(7), 5990–5998 (2018). https://doi.org/10.1109/TIE.2017.2774777
3. He, Y., Zhao, J.: Temporal convolutional networks for anomaly detection in time series. J. Phys. Conf. Ser. **1213**(4), 042050 (2019). https://doi.org/10.1088/1742-6596/1213/4/042050
4. Xu, J., Wu, H., Wang, J., Long, M.: Anomaly transformer: time series anomaly detection with association discrepancy. In: International Conference on Learning Representations (2022). https://openreview.net/forum?id=LzQQ89U1qm_
5. Lea, C.S., Vidal, R., Reiter, A., Hager, G.: Temporal convolutional networks: a unified approach to action segmentation. In: ECCV Workshops (2016). https://api.semanticscholar.org/CorpusID:12414640
6. Wu, H., Hu, T., Liu, Y., Zhou, H., Wang, J., Long, M.: TimesNet: temporal 2D-variation modeling for general time series analysis. In: The Eleventh International Conference on Learning Representations (2023). https://openreview.net/forum?id=ju_Uqw384Oq
7. He, K., Zhang, X., Ren, S., Sun, J.: Deep residual learning for image recognition. In: Proceedings of the IEEE Conference on Computer Vision and Pattern Recognition, pp. 770–778 (2016)
8. Szegedy, C., et al.: Going deeper with convolutions. In: 2015 IEEE Conference on Computer Vision and Pattern Recognition (CVPR), IEEE Computer Society, Los Alamitos, CA, USA, pp. 1–9 (2015). https://doi.org/10.1109/CVPR.2015.7298594, https://doi.ieeecomputersociety.org/10.1109/CVPR.2015.7298594

An Explainable AI Framework for Comparative Analysis of the Model Explanations in Breast Cancer Prediction

Ghazaleh Emadi[1] and Ana-Belén Gil-González[1,2](✉)

[1] University of Salamanca, Salamanca, Spain
{Idu030708,abg}@usal.es
[2] BISITE Research group. University of Salamanca, Salamanca, Spain

Abstract. This work proposes a systematic framework that employs eXplainable Artificial Intelligence (XAI) to conduct an in-depth comparative analysis of model explanations applied to breast cancer prediction. As the need for transparency and interpretability within Artificial Intelligence grows, our study delves into critical inquiries about the assessment of explanation quality, a comparative study of both direct and post-hoc explanation methods, and the trade-off between interpretability and explainability. The evaluation assists in carrying out well-rounded analyses of the explanations provided by the models. Our research not only enhances the understanding of explanation quality assessment but also underlines the required balance between interpretability and model efficiency, thereby revealing the correlation between model complexity and interpretability. Consequently, this framework serves as an efficient device for evaluating and selecting algorithms for breast cancer prediction, thereby encouraging more comprehensive utilization of Artificial Intelligence in clinical oncology.

Keywords: Explainable AI · Interpretability · Transparency · Transparency Performance Trade-off · XAI Techniques · Direct Explanations · PostHoc Explanations · Local Explanations

1 Introduction

Artificial Intelligence (AI) has been rapidly advancing, leading to the creation of various models with the capability to transform multiple domains, including the crucial field of healthcare, more pointedly, oncology [2,8]. Such promising developments, however, encounter a roadblock owing to the opaque 'black-box' nature of several machine learning models that restrict their full application, particularly in healthcare settings where real patient lives can be influenced by these advancements [4].

This research addresses the paradox of powerful yet opaque AI models in healthcare by utilizing eXplainable AI (XAI) to enhance interpretability, focusing specifically on breast cancer prediction. Various model explanations are systematically compared, delving into evaluation of explanation quality, contrasting

R. Chinthaginjala et al. (Eds.): DCAI 2024, LNNS 1259, pp. 21–30, 2025.
https://doi.org/10.1007/978-3-031-82073-1_3

post-hoc and direct explanation methods, and understanding 'white-box' models [13]. The methodology involves applying diverse explanation methods to a breast cancer dataset to provide comprehensive insights [7,16]. The research findings aim to guide the selection of the most suitable explainability algorithm, paving the way for wider application of XAI in clinical oncology. The subsequent sections elaborate on related works, our methodology, results, and discussion, leading to a conclusion that summarizes the study findings and their potential impact on ongoing and future breast cancer research.

The rest of this paper is organized as follows: Sect. 2 presents a review of related work, and reviews the advances, algorithms, and challenges in evaluating AI explanations. This provides the necessary background and understanding of the various techniques applied. Section 3 outlines our proposed framework's methodology, detailing how we apply each analysis method, from Lime to MLP models, to the breast cancer dataset, and how we evaluate their explanations. Section 4 discusses the results obtained from our analysis, with a focus on the implications of these outcomes. We deciphered the various insights gathered, allowing us to compare and contrast the different techniques. Section 5 concludes the study with a summary of our findings and their potential impact on breast cancer research. In this final section, we also look ahead to future work, discussing potential improvements, novel approaches, and how our current research can inform and shape the continuing exploration into Explainable AI in clinical oncology.

2 Related Work

Explaining the intricacies and decisions of AI models presents an array of substantial challenges rooted in both the technical complexities of models and human ability to understand these complex concepts. The main challenge arises from the infamous blackbox nature of many AI models. These "black boxes" are obscured by algorithmic complexity, high dimensionality, and nonlinear characteristics that are commonly embedded within a plethora of machine learning models. This creates an opaque system, while their output can be straightforwardly measured, the internal decision-making structure remains a labyrinth that is hard to navigate [5,19].

A correlated challenge lies in balancing the trade-offs between the interpretability of the model decisions and their performance [6]. High-end, complex models often exhibit unparalleled performance, but their complexity inversely affects their interpretability. By contrast, simpler models offer greater transparency, but their performance is typically less precise.

The subjective nature of explanations implies that what one person sees as a sufficient explanation and another might find lacking. Hence, establishing a universally acceptable method for measuring or quantifying the explanation quality is a daunting task. Furthermore, even with a perfect explanation, comprehension is often a challenge because of the technical nature of AI and machine learning, which makes it difficult for non-experts to understand. The fact that the AI models learn and adapt over time presents an additional challenge. As the model evolves, maintaining consistency and coherency in explanations, particularly for models used in critical decision-making scenarios, can be problematic. Adding to these complexities is the difficulty in choosing an appropriate explanation method for specific use cases, making the field of explainable AI a challenging terrain for navigation [3].

Along with the above, it is necessary to add the concept of the level of explainability. Explainability techniques are mainly divided into two broad categories: global and local. Global, where they explain the model in a general way, while pointing out its generic rules of operation. The local level explains, for each specific piece of data, how the model reasons and the rules that lead to a given result. The correct granularity and relationship of the two levels is also an important aspect of tuning the explainability of the results. Some of the main methods for explainability are considered below.

2.1 Post-Hoc Explanation Methods

Local Interpretable Model-Agnostic Explanations (LIME) [17] approach the issue of explainability by approximating the decision boundary of a complex model with a simpler, locally fitted model. LIME creates easy-to-understand visualizations detailing the extent to which each feature contributes to a model's prediction for a single instance. LIME, then, has been designed to be applied locally.

SHapley Additive exPlanations (SHAP) [10] utilizes concepts from cooperative game theory to allocate a 'fair' contribution of each feature to the model's prediction. It provides an interpretative understanding of feature interactions and dependencies in driving a prediction of an individual data instance [11]. SHAP can be used both globally and locally.

KernelSHAP[18] is a variant of the SHAP that can be applied to any model. Unlike SHAP, which expounds on feature contributions specifically in tree-based models, KernelSHAP can extract explanations from other types of models. KernelSHAP calculates an approximation of the SHAP values using weighted linear regression.

Counterfactual Explanations operate on the principle of identifying which features of a sample would need to be minimally and realistically changed to alter the model's prediction. Thus, a counterfactual explanation reveals how a specific prediction can differ [1].

The Permutation Feature Importance (PFI) assesses the impact of each feature on a model's predictions by permuting the values of a feature and measuring the decrease in the model's performance [14]. The features that caused the most significant performance degradation were deemed to be the most important. Obviously, permutation importance can only be applied globally.

2.2 Direct Explanation Methods

Explainable Boosting Machine (EBM) [9,20] is a model type, developed at Microsoft Research, that incorporates transparency into its design from the ground up. It uses a modern take on traditional generalized additive models (GAMs), strengthening their interpretability while preserving strong performance. The entire model can be explained and visualized, giving insight into the influence of each feature [12]. EBM can be used both globally and locally.

The linear model assumes a linear relationship between input variables and a single output variable. It provides a simple interpretable model that explains how each feature influences the prediction via its coefficients. Linear models are transparent, but may lack the accuracy of more complex models [15]. Table 1 compares the cons and pros of described direct and post-hoc methods.

Table 1. Comparison of AI Explainability Methods [3,5,6,19]

Method	Pros	Cons
LIME	Instance-level explanations, works with complex models	Unstable explanations, contingent on hyperparameters
SHAP	'Fair' feature contribution, provides global and local interpretability	Computationally expensive
KernelSHAP	Extends SHAP to any model, both local and global interpretability	Computationally expensive, not suited for large datasets
Counterfactual Explanations	Offers potential actionable insights	Computationally complex, multiple solutions possible
PFI	Model-agnostic, simple to understand and implement, provides global importance	Doesn't consider feature interactions, inaccurate when features are correlated
EBM	High interpretability without compromising performance, provides global importance and pairwise interactions	May not perform as well as more complex models on certain tasks
Linear Model	Direct, interpretable relationship between features and predictions, simple computations	May lack the performance of complex models, linear relationships might not reflect reality

3 Methodology

Breast Cancer Dataset. The Wisconsin Breast Cancer dataset is a well-recognized benchmark dataset that is extensively utilized in Machine Learning and AI for predictive modeling studies. The dataset was derived from digitized images of breast mass aspirates, with each of the 569 instances having 32 attributes encompassing key factors such as radius, texture, and smoothness. The dataset details 31 features describing the characteristics of the breast mass, including the mean values, standard errors, and worst-case scenarios for each attribute. The 'target' attribute presents the diagnosis as malignant or benign. Figure 1 represents the frequency distribution of the 'target' variable in the breast cancer dataset. The 'target' variable has two categories: Malignant, denoted by 0, and Benign, denoted by 1. The height of each bar corresponded to the total number of observations for each class. The plot shows the distribution of malignant and benign tumors in the dataset. Analyzing the height of the bars can reveal whether there is an imbalance in the classes, which is an important factor to consider during the process of building predictive models, as it could affect their performance. Reviewing this plot provides a clear understanding of the proportion of malignant to benign samples, supporting informed decision making in further data processing and modeling.

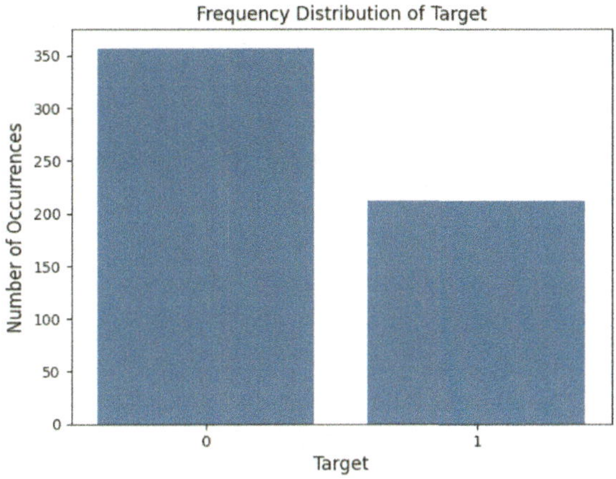

Fig. 1. Target distribution of data points.

Figure 2 shows the correlation heatmap, providing visual insight into the relationship between the various attributes in our dataset. Colors in the heatmap correspond to the degree and direction of correlation between pairs of variables. Darker shades indicate high correlation, with red signifying positive relationships and blue illustrating negative relationships. Light or near-white squares signify weak or no correlation, indicating that the corresponding features do not have much influence on each other. By revealing how closely the features are related, this heatmap helps in identifying potential collinearity where two or more variables are highly correlated.

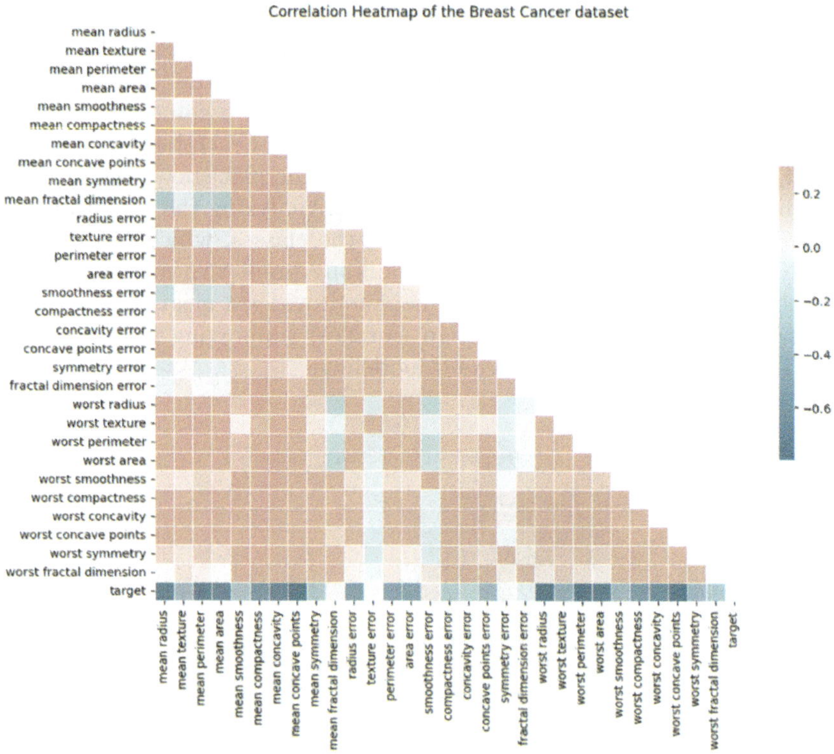

Fig. 2. Correlation Heatmap of the Breast Cancer dataset

In this study, our methodology, as illustrated in Fig. 3, kicks off with the Breast Cancer dataset, an esteemed dataset regularly employed in classification tasks to differentiate between malignant and benign tumor cases based on a comprehensive range of cellular attributes.

The research methodology begins by using a renowned Breast Cancer dataset for classification tasks. Exploratory Data Analysis (EDA) was applied to understand the essential characteristics of the data and test the initial assumptions. The dataset then undergoes preprocessing techniques, such as data cleaning, transformation, feature selection, encoding categorical variables, and feature scaling, preparing it for machine learning algorithms. The model training phase employs both Black-box and White-box models to balance performance and transparency. After training, explainable AI algorithms, such as LIME, KernelSHAP, EBM, and Logistic Regression, were used to interpret the decision-making of the models. Finally, the performance of the models and algorithms was analyzed to select the most suitable model that fulfills the research goal.

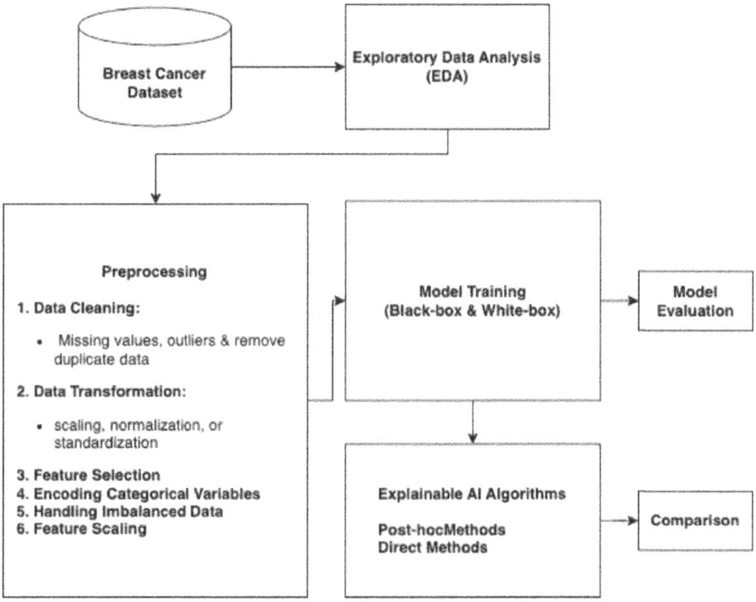

Fig. 3. The Architecture flow of the proposed Explainable framework for model predictions on the Breast Cancer dataset

4 Results

Some results are shown at Table 2, that illustrates the performance comparison of the four distinct machine learning models, each measured in terms of accuracy, precision, and recall. The Explainable Boosting Machine (EBM) classifier demonstrated a well-rounded performance, with a 97% score across all three metrics, indicating reliable prediction capabilities.

Table 2. Performance Comparison of the trained models

Model	Accuracy	Precision	Recall
Explainable Boosting Machine Classifier	0.97	0.97	0.97
Random Forest Classifier	0.96	0.95	0.98
Logistic Regression	0.96	0.96	0.99
Multi-layer Perceptron	0.98	0.97	0.99

The Random Forest Classifier shows a slightly lower precision of 95%, which means that it might classify some instances incorrectly as positive. However, its recall is 98%, suggesting that it is excellent for detecting actual positive cases. The Logistic Regression model exhibits the same overall accuracy as the Random

Forest Classifier but achieves a higher precision of 96%, pointing to its acute ability to correctly identify positive instances. Its recall is 99%, the best among all the models, demonstrating its exceptional identification of actual positive cases, and the multilayer perceptron model exhibits the highest accuracy among all models, at 98%. Its precision is slightly lower than its accuracy at 97%, but its recall is exceptionally high, equal to that of the Logistic Regression model at 99%. Although these models exhibit exceptional operational performance, they highlight a common trade-off encountered in machine learning: the balance between model interpretability (transparency) and model performance.

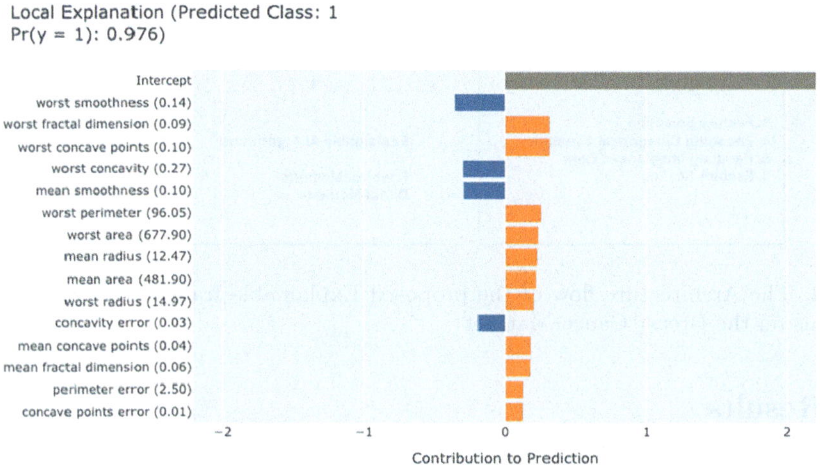

Fig. 4. Feature importance for the Explainable Boosting Machine (EBM) model

Figure 4 shows the feature importance plot for the Explainable Boosting Machine (EBM) model, which describes the contribution of each feature to the final prediction for a given data point. Each bar in the plot signifies a specific feature, and its length indicates the degree to which that feature influences the prediction, enabling a visual understanding of the feature effects. By examining this plot, we can discern the key features that have a significant impact, thus offering insights into the model's decision-making process.

5 Conclusion and Future Work

In conclusion, this study highlights the trade-off between model performance and interpretability, examined through the lens of four distinct machine learning models applied to a breast cancer dataset. While the Explainable Boosting Machine Classifier offers robust performance with increased transparency, more complex models, such as the random forest classifier and multilayer perceptron, present slightly superior performance at the cost of less interpretability. Therefore, choosing an appropriate model is a nuanced decision that hinges on the unique demands of the task, underscoring the need to strike a balance between accurate predictions and the capacity to understand and explain the underlying decision-making process.

Additional datasets will be worked on in the future. Work will be done on practical methodologies for the deployment model of applications in the health sector that incorporate the studies in operational environments.

Acknowledgments. This research has been supported by the project "COordinated intelligent Services for Adaptive Smart areaS (COSASS), Reference: PID2021-123673OB-C33, financed by MCIN /AEI /10.13039/501100011033/FEDER, UE.

References

1. Alfeo, A.L., Zippo, A.G., Catrambone, V., Cimino, M.G.C.A., Toschi, N., Valenza, G.: From local counterfactuals to global feature importance: efficient, robust, and model-agnostic explanations for brain connectivity networks. Comput. Methods Programs Biomed. **236**, 107550 (2023)
2. Baryannis, G., Dani, S., Antoniou, G.: Predicting supply chain risks using machine learning: the trade-off between performance and interpretability. Futur. Gener. Comput. Syst. **101**, 993–1004 (2019)
3. Chaddad, A., Peng, J., Jian, X., Bouridane, A.: Survey of explainable AI techniques in healthcare. Sensors **23**(2), 634 (2023)
4. Dziugaite, G.K., Ben-David, S., Roy, D.M.: Enforcing interpretability and its statistical impacts: Trade-offs between accuracy and interpretability. arXiv preprint arXiv:2010.13764 (2020)
5. Hassija, V., et al.: Interpreting black-box models: a review on explainable artificial intelligence. Cogn. Comput. **16**(1), 45–74 (2024)
6. Javed, A.R., Ahmed, W., Pandya, S., Maddikunta, P.K.R., Alazab, M., Gadekallu, T.R.: A survey of explainable artificial intelligence for smart cities. Electronics **12**(4), 1020 (2023)
7. Khater, T., et al.: An explainable artificial intelligence model for the classification of breast cancer. In: IEEE Access (2023)
8. Kucklick, J.-P., Müller, O.: Tackling the accuracy-interpretability trade-off: interpretable deep learning models for satellite image-based real estate appraisal. ACM Trans. Manag. Inf. Syst. **14**(1), 1–24 (2023)
9. Liu, G., Sun, B.: Concrete compressive strength prediction using an explainable boosting machine model. Case Stud. Constr. Mater. **18**, e01845 (2023)
10. Lundberg, S.M., Lee, S.-I.: A unified approach to interpreting model predictions. In: Guyon, I., et al. (eds.) Advances in Neural Information Processing Systems. vol. 30, pp. 4765–4774. Curran Associates, Inc. (2017). http://papers.nips.cc/paper/7062-a-unified-approach-to-interpreting-model-predictions.pdf
11. Lundberg, S.M., et al.: Explainable machine-learning predictions for the prevention of hypoxaemia during surgery. Nat. Biomed. Eng. **2**(10), 749 (2018)
12. Lundberg, S.M., et al.: From local explanations to global understanding with explainable AI for trees. Nat. Mach. Intell. **2**(1), 56–67 (2020)

13. Ma, X., Jian, L., Liu, X., Weibin, Q.: A genetic programming approach for real-time crash prediction to solve trade-off between interpretability and accuracy. J. Transp. Saf. Secur. **15**(4), 421–443 (2023)
14. Molnar, C., et al.: Relating the partial dependence plot and permutation feature importance to the data generating process. In: World Conference on Explainable Artificial Intelligence, pp. 456–479. Springer (2023)
15. Rahman, M.A., Ibrahim, A.A., Tumian, A.: Feature selection using generalized linear model for machine learning-based sepsis prediction. In: 2023 International Conference on Advances in Intelligent Computing and Applications (AICAPS), pp. 1–6. IEEE (2023)
16. Rajpal, S., et al.: XAI-MethylMarker: explainable AI approach for biomarker discovery for breast cancer subtype classification using methylation data. Expert Syst. Appl. **225**, 120130 (2023)
17. Ribeiro, M.T., Singh, S., Guestrin, C.: "why should i trust you?" Explaining the predictions of any classifier. In: Proceedings of the 22nd ACM SIGKDD International Conference on Knowledge discovery and Data Mining, pp. 1135–1144 (2016)
18. Roshan, K., Zafar, A.: Using kernel SHAP XAI method to optimize the network anomaly detection model. In: 2022 9th International Conference on Computing for Sustainable Global Development (INDIACom), pp. 74–80. IEEE (2022)
19. Saeed, W., Omlin, C.: Explainable AI (XAI): a systematic meta-survey of current challenges and future opportunities. Knowl.-Based Syst. **263**, 110273 (2023)
20. Tasin, I., Nabil, T.U., Islam, S., Khan, R.: Diabetes prediction using machine learning and explainable AI techniques. Healthc. Technol. Lett. **10**(1-2), 1–10 (2023)

Enhancing Power Forecasting Through Contextual Awareness with C-Means and K-Means Approaches

Brígida Teixeira[1](✉), Tiago Pinto[2], and Zita Vale[1]

[1] GECAD – Research Group on Intelligent Engineering and Computing for Advanced Innovation and Development, LASI - Intelligent Systems Associate LAboratory, Polytechnic of Porto, R. Dr. António Bernardino de Almeida, 431, 4249-015 Porto, Portugal
{bct,zav}@isep.ipp.pt

[2] Universidade de Trás-os-Montes e Alto Douro e INESC-TEC, Quinta de Prados, 5000-801 Vila Real, Portugal
tiagopinto@utad.pt

Abstract. This study delves into enhancing energy consumption forecasting in smart buildings by analyzing the implications of different contextual detection strategies, employing both K-Means and Fuzzy C-Means clustering techniques. Utilizing a comprehensive dataset from various sensors, the research segments power usage data to uncover distinct consumption patterns, extracted from user's behavior. A suite of seven machine learning forecasting algorithms undergoes evaluation, guided by two distinct methodologies: crafting specialized models for individual clusters and embedding contextual data as features within an integrated model. To perform the selection of the most suitable algorithm for each approach and cluster, an automated machine learning process is utilized, streamlining the optimization and comparison process. Moreover, the study explores the influence of soft clustering in providing nuanced, probabilistic cluster memberships and their impact on forecasting accuracy. A baseline scenario, devoid of contextual considerations, serves as a benchmark for comparison, allowing for a clear evaluation of the benefits introduced by integrating context into predictive models. The research aims to identify the optimal contextual detection strategy, considering the unique advantages of soft and hard clustering, to significantly improve energy consumption forecasts.

Keywords: Contextual Awareness · Fuzzy C-Means (FCM) · K-Means · Power Consumption Forecasting

1 Introduction

In intelligent building energy management, context-awareness emerges as a fundamental element in achieving precise load forecasting, which stands as a cornerstone for effective system planning and real-time grid operations [1]. Leveraging a comprehensive analysis of environmental data (such as weather conditions and seasonality), user behavior (including occupancy patterns and activities), and building systems (encompassing

R. Chinthaginjala et al. (Eds.): DCAI 2024, LNNS 1259, pp. 31–42, 2025.
https://doi.org/10.1007/978-3-031-82073-1_4

equipment status and layout), energy management systems gain significant advantages [2]. These include the development of improved forecasting models capable of accurately predicting energy needs based on contextual factors, adaptive optimization strategies enabling real-time response to demand fluctuations, implementation of preventative maintenance measures to detect potential issues and optimize equipment health, and customization of energy usage to align with the unique characteristics of each building, among others [3].

In this regard, clustering techniques enhance these benefits by discerning patterns and groupings within the complex contextual data landscape. By categorizing users with similar energy profiles, clustering enables the development of targeted strategies, identification of outliers, and refinement of forecasting processes to achieve even greater precision [4]. Consequently, the integration of clustering methodologies unlocks new levels of efficiency within intelligent energy management systems, facilitating optimized resource allocation and enhanced operational performance.

The K-Means clustering algorithm [5] categorizes users by weather conditions or consumption levels to unveil similarities among load patterns, a method substantiated by numerous studies such as [6]. This approach identifies distinct consumer groups and facilitates the creation of tailored forecasting models for each cluster, thereby enhancing overall model accuracy. An alternative strategy involves using K-Means to extract novel features based on consumption similarities. This leads to a unified forecasting model that spans all clusters [7]. Moreover, K-Means has been instrumental in load forecasting models by identifying anomalies such as noise and outliers, thereby streamlining consumption scenarios for improved forecasting accuracy [8].

The Fuzzy C-Means (FCM) [9], a derivative of K-Means, allows data points to belong to multiple clusters with varying degrees of membership. This capability is particularly beneficial in scenarios with ambiguous cluster boundaries. The flexibility of FCM is leveraged for load profiling [10], accommodating the complex and overlapping consumption patterns typical among consumers.

This study delves into applying clustering techniques, specifically K-Means and FCM, to discern contextual patterns emerging from user behavior within smart building environment. It aims to determine the optimal approach for forecasting energy consumption by amalgamating hard and soft clustering methodologies. The Halving Grid SearchCV AutoML process systematically evaluates seven forecasting algorithms to tailor models to each unique scenario and context identified within the data, ensuring the models' robustness and specificity. Moreover, the research emphasizes the importance of feature selection, highlighting the correlation between various data features and power consumption. By prioritizing relevant variables, the models are enriched, thereby enhancing predictive accuracy and seeking the most pertinent data for context definition. The comparison between K-Means and FCM's soft clustering capabilities offers a nuanced understanding of energy consumption behaviors, facilitating flexible cluster memberships and providing insights into consumption patterns that may not neatly align with distinct categories.

The remainder of the paper is structured as follows: Sect. 2 delves into the methodology, detailing the data selection, clustering techniques, the machine learning algorithms evaluated, and the AutoML process employed. Section 3 presents a case study applying the proposed approaches to a real-world dataset. Section 4 presents the results and discusses the findings' implications, outlining the research's potential applications and limitations. Finally, Sect. 5 concludes the paper by summarizing the key contributions and suggesting avenues for future research in the energy management domain within smart buildings.

2 Methodology

This section describes the methodology adopted in this work, aimed at enhancing the accuracy of power consumption forecasting in smart buildings through an integrated approach combining the strengths of clustering techniques, such as K-Means and FCM, with the predictive power of advanced machine learning algorithms, all within the framework of an AutoML process for model selection. Figure 1 illustrates the comprehensive forecasting framework from data profiling to performance evaluation.

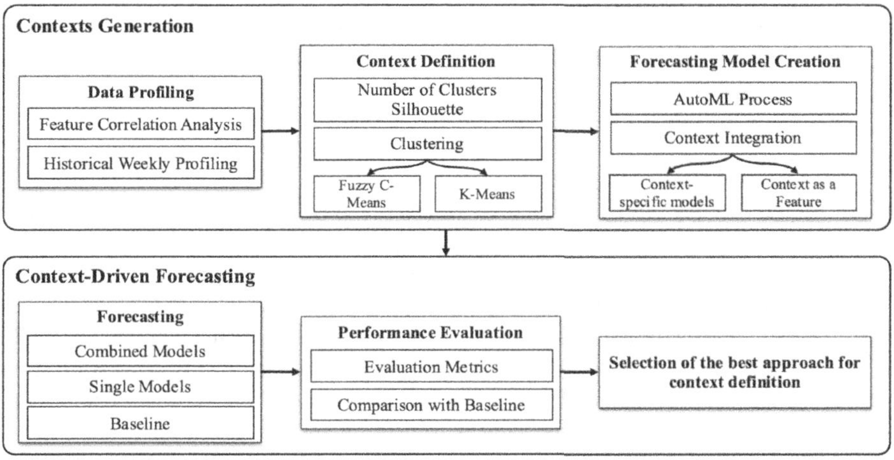

Fig. 1. Proposed methodology.

2.1 Features Correlation

A pivotal step in the methodology is the correlation analysis between different features and energy consumption, which aims to refine the dataset for enhanced model accuracy. This analysis is based on the premise that not all data points significantly influence energy consumption patterns. By calculating the correlation coefficient for each feature and retaining only those with strong correlations, the forecasting models are streamlined to focus on the most impactful predictors.

Features falling below a user-defined correlation threshold are deemed to have minimal impact on energy consumption and are excluded. This threshold is carefully selected to balance the breadth of predictors against the need to minimize noise in the data. The importance of this feature selection process lies in its ability to simplify model complexity, enhance interpretability, and improve computational efficiency. By concentrating on key features, the models become more straightforward and less prone to overfitting while also speeding up the training and evaluation phases.

2.2 Profiling

This methodology utilizes four-month historical data segments to grasp seasonal influences on power consumption in smart buildings, acknowledging how changes in weather, operations, and occupancy drive power usage patterns. The forecasting models become more sensitive to seasonal power consumption fluctuations by dissecting the year into these distinct periods. The choice of these quartiles aims to dissect how shifts in climate, heating and cooling needs, and occupancy influence power usage, crafting models that align with these variables and adapt to energy demand's cyclical nature. This approach underscores the importance of weaving temporal context into forecasting, recognizing the strong tie between power usage and seasons. Subsequently, the data is refined to create a weekly power consumption profile, averaging usage every 15 min to delineate consumption contexts. This step is key in capturing the diverse power usage patterns, providing a foundation for models to predict power consumption dynamics in smart buildings accurately.

2.3 Contexts Definition

Context definition leverages FCM and K-Means clustering, guided by silhouette scores [11], to determine the optimal cluster number. The silhouette score assesses an object's similarity to its cluster versus others, indicating clearer cluster definitions with higher scores. K-Means efficiently creates distinct, non-overlapping clusters by assigning points to the nearest center, which is ideal for clear consumption patterns. Conversely, FCM's soft clustering allows points to belong to multiple clusters, capturing overlapping consumption contexts with nuanced classification. This methodology maps each 15-min interval to a context label via clustering, simplifying future data context applications. Models then use context labels to predict consumption based on historical patterns.

Contextual insights are integrated into models in two ways. In the context-specific approach, the dataset is divided by contexts into separate datasets for each cluster. For FCM, datasets include data from the primary and secondary closest clusters. This approach ensures models are trained on context-relevant data, enhancing forecasting accuracy for distinct scenarios. In the approach using the context as a feature, the context labels are added as a dataset feature (with application of one hot encoding), allowing models to adjust predictions based on context. These strategies aim to enhance forecasting models with contextual insights, exploring the potential of both context-specific and unified models in capturing energy consumption dynamics.

2.4 Forecasting Model Creation

The selection and fine-tuning of forecasting algorithms and their parameters utilize the HalvingGridSearchCV AutoML technique [12]. This approach evaluates seven key machine learning algorithms, from [13] and [14]: Adaptive Boosting (AdaBoost), Gradient Boosting (GBoost), K-Nearest Neighbors (KNN), Multilayer Perceptron (MLP), Random Forest (RF), Support Vector Regressor (SVR), Extreme Gradient Boosting (XGB), each applied within specified contexts. The halving strategy reduces computational effort by iteratively concentrating on the most promising configurations, enabling quick and extensive exploration of the algorithms. This ensures an efficient and effective determination of the best forecasting models.

2.5 Results Evaluation

The methodology concludes with an analysis comparing the forecasting models' effectiveness. Key metrics, namely Mean Absolute Error (MAE) to measure the average error magnitude, Mean Squared Error (MSE) for the average squared differences, and Coefficient of Determination (R2) to assess model fit, provide an analysis of each model's accuracy in predicting power consumption. Results are then contrasted with a baseline scenario that does not feature the advanced clustering and contextual strategies, highlighting the superior performance and added value these methodologies bring to the forecasting process.

3 Case Study

In this section, the case study focuses on the role of different clustering strategies in defining contexts for power consumption forecasting within a smart building. The aim is to evaluate how distinct clustering methodologies, specifically K-Means and FCM, influence the formation of these contexts and, subsequently, the accuracy of the forecasting models derived from them.

The case unfolds within a smart building equipped with an array of sensors capturing data points such as CO_2 levels, air quality, temperature, and humidity. Historical data from 2019 spanning four months is dissected into 15-min intervals, resulting in approximately 11425 samples. Table 1 provides the descriptive statistics for the extracted features used in this study, including mean, standard deviation, minimum, and maximum values. From pre-processing and feature engineering, the base features included data from light intensity sensors, humidity sensors, temperature sensors, VOC air quality, CO_2 levels, as well as time-related features such as day, month, and year. Lag values were also considered to capture temporal dependencies in the data. These features are

Table 1. Descriptive Statistics of Features.

Feature	Light (%)	Humidity (%)	Temperature (°C)	VOC (%)	CO_2 (ppm)
Mean	18.33	0.00	28.52	4.69	3.16
Std	34.82	0.00	2.63	0.58	6.66
Min	0.00	0.00	21.38	2.20	0.00
Max	100.00	0.00	40.90	6.59	20.00

vital for defining consumption profiles as they capture both the environmental conditions and occupancy patterns that influence power usage.

The forecasting timeline is set from October 7, 2019, over 7, 14, and 21 days to assess the models' performance under typical operational conditions. The historical dataset, along with the features, is used to create the weekly power consumption profile by calculating the average usage for each 15-min period. This weekly profile serves as the foundation for employing clustering methods to identify the contexts.

By applying K-Means and FCM, the historical contextual data is grouped to identify distinctive consumption patterns. This clustering lays the groundwork for feature selection. A subsequent correlation analysis filters features based on a threshold of 0.3, ensuring only significant information is used by the model. The most correlated features with consumption were selected, with light intensity and CO_2 levels exhibiting the highest correlation. These features were critical as they provided insights into the factors most strongly influencing power usage, thereby enhancing the interpretation and analysis of the results.

The study uses the HalvingGridSearchCV AutoML process to evaluate seven forecasting algorithms, exploring two main methodologies: building custom models for each cluster and creating a unified model that incorporates cluster-informed features. To avoid overfitting, cross-validation with 3 folds was used. Silhouette scores, ranging from 2 to 6, were used to determine the optimal number of clusters. This process is applied to both subsets of the data: one filtered based on the best features identified through correlation analysis, and another encompassing the entire dataset.

4 Results and Discussion

This section presents the results of the scenarios executed to understand the efficacy of different context definition strategies in power consumption forecasting. The results illustrate the improvements achieved in forecasting accuracy when applying the Best Features (BF) approach, across different forecasting days (FD), and by employing distinct clustering methods (FCM and K-Means) for various contexts.

Figure 2 and Fig. 3 offer a visual representation of the contexts for power consumption, as delineated by two distinct clustering methods, where considering the complete dataset (without BF approach): FCM with membership degrees, and K-Means. These graphical representations serve as weekly profiles, displaying the variation in context membership and consumption patterns across different period of the week.

For the FCM approach, the first set of plots illustrates the membership degrees for each cluster, indicating the degree to which each time interval belongs to a given context. These degrees range from low to high, signifying the strength of association between time intervals and the clusters. The gradations of color intensity reflect the variability within the week, with darker shades representing higher degrees of membership and thus a stronger relationship to the specific context.

In contrast, the K-Means weekly profiles exhibit a more segmented structure, with time periods being distinctly classified into one of the clusters. This is indicative of the 'hard clustering' nature of K-Means, where each time interval is assigned to only one cluster without overlap, as opposed to the 'soft clustering' technique of FCM that allows for shared membership among clusters.

Table 2 delineates the performance of baseline models, serving as a benchmark to assess the impact of the context-driven clustering strategies. The baseline, devoid of context-specific clustering and the refined BF approach, delivered an MAE of 0.0183, an MSE of 0.0008, and an R2 of 0.8695. The scenarios utilizing the BF approach did not fare as well, primarily due to the reduced dataset complexity. By selecting only the features with the highest correlation, the dataset was reduced. Consequently, the BF-inclusive scenario registered an increased MAE of 0.0265 and MSE of 0.0012, accompanied by a lower R2 of 0.7966, suggesting that the omission of less-correlated features might have inadvertently removed subtle yet valuable predictive information.

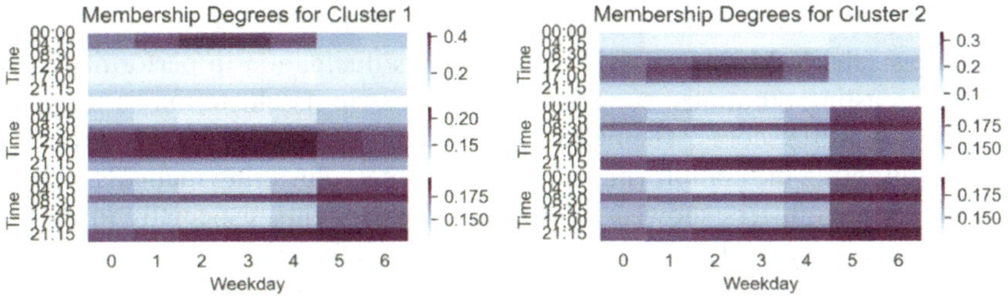

Fig. 2. Example of Weekly Contextual Profiles Based on FCM with Membership Degrees

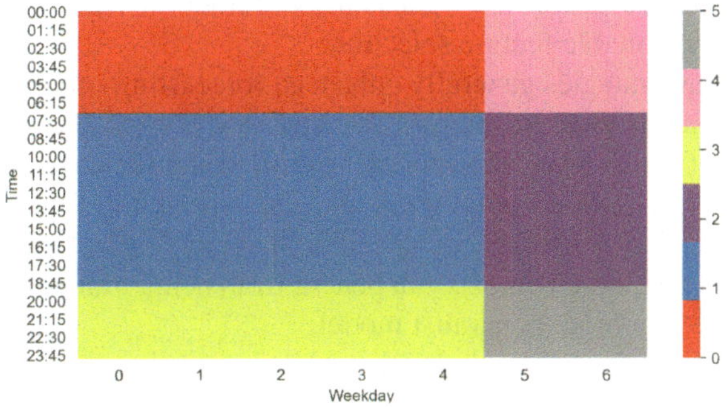

Fig. 3. Example of Weekly Contextual Profiles Based on K-Means Clustering

Table 3 presents the results of forecasting accuracy across different contexts defined by clustering methodologies. Silhouette analysis determines the optimal number of clusters or contexts. For both the FCM and K-Means methods, the silhouette analysis was pivotal in determining the optimal cluster count. The analysis varied depending on whether the Best Features (BF) were incorporated. With BF included, 6 clusters were identified as the optimal configuration, whereas using the complete dataset suggested a more streamlined approach, with 3 clusters being the most suitable. This variation in cluster number directly influences the construction of the forecasting models, as the silhouette scores provide insights into the best cluster configurations by balancing within-cluster cohesion against between-cluster separation.

However, the relationship between the optimal number of clusters and forecasting accuracy proved to be complex. For instance, without BF, the FCM approach generally

Table 2. Baselines results.

Best Features	MAE	MSE	R^2
Yes	0.0265	0.0012	0.7966
No	0.0183	0.0008	0.8695

improved R2 values for a 7-day forecast with 6 clusters. Still, it decreased performance with 3 clusters over longer forecasting periods. This illustrates the nuanced interaction between the granularity of context definition and forecasting performance, where a greater number of clusters does not necessarily equate to higher accuracy.

In contrast, the K-Means method exhibited less fluctuation in performance across different forecasting durations, suggesting robustness in its clustering strategy as dictated by the silhouette scores. The FCM combined model displayed many improvement metrics, highlighting the complexities of employing soft clustering methods in forecasting models. Some clusters showed significant enhancements, especially in R2 values, while others experienced declines, indicating variable efficacy.

The analysis also revealed mixed results for models that incorporate BF. In some instances, these models outperformed the baseline, particularly where the reduction of features did not omit critical predictors. However, excluding features based on lower correlation values decreased forecasting accuracy in other cases, underscoring the importance of reasonable feature selection.

While clustering strategies generally enhanced forecasting capabilities compared to the baseline, the benefits were not uniformly distributed across all contexts or scenarios. The effectiveness of context-specific approaches varied significantly, with some contexts showing marked improvements and others not. This variability highlights the need for a tailored approach that considers the specific characteristics of each cluster and the forecasting timeframe. Some contexts can benefit from being trained along with the data from other contexts in a more generalist model.

From this study can also be concluded that while integrating clustering and contextual information into forecasting models often leads to superior outcomes than baseline models, success is highly dependent on the specifics of each scenario. Careful analysis of silhouette scores, clustering methodologies, feature selection, and model performance across different horizons is essential to optimize the use of these advanced techniques in practical applications.

Furthermore, while the results demonstrate improvements in energy consumption forecasting within the studied smart building, it is essential to consider the generalizability of the proposed method. The methodology is designed to be applicable to other datasets from different smart buildings, accommodating various characteristics and usage profiles. However, further studies are needed to evaluate and validate the performance of the proposed approach across diverse datasets.

Table 3. Scenarios results.

B.F	F.D	Method	A	Contexts	MAE				MSE				R^2			
					General	Improvement			General	Improvement			General	Improvement		
						Average	Min	Max		Average	Min	Max		Average	Min	Max
No	7	FCM	C	3	0.0184	39.6577	17.3285	70.4146	0.0010	43.1846	14.7998	82.2132	0.8419	244.5222	6.5464	634.7824
			S	3	0.0258	2.8082	–	–	0.0012	3.0304	–	–	0.8027	0.7739	–	–
		K-Means	C	3	0.0179	36.5791	14.2260	71.4037	0.0008	47.7203	21.2348	86.3967	0.8736	175.6259	15.7157	414.2308
			S	3	0.0183	30.9312	–	–	0.0008	35.7541	–	–	0.8693	9.1307	–	–
No	14	FCM	C	3	0.0251	25.8657	–21.2802	67.6474	0.0017	18.6715	–58.5503	82.0238	0.7223	233.6364	–25.8984	634.7824
			S	3	0.0267	–0.8782	–	–	0.0013	–0.7214	–	–	0.7951	–0.1842	–	–
		K-Means	C	3	0.0176	37.3820	15.5707	71.4037	0.0008	48.6352	22.3348	86.3967	0.8765	183.0211	16.4431	435.6888
			S	3	0.0183	30.9312	–	–	0.0008	35.7541	–	–	0.8693	9.1307	–	–
No	21	FCM	C	3	0.0181	40.4791	18.2100	71.9973	0.0010	43.4643	12.5700	85.2821	0.8387	245.3411	5.5601	634.7824
			S	3	0.0288	–8.5304	–	–	0.0013	-6.9622	–	–	0.7824	–1.7780	–	–
		K-Means	C	3	0.0179	34.8237	8.3552	70.2573	0.0007	49.2265	19.2420	89.0461	0.8798	164.2287	17.4239	375.3584
			S	3	0.0183	30.9312	–	–	0.0008	35.7541	–	–	0.8693	9.1307	–	–
Yes	7	FCM	C	6	0.0176	5.0666	-20.7661	35.1859	0.0008	17.4654	-19.0968	65.1215	0.8760	26.0364	-31.4976	98.6628
			S	6	0.0182	0.5629	–	–	0.0008	1.3900	–	–	0.8714	0.2085	–	–
		K-Means	C	6	0.0174	3.1101	–13.0775	25.3018	0.0007	15.1031	-14.8211	53.9639	0.8913	25.8018	–15.0719	96.2733
			S	6	0.0183	–0.1200	–	–	0.0008	0.0082	–	–	0.8696	0.0012	–	–

(*continued*)

Table 3. *(continued)*

B.F	F.D	Method	A	Contexts	MAE				MSE				R^2			
					General	Improvement			General	Improvement			General	Improvement		
						Average	Min	Max		Average	Min	Max		Average	Min	Max
Yes	14	FCM	C	6	0.0176	0.6908	-22.4564	34.2002	0.0007	18.8032	-6.8662	64.2985	0.8782	49.1273	-9.7222	196.7557
			S	6	0.0182	0.5629	–	–	0.0008	1.3900	–	–	0.8714	0.2085	–	–
		K-Means	C	6	0.0186	–0.5276	-16.9285	25.3018	0.0008	10.8736	–14.8211	53.9639	0.8762	20.4250	–15.0719	96.2733
			S	6	0.0183	–0.1200	–	–	0.0008	0.0082	–	–	0.8696	0.0012	–	–
Yes	21	FCM	C	6	0.0171	0.3752	-35.7853	44.6774	0.0007	11.5802	-39.9674	62.1459	0.8804	40.0197	-56.5914	213.4748
			S	6	0.0182	0.5629	–	–	0.0008	1.3900	–	–	0.8714	0.2085	–	–
		K-Means	C	6	0.0174	2.8729	-13.0775	24.2522	0.0007	18.0812	-14.8211	53.5044	0.8860	30.2799	–15.0719	95.4536
			S	6	0.0183	–0.1200	–	–	0.0008	0.0082	–	–	0.8696	0.0012	–	–

5 Conclusion

This paper proposes a methodology to assess the potential of clustering methodologies to refine power consumption forecasting in smart buildings. The case study revealed that FCM and K-Means clustering can improve forecasting accuracy compared to baseline models, though benefits vary across different contexts and are not uniformly distributed. Employing Best Features (extracted from correlation degree) and silhouette analysis was critical in defining the optimal number of clusters, showing a preference for fewer clusters with the entire dataset and more clusters with only best features. The results suggest that a tailored approach to feature selection and cluster determination can enhance model performance, but for some clusters only. The combined model (one model per context) often outperformed the single model strategy (including contextual information as new features).

Future work will explore additional timeframes (e.g., 30, 60, or 90 days), considering seasonal factors and changes in user behavior, datasets from other smart buildings to validate and refine the approaches discussed, and alternative clustering methodologies that can offer improved performance. These efforts will examine the methodology's applicability to energy consumption and other characteristics of smart buildings.

Acknowledgments. This work has received funding from FEDER Funds through COMPETE program and from National Funds through (FCT) under the project PRECISE (PTDC/EEI-EEE/6277/2020), DOI 10.54499/PTDC/EEI-EEE/6277/2020. The authors acknowledge the work facilities and equipment provided by GECAD research center (UIDB/00760/2020), DOI:10.54499/UIDB/00760/2020 to the project team. Brígida Teixeira is supported by FCT with Ph.D. Grant 2020.08174.BD, DOI:10.54499/2020.08174.BD.

References

1. Miraftabzadeh, S.M., et al.: Advances in the application of machine learning techniques for power system analytics: a survey. Energies **14**, 4776 (2021). https://doi.org/10.3390/en14164776
2. Trotta, G.: An empirical analysis of domestic electricity load profiles: who consumes how much and when? Appl. Energy **275**, 115399 (2020). https://doi.org/10.1016/j.apenergy.2020.115399
3. Miraftabzadeh, S.M., et al.: K-Means and alternative clustering methods in modern power systems. IEEE Access **11**, 119596–119633 (2023). https://doi.org/10.1109/ACCESS.2023.3327640
4. Jozi, A., et al.: Contextual learning for energy forecasting in buildings. Int. J. Electr. Power Energy Syst. **136**, 107707 (2022). https://doi.org/10.1016/j.ijepes.2021.107707
5. MacQueen, J.: Some methods for classification and analysis of multivariate observations. In: Proceedings of the Fifth Berkeley Symposium on Mathematical Statistics and Probability, Volume 1: Statistics, pp. 281–298. University of California Press (1967)
6. Acquah, M.A., et al.: Spatiotemporal sequence-to-sequence clustering for electric load forecasting. IEEE Access **11**, 5850–5863 (2023). https://doi.org/10.1109/ACCESS.2023.3235724

7. Huang, S., et al.: Multi-scene design analysis of integrated energy system based on feature extraction algorithm. Energy Rep. **8**, 466–476 (2022). https://doi.org/10.1016/j.egyr.2022.03.161
8. Cui, M., Wang, J., Yue, M.: Machine learning-based anomaly detection for load forecasting under cyberattacks. IEEE Trans. Smart Grid. **10**, 5724–5734 (2019). https://doi.org/10.1109/TSG.2018.2890809
9. Dunn, J.C.: A fuzzy relative of the ISODATA process and its use in detecting compact well-separated clusters. J. Cybern. **3**, 32–57 (1973). https://doi.org/10.1080/01969727308546046
10. Liu, F., et al.: A hybrid short-term load forecasting model based on improved fuzzy C-Means clustering, random forest and deep neural networks. IEEE Access **9**, 59754–59765 (2021). https://doi.org/10.1109/ACCESS.2021.3063123
11. Rousseeuw, P.J.: Silhouettes: a graphical aid to the interpretation and validation of cluster analysis. J. Comput. Appl. Math. **20**, 53–65 (1987). https://doi.org/10.1016/0377-0427(87)90125-7
12. Li, L., et al.: Hyperband: A Novel Bandit-Based Approach to Hyperparameter Optimization http://arxiv.org/abs/1603.06560 (2018). https://doi.org/10.48550/arXiv.1603.06560
13. Pedregosa, F., et al.: Scikit-learn: machine learning in python. J. Mach. Learn. Res. **12**, 2825–2830 (2011)
14. Chen, T., Guestrin, C.: XGBoost: a scalable tree boosting system. In: Proceedings of the 22nd ACM SIGKDD International Conference on Knowledge Discovery and Data Mining, pp. 785–794 (2016). https://doi.org/10.1145/2939672.2939785

Development and Assessment of a System to Help Students Improve Self-compassion

Masashi Koizumi[1], Mieko Ohsuga[2](✉), and Juan Manuel Corchado[3]

[1] Graduate School of Robotics and Design, Osaka Institute of Technology, Osaka, Japan
[2] Faculty of Robotics and Design, Osaka Institute of Technology, Osaka, Japan
mieko.ohsuga@oit.ac.jp
[3] Bisite Research Group, University of Salamanca, Salamanca, Spain
corchado@usal.es

Abstract. Mental health issues are becoming more prevalent among university students. The mindful self-compassion (MSC) training program, which was introduced to address this issue, has shown some efficacy. However, many people, particularly Japanese people, have difficulty recalling visual imagery or feel uncomfortable or resistant to treating themselves with compassion. This study proposes and develops a system that uses virtual space and avatars to help individuals improve their self-compassion. In the proposed system, the user first selects an avatar of a person with whom to talk (hereafter referred to as "partner"), and then talks about the problem to the avatar of his/her choice. Next, the user changes viewpoints and listens to the problem as the partner's avatar and responds with compassion. Finally, the user returns to his/her own avatar and listens to the compassionate response spoken as the partner avatar. We first conducted surveys to understand the important system components, and then developed prototypes. In light of the results of the experiments, we improved the prototype by introducing a generative AI. The first prototype used the user's spoken voice as it was, but the improved system uses the generative AI to organize and convert the voice and present it. In addition, we added a function to generate and add advice with compression. The proposed system is expected to contribute to the improvement of students' self-compassion.

Keywords: Self-compassion · Virtual Space · Avatar · Generative AI

1 Introduction

Recently, student mental health has become a growing concern [1, 2]. In particular, studies have shown that the prevalence of depressive symptoms is higher in Asian countries [3], the rate of negative self-evaluation [4] and self-esteem is higher in Western countries and lower in Japanese people [5]. High and low self-esteem have a significant impact on mental health, but excessively high self-esteem has been linked to narcissism, prejudice, and discrimination [6]. As an alternative measure, the psychological concept of self-compassion (SC) has recently attracted attention [7]. To strengthen SC, a training program called mindful self-compassion (MSC), which uses meditation and visual

R. Chinthaginjala et al. (Eds.): DCAI 2024, LNNS 1259, pp. 43–52, 2025.
https://doi.org/10.1007/978-3-031-82073-1_5

imagery, was held [8]. In MSC training, individuals engage in self-talk by imagining a kindly person and talking to that person about their concerns and then responding with compassion as that person. However, many people struggle to recall images or find compassion as uncomfortable. In this study, we propose a system that enables such individuals to easily engage in MSC by simulating an MSC session, wherein they have conversations with themselves by visually presenting themselves and their partner's avatars in a virtual space. In particular, we intend to contribute to the improvement of students' SC and make the system more user-friendly in everyday life by comparing and implementing elemental technologies into a prototype, and repeatedly evaluating and refining the system to make it simpler and more versatile. There have been many studies on VR and AI for enhancing self-compassion [9, 10]. However, most of them are related to coaching from the perspective of others using chatbots or avatars, and those applied to support self-compassion are still few. One study with a concept similar to our proposed system is a work based not on MSC but on cognitive-behavioral therapy, using avatars [11]. It emphasizes the diversity of individual characteristics and the importance of changing avatars and VR environments to adapt to these characteristics.

2 Component Survey

To develop the prototype, we selected the system's components based on a survey of 23 university students in their 20s and 30s regarding the devices they used in their daily lives, their avatar preferences, and preferred secure locations (Ethical Approval No. 2022-31). Therefore, "(c) Notebook PC" was selected from among (a) smartphones, (b) tablets, (c) notebook PCs, (d) home game consoles, and (e) HMDs. The avatars presented in the virtual space were: (A) a real avatar made from a photograph, (B) a deformed 3D avatar, (C) an animated avatar, (D) an illustrated avatar, and (E) an animal avatar; among these we chose "(C) an animated avatar" (see Fig. 1). Furthermore, we prepared three types of virtual spaces for self-talk: (1) a room with furniture, (2) a simple room, and (3) a natural environment. This allowed the user to choose the environment according to their psychological state when using the system (see Fig. 2). Next, to investigate the user's resistance to interacting with an animated avatar, we considered demonstrating the system on our university's open campus and conducted a survey of 25 high school students (Ethical Approval No. 2022-33). We conducted a free discussion session with high school students by presenting (I) both avatars and real people on a projector, (II) only avatars on the screen, and (III) only people on the screen (see Fig. 3). Finally, we surveyed the participants to determine whether they were more comfortable conversing with avatars or people. The survey results demonstrated that there was no difference in the conversations between the avatars and real people, and it was confirmed that using avatars for the prototype posed no problems.

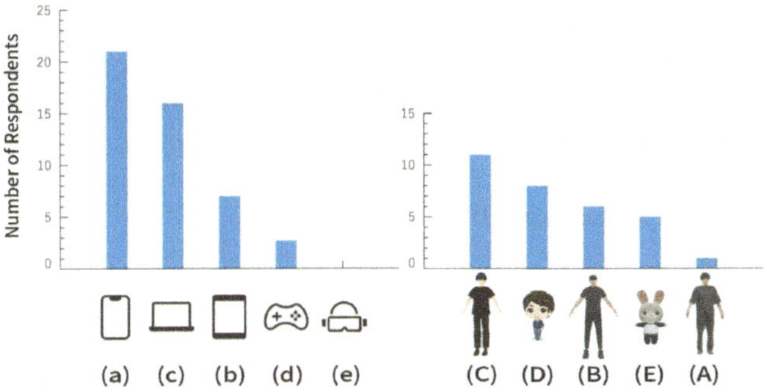

Fig. 1. Devices and avatar survey results.

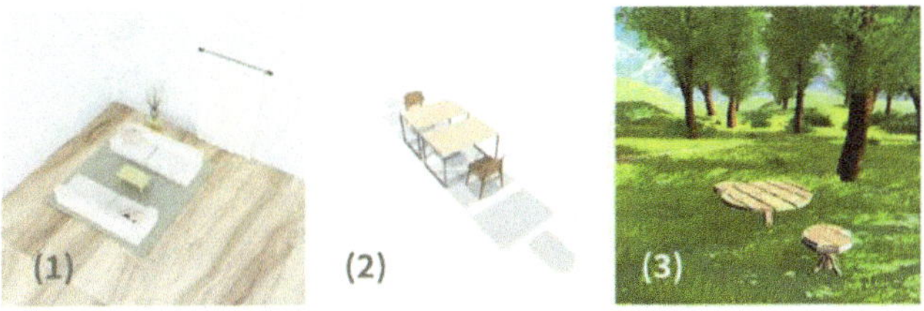

Fig. 2. Three types of virtual spaces.

Fig. 3. Survey procedure for resistance to talking to avatars.

3 System Development

We identified the components based on the survey results and developed Prototype 1 based on a work scenario using the following visual images: (1) talk one's own problems to the avatar of the intimate someone (Hereafter referred to as "partner") displayed on the screen; (2) switching viewpoints and assuming the role of the partner to listen to own problems; (3) taking the role of the partner, to own avatar displayed on the screen, and listening to one's own problems; and taking on the role of the partner and respond to own avatar on the screen with compassion about own concerns, (4) switching viewpoints and assuming entering in one's own avatar, the user listens to the partner avatar's answers given with compassion. The self-talk systems (1)–(4) were implemented using Unity software (see Fig. 4). The motion capture tool for moving avatars on Unity software was selected based on the following criteria: open sound control (OSC) communication, ability to control multiple avatars, and ability to operate the system using only a laptop PC to simplify the system. VSeeFace was used as the motion capture tool based on the following selection criteria. To realize this scenario, the user must hear themselves; therefore, we designed the system to record the screen when the user speaks. The recording data were saved in MP4 format in a local folder. The recording can be started, stopped, and played back at the touch of a button (see Fig. 5). The virtual space was created using Unity assets and Blender software.

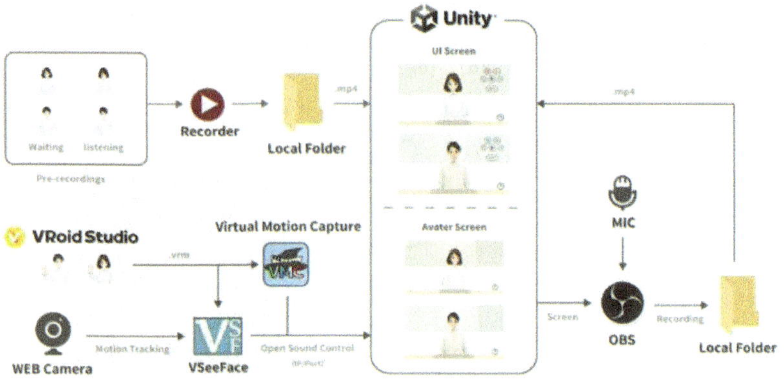

Fig. 4. System Configuration.

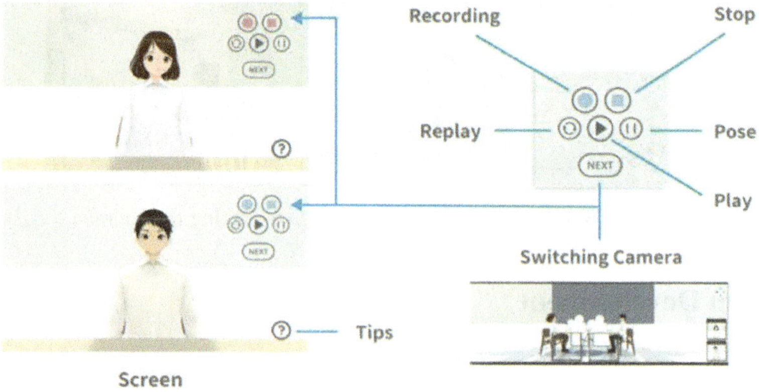

Fig. 5. Operation buttons on the game screen.

4 Preliminary Experiments on Avatars

Using Prototype 1, we conducted a preliminary experiment to confirm the appropriateness of the animated avatars presented to seven university students by comparing (A) real, (B) deformed, and (C) animated avatars (Ethical Approval No. 2022-54). The results showed that, for the user's avatar, the order of resistance was (C), (B), and (A), and it was (C), (A), and (B) for the partner's avatar (see Table 1). However, because some participants showed different preferences for their own and their partner's avatars, we decided to allow them to create their own character instead of an identical avatar.

Table 1. User's subjective resistance to avatars.

User's avatar		Low ← Resistance → High			Partner's avatar		Low ← Resistance → High		
		1	2	3			1	2	3
(A)		1	3	6	(A)		2	7	1
(B)		0	6	4	(B)		0	1	9
(C)		9	1	0	(C)		8	2	0

5 Experiment 1

Experiment 1 was conducted with 10 students from our campus to assess the prototype (Ethics Approval No. 2022-54). The experiment investigated the acceptability of self-talk using avatars and the possibility of its continued use and identified areas for improvement. Scores were obtained on the self-compassion scale short form [12], which is a measure of SC, before the experiment began, and the experiment was conducted in the following steps: (1) character making, (2) explanation of how to use the prototype, and (3) practice of the Prototype 1 work (see Fig. 6). Following the experiment, the participants were asked to evaluate each scenario on four levels in terms of resistance, desire for continued use, and expected usefulness. Free comments were also collected. This experiment produced a subjective assessment of resistance and physiological indices that change with emotion, yielding a multifaceted assessment. In particular, the participants were asked to wear a wristband-type wearable device (E4, Empatica) that is commercially available for research to measure skin electrical activity and pulse rate. The pulse rate increased in scenarios (1) and (3) and decreased in scenarios (2) and (4). The increase in pulse rate reflects the physical load caused by speech and emotional changes (mainly tension), whereas the decrease in pulse rate reflects the attentional focus caused by listening. The skin conductance level increased monotonically in most cases, but after the scenario began, it increased rapidly and the skin conductance response also increased, indicating an emotional change caused by increased tension (see Fig. 7). The participants who experienced emotional change in (3) commented that they felt uncomfortable because their speech was not coherent, and in (4) they felt a sense of resistance because they disliked their own voices. This implies that there is potential for improvement, such as modifying sentences according to context, even when the user does not speak well or making the system output in a voice that is not their own. In the free description, positive comments regarding acceptability were obtained, such as "I feel positive by verbalizing my problems" and "I calmly think about solutions by talking to myself when I cannot talk to others about my problems." Furthermore, nine participants expressed a desire for continued use of the system and eight expressed expectations for its efficacy, confirming a certain level of acceptance of the system.

Fig. 6. Experimental Scene.

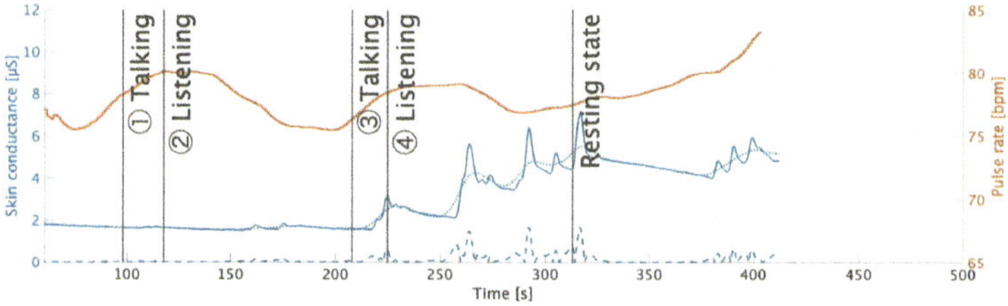

Fig. 7. Changes in skin conductance level and pulse rate.

6 System Improvements

Based on the improvements identified in Experiment 1, we developed Prototype 2 using a generative AI (ChatGPT) to organize speech context and convert voices using a synthetic voice tool (VOICEVOX) (see Fig. 8). The user can feel less resistance and discomfort when using the system by using ChatGPT to "organize" the speech content when the user fails to speak properly or includes dog words in the speech in scenarios (1) and (3). Furthermore, in (4), we decided to add a bit of advice created by the generated AI. This is because we thought that the message heard in (4) would sound more like a message with compassion from others. The above system was implemented using Unity. By clicking the "record" button, the speech in (1) and (3) is saved in WAV format, the recorded audio file is sent to Whisper, a speech-to-text model, and the transcribed text data are sent to ChatGPT. The text data processed by the ChatGPT prompt were sent to VOICEVOX, a synthetic voice tool, causing the avatar to perform lip-sync when reading aloud. The avatar can also be attached to the program using Auto Blink and Mixamo for a more natural avatar expression. Having the avatar speak with a synthesized voice eliminates the problem of people who are uncomfortable with their own voice or hearing themselves from another avatar. Furthermore, the system becomes simpler because tracking and recording are no longer necessary. Additional features include a tutorial and a guide function that presents the next scenario on the screen to simplify the system for first-time users (see Fig. 9).

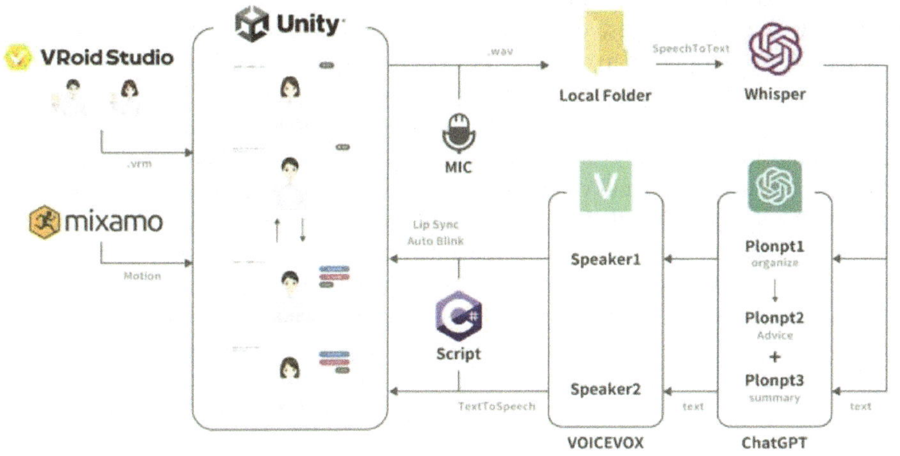

Fig. 8. System Configuration.

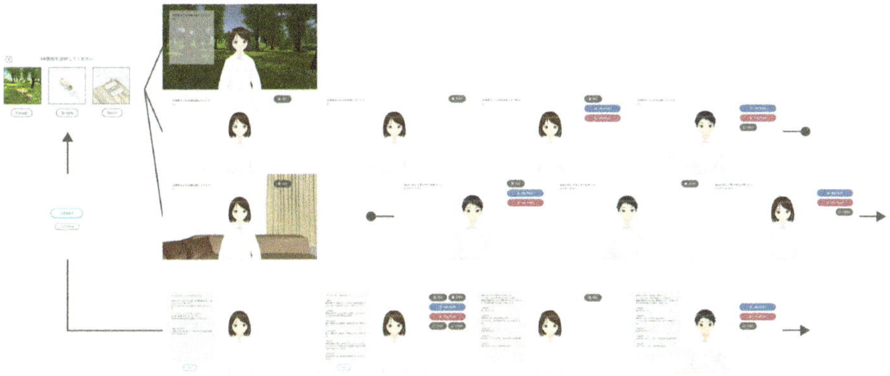

Fig. 9. Screen Transition.

7 Experiment 2

We conducted Experiment 2 to assess Prototype 2 using ten university students (Ethics Approval No. 2022-54). The experimental procedure and evaluation methods were almost the same as in Experiment 1, with the only addition of a procedure in which participants were asked to select character voices for themselves and their partner's avatar. The results of the experiment showed that all participants wanted to continue using the system and found it useful after the modification, although not all of them did before the modification. The participants with low SC commented the following in the free description: "I would like to use this system because I am to do my best if someone speaks kindly to me about my problems when I am working hard" and "I would like to use this system because I have a personality that makes it difficult to talk about my problems with others." These comments indicate that the system has a certain level of acceptability even for participants who are not comfortable talking to others with a sense of compassion. The comments were "Using the system is not complicated," "I felt like I was talking to by other person because the tone and phrasing were slightly different from what I said," "I felt that the content added was more persuasive than my own words," "I felt that my chosen avatar and voice were more convincing than my own," "I felt that the system was easy to use," and "I felt that I could interact with the avatar and voice I chose without any resistance." These results indicate that system improvements, regardless of SC level, improved system acceptability. However, in terms of user interface, some users commented that the buttons were easily mishandled, it was difficult to judge whether a button was pressed during the system processing, the AI processing time was too long, and the advice in (4) felt a bit too much.

8 System Re-improvement

The ChatGPT prompts were modified, and Prototype 3 was created as the final stage of the proposed system. The UI was improved by displaying only the buttons necessary for the progress of the scenario and by presenting a waiting screen while ChatG- PT is processing (see Fig. 10). Furthermore, the total output characters were limited to reduce the processing time of ChatGPT. The processing time required to start speaking was calculated to be "approximately 10 – 15 s per 100 characters." Assuming a maximum waiting time of "30 s" we set the limit at "200 characters". To address the issue of the advice feeling like too much, we changed the prompt from "advice" to "compassionate" words. This is because we believe that the core of this system is not an advice system but the ability to think and speak compassionate words on one's own. By changing prompts, it is expected that even in situations where people who are not accustomed to interacting with compassionate people will be able to express more compassion than they have expressed verbally.

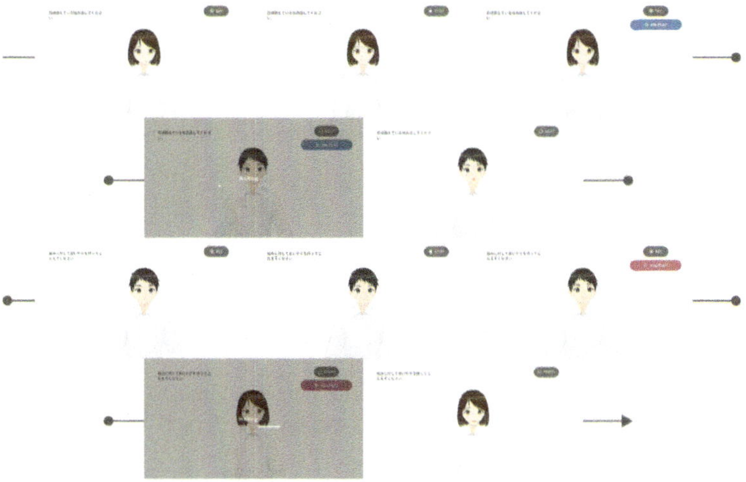

Fig. 10. Screen Transition after Improvement.

9 Limitation

Due to the small number of participants in the experiment and the biased sample, which consisted solely of students from a specific university, it was not possible to examine factors such as personal characteristics (age, education, income, etc.) or psychological profiles (personality, sociability, etc.). Furthermore, the results obtained may have been influenced by the participants' physical and emotional states at the time of the experiment. Because of these limitations, we chose not to conduct well-designed quantitative tests but instead gathered insights from free-form comments. However, these insights cannot be considered universal. Additionally, since each participant only took part in the experiment once, we were unable to assess the medium- and long-term effects, which is another issue to be addressed in future research.

10 Conclusion

The self-talk system proposed and developed in this study can be easily used without special equipment and has a certain level of acceptability regardless of the user's SC level, indicating the possibility of reducing the user's resistance to comforting themselves with others. We believe that this system will contribute to the improvement of students' SC because it is expected to be used continuously and will be highly usefull. In order to increase the acceptability of the system, it is important to adapt to individual differences. Specifically, the following three measures can be considered. First, users should have the option to choose a realistic avatar. Second, the system should be flexible enough to accommodate various devices, such as smartphones and HMD. Third, the responses from ChatGPT should be tailored to the user's personality to enhance acceptability. In the future, we expect all operations such as avatar creation and character voice settings to be completed within Unity, increasing its versatility and enabling users' daily use.

Acknowledgment. The authors would like to thank all those who participated in the experiments of this study. The second prototype of this study was developed during the first author's short stay at the University of Salamanca. We thank the BISITE research group for their helpful advice and warm encouragement. We thank Prof. Kenji Matsui for suggesting the opportunity to study at the University of Salamanca and for reading and commenting on the manuscript of the paper.

References

1. Shorey, S., Ng, E.D., et al.: Global prevalence of depression and elevated depressive symptoms among adolescents: a systematic review and meta-analysis. Br. J. Clin. Psychol. **61**(2), 287–305 (2022)
2. Auerbach, R.P., Mortier, P., et al.: WHO world mental health surveys international college student project: prevalence and distribution of mental disorders. J. Abnorm. Psychol. **127**(7), 623 (2018)
3. Steptoe, A., Ardle, J., et al.: Depressive symptoms, socio-economic background, sense of control, and cultural factors in university students from 23 countries. Int. J. Behav. Med. **14**, 97–107 (2007)
4. Heine, S.J., Lehman, D.R., et al.: Is there a universal need for positive self-regard? Psychol. Rev. **106**(4), 766 (1999)
5. Schmitt, D.P., Allik, J.: Simultaneous administration of the Rosenberg self-esteem scale in 53 nations: exploring the universal and culture-specific features of global self-esteem. J. Pers. Soc. Psychol. **89**(4), 623 (2005)
6. Neff, K.D.: Self-Compassion [Translated from Japanese.], Kongo Publishing (2019)
7. MacBeth, A., Gumley, A.: Exploring compassion: a meta-analysis of the association between self-compassion and psychopathology. Clin. Psychol. Rev. **32**(6), 545–552 (2012)
8. Neff, K.D., Germer, C.K.: Mindful Self-Compassion Workbook [Translated from Japanese.], Seiwa Shoten. Kohki Arimitsu, Yasuhiko Aoki (2021)
9. Žilinský, T., Halamová, J.: The effects of virtual reality on enhancement of self-compassion and self-protection, and reduction of self-criticism: a systematic review. Int. J. Environ. Res. Public Healthnt. J. Environ. Res. Public Health **20**(3), 2592 (2023)
10. Morrow, E., Zidaru, T., et al.: Artificial intelligence technologies and compassion in healthcare: a systematic scoping review. Front. Psychol. **13**, 971044 (2023)
11. Halim, I., Stemmet, L., et al.: Individualized virtual reality for increasing self-compassion: evaluation study. JMIR Mental Health **10**, e47617 (2023)
12. Arimitsu, K., Aoki, Y., et al.: Construction and validation of a short form of the Japanese version of the self-compassion scale. Komazawa Ann. Reports Psychol. **18**, 1–9 (2016)

Leveraging Machine Learning and Deep Learning Models for Enhanced Stock Price Prediction: A State-of-the-Art Analysis

Safae Belamfedel Alaoui[1(✉)], Abdelatif Hafid[2], Mhamed Sayyouri[1], and Mohamed Rahouti[3]

[1] LISA, Ecole Nationale des Sciences Appliquées (ENSA), 72 Fez, Morocco
safae.belamfedelalaoui@usmba.ac.ma

[2] ESISA Analytica, Ecole Supérieure d'Ingénierie En Sciences Appliquées (ESISA), 505 Fez, Morocco

[3] Computer and Information Science Department, Fordham University, New York, USA

Abstract. The prediction of stock prices has recently gained considerable attention as a complex and challenging issue within the realms of economics and finance. Stock prices are affected by various factors, such as the business environment, stock market operations, inflation, and unexpected events. Since the stock market is volatile and nonlinear, finding the most effective model to forecast stock prices is one of the most challenging problems. Researchers have increasingly explored various Machine Learning (ML) and Deep Learning (DL) models to address this issue due to their capacity to handle time series data and nonlinear patterns. These models often outperform traditional approaches in predicting stock prices with high accuracy and lower root mean square error (RMSE). This paper reviews various works that have utilized ML approaches for stock price prediction, covering research published between 2017 and 2023. This literature review discusses various techniques, their performance, limitations, and future work. We assess the latest techniques in many studies, including ML and DL models. The findings of this review conclude that Neural Networks (NNs) are the most commonly used approaches in predicting stock prices due to their effectiveness in detecting complex patterns in financial data.

Keywords: Stock price prediction · Machine Learning · Deep Learning · Neural Networks · Sentiment Analysis

1 Introduction

Nowadays, stock market prediction is becoming an interesting field thanks to its impact on the economy of a country and social organizations. There are two approaches to predicting the stock market price: technical and fundamental. Predicting stock prices helps in managing risks and influencing investment decisions, financial planning, and market analysis. However, developing an effective

R. Chinthaginjala et al. (Eds.): DCAI 2024, LNNS 1259, pp. 53–64, 2025.
https://doi.org/10.1007/978-3-031-82073-1_6

prediction model is challenging because the market is dynamic and nonlinear. [1] illustrated in their research that various factors, such as economic, political, psychological, and company-specific variables, impact stock markets. In the literature various methods have been explored, including traditional approaches (e.g., statistical arbitrage, technical indicators) and advanced technologies (e.g., Machine Learning models), to predict accurate stock prices. [2], compared traditional statistical methods including Simple Moving Average (SMA), Weighted Moving Average (WMA), Exponential Smoothing, and Naive Approach, with ML techniques for stock price prediction. They conclude that Multi-Layer Perceptron (MLP) and Long Short Term Memory (LSTM) models, significantly outperform traditional approaches.

Over the past few years, review studies that address the application of the most advanced ML algorithms in the stock price prediction and classification field have gained considerable attention [3]. [4], provided a recent assessment of ML approaches used in financial market prediction. Their study performs a literature review of various selected articles. The researchers reveal that Support Vector Machine (SVM) and NNs are the most widely used models for stock price prediction thanks to their ability to handle non-linear and complex patterns characteristic of financial data.

Similarly, [5] reviewed more than 50 research articles on stock market prediction. The authors categorize the methodologies into various prediction and clustering techniques, including Bayesian models, fuzzy classifiers, artificial neural networks (ANN), SVM classifiers, and other neural network approaches. They identify ANNs and fuzzy-based techniques as the most commonly utilized methods for achieving effective prediction. [6], offered a brief review and categorization of stock market forecasting approaches, including deep learning (e.g., LSTM, RNN, and CNN), statistics (e.g., ARIMA), and their hybrid combinations. The authors highlighted the potential of deep learning models, particularly when combined in hybrid forms with sentiment analysis to capture dynamic changes in stock performance, compared to the classic ARIMA, [7], is another recent work that presents a critical analysis of state-of-the-art ML techniques in stock price prediction, ranging from traditional techniques to advanced NNs. The authors focus in their study on analyzing the effectiveness and challenges of each method in the context of stock market prediction, to offer a comprehensive overview of the current state of research in this field.

In this review, we present a summary of advanced approaches and commonly employed ML techniques for stock price prediction, such as Random Forest (RF), SVM, k-Nearest Neighbor (KNN), Naïve Bayes(NB), LSTM, and Gated Recurrent Network (GRU), as well as a combination of two or more hybrid approaches, statistical approaches (e.g., FB Prophet (Facebook Prophet), and ARIMA (autoregressive integrated moving average), and sentiment analysis. Moreover, a list of the most used datasets (e.g., S&P500 and Yahoo Finance data) and evaluation metrics (e.g., RMSE and the mean absolute error (MAE)) is described. Finally, we discuss the main open challenges within this field and promising research directions for future studies on stock price predictions.

Figure 1 presents a taxonomy of various approaches used in stock price prediction, classifying various methodologies based on key characteristics and strategies into three main categories: articles using ML and DL algorithms, novel hybrid approaches, and sentiment analysis.

The reminder of the paper is organized as follows: Sect. 2 presents a summary of the state-of-the-art ML and DL approaches for stock price prediction. Section 3 compares our review with seminal works in literature, and discusses the limitations and future research directions of this review. Finally, Sect. 4 concludes the paper.

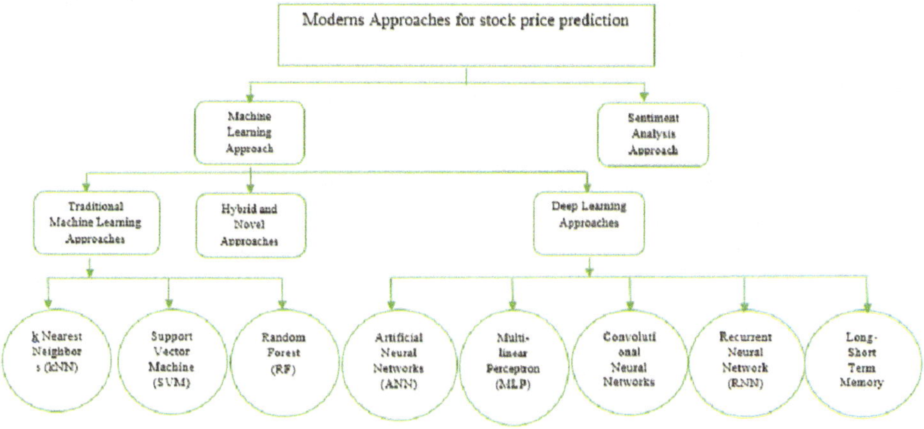

Fig. 1. Taxonomy of State-of-the-art Approaches in Stock Price Prediction.

2 Literature Review

This section presents a comprehensive examination of the most commonly utilized machine learning approaches for stock price prediction in the literature, as depicted in Fig. 1.

2.1 Sentiment Analysis Approach

Sentiment analysis consists of interpreting human emotions and sentiments, especially in the context of stock markets, as it directly affects stock prices. The authors have examined different data sources, including StockTwits, Twitter, and Yahoo Finance, to identify their impact on the markets.

[8] proposed models based on sentiment analysis and deep learning models, the authors used various methods including the authors used various methods including RNN and LSTM. The dataset used contains five years of daily stock prices for S&P500 companies and over 265,000 related financial news articles.

To evaluate the model's performance, the authors employed the Mean Absolute Percentage Error (MAPE) metric, and they concluded that the RNN, incorporating news sentiment, achieved better accuracy. The paper demonstrates that incorporating news sentiment with historical stock price data enhances the accuracy of stock price predictions.

[9] proposed a hybrid model that combines sentiment analysis with LSTM to predict stock market prices. The authors extracted the sentiment data by analyzing comments on the stock market and then integrated this data with Apple Inc.'s historical stock data. To evaluate the model's performance, the researchers used various metrics, including MSE, MAPE, MAE, and R2. This study highlights the significant role of sentiment analysis in enhancing the prediction performance of models.

[10] explored the StockTwits data to integrate sentiment analysis into the stock price prediction field. The authors used various machine learning models, including NB, SVM, and Logistic Regression (LR), with text processing techniques such as a bag of words, bigram, Term Frequency-Inverse Document Frequency (TF-IDF), and Latent Semantic Analysis (LSA) for sentiment classification. The dataset was collected from five different companies. The evaluation of the models used demonstrates that Logistic Regression (LR) with TF-IDF and SVM with LSA performed better than other combinations, with an accuracy between 75% and 85% for all five companies. The researchers conclude that integrating sentiment analysis with traditional stock time series data can significantly enhance prediction results.

[11] highlighted the potential of machine learning and sentiment analysis in predicting cryptocurrency price movements. They combined sentiment analysis with machine learning models, including MLP, SVM, and RF, to predict trends prices of Litecoin, Ethereum, Bitcoin, and Ripple. The researchers used Twitter data and market trends as inputs. They evaluated model performance using precision, accuracy, recall, and the F1 score. The results conclude that the MLP model outperforms other models in predicting the direction of market movements for each cryptocurrency.

Table 1 presents a brief comparative of sentiment analysis methods, providing a quick overview of the various approaches used in this analysis.

Table 1. Comparative Analysis of Various Methods Utilized in Sentiment Analysis Research

References	Methods								
	RNN	LSTM	ARIMA	F. Prophet	NB	SVM	RF	LR	MLP
[8]	✓	✓	✓	✓					
[9]		✓							
[10]					✓	✓		✓	
[11]							✓	✓	✓

2.2 Machine Learning Approaches

Traditional Machine Approach. In the last few years, ML has significantly impacted the financial markets, particularly in predicting stock prices. The widely traditional machine learning algorithms used in the literature include Random Forest (RF), KNN, and SVM. A literature survey of some traditional ML methods applied in stock price prediction is presented next.

[12] have developed a machine learning-based classification methodology to predict whether market trends will be upward or downward. The authors utilize key features from historical data, such as volume, and technical metrics as input for the RF classifier. The proposed methodology is validated through a comprehensive analysis of Bitcoin's closing prices. The authors evaluate the process through various simulations, including a backtesting strategy, and conclude that their approach achieved an accuracy of over 86 % in providing buy and sell signals.

[13] demonstrate the effectiveness of combining traditional technical indicators with traditional machine learning algorithms, such as SVM and LR, and advanced ML algorithms (e.g., ANN), within a time series framework in stock market prediction. These algorithms have been trained on fifty stocks of India's National Stock Exchange (NSE) from 1 January 2013 to 31 December 2018. The suggested combination enhances prediction accuracy with an average of 89.93 % for SVM, 76.68 % for ANN, and 86.98% for LR.

[14] used RF and ANN models to predict the next day's closing price of the stock. The authors explored 10-year datasets from Yahoo Finance for five different companies. To assess the effectiveness of these models, the authors employed various metrics, including RMSE, MAPE, and Mean Bias Error (MBE). The obtained results indicate that ANN achieves the highest performance, yielding values of 0.42, 0.77, and 0.013 for RMSE, MAPE, and MBE, respectively.

In their study, [15] used five supervised ML models to predict stock price movements, including KNN, RF, SVM, Softmax algorithms, and Naive Bayes. These models were trained on datasets from Yahoo Finance and Quandl. Moreover, the researchers used twelve different technical indicators, such as the Moving Average and Relative Strength Index, to enhance prediction precision. The authors employed accuracy and F1 score metrics to evaluate the models, and the results indicate that the RF approach, with a reduced size of the dataset, outperformed other models in predicting stock market prices.

Table 2 shows a brief comparative of various methods employed in research utilizing traditional machine learning models.

Deep Learning Approach. DL models demonstrate their performance in the prediction of stock price, by capturing the complex and nonlinear character of trends in the stock market. [16] conducted an extensive examination of DL models, including LSTM, GRU, CNN, and ELM, to forecast both one and four-step stock prices of the Bombay Stock Exchange Index (S&P BSE-BANKEX) using data from Yahoo Finance between July 12, 2005, and November 3, 2017. The authors evaluated prediction accuracy through measures such as RMSE, Median Absolute Percentage Error (MdAPE), and Directional Accuracy (DA).

Table 2. Comparative Analysis of Various Methods Utilizing Traditional Machine Learning Models.

References	Methods					
	RF	SVM	KNN	LR	ANN	NB
[12]	✓	✓		✓		
[13]		✓		✓	✓	
[14]	✓				✓	
[15]	✓	✓	✓			✓

[17] used an advanced LSTM approach to forecast stock market prices. They trained the model on three historical datasets from the Shanghai Composite Index and stocks including Petro China and ZTE. In the evaluation process, the authors used RMSE and MAE metrics. The proposed model achieves over 95 % prediction accuracy, outperforming LSTM and deep current network models.

In [18], researchers introduced a new method for forecasting stock market prices using DL algorithms. They propose two models: ANN and CNN, trained on India's NSE dataset for the period from April 2008 to April 2018, focusing on the NIFTY price index. The performance of the proposed approach was validated using diverse metrics such as accuracy, MSE, and MAE. The ANN model achieved an accuracy score of 97.66 %, while the CNN model outperformed the ANN model with an accuracy of 98.92 %.

[19], intended to illustrate the efficacy of using LSTM, and RNN, for stock market trend prediction. The authors used Google (GOOGLE) opening prices and Nike (NKE) stocks from the stock exchange in New York (NYSE). The study demonstrates that training the model with fewer datasets and more epochs can enhance the model's performance. For 100 epochs, the loss rate for GOOGLE was recorded at 4.97E-04, while for NKE, it was 8.74E-04.

[3] trained four ML models, including ANN, LSTM, SVM, and RF, on the iShares MSCI ETF United Kingdom exchange datasets between January 2015 and June 2018 to predict stock prices. They employed metrics such as MAE, MSE, and RMSE to assess the model's performance. The results demonstrated that LSTM showed higher performance with scores of 0.210350, 0.093969, and 0.306543 for MAE, MSE, and RMSE, respectively. This study demonstrates the capacity of DL models to provide robust predictions for complex financial time series data.

[20] presented a hybrid model combining LSTM and the Bi-Directional LSTM (BI-LSTM) techniques to predict stock market prices. They trained these models on Google stock market data from August 19, 2004, to October 4, 2019, and found that the BI-LSTM model outperforms the LSTM model in predicting stock prices, achieving the highest level of accuracy based on RMSE metric evaluations.

[21] introduced an improved LSTM network to predict Indian market stock prices, comparing stateful and stateless LSTM models. For evaluating the models, the authors used the RMSE metrics and found that using stateless LSTM

with only a single hidden layer achieves a higher RMSE value for the time series prediction.

[22] proposed a hybrid prediction approach using LSTM and GRU models. They trained the model on the time series dataset S&P 500. The LSTM generates the first prediction, which is passed to the GRU to obtain the final prediction. The proposed approach outperformed previous neural network approaches with an MSE value of 0.00098.

[23] employed four deep learning models, including MLP, RNN, LSTM, and CNN, for predicting stock market prices. They trained the models on data from Tata Motors and tested them on NSE and NYSE stocks. The evaluation results conclude that the CNN model outperforms other models, with MAPE values of 0.043, 0.0433, and 0.0532 for different stocks, demonstrating its superior ability to capture stock market trends.

Table 3, shows the various methods employed in research using traditional Deep Learning models. More specifically, we classify research papers based on the use of ANN, LSTM, GRU, CNN, ELM, RNN, or Bi-LSTM.

Table 3. Comparative Analysis of Various Deep Learning Models

References	Methods					
	ANN	LSTM	GRU	CNN	ELM	RNN	Bi-LSTM
[16]		✓	✓	✓	✓		
[17]		✓					
[18]	✓			✓			
[19]		✓				✓	
[3]	✓	✓					
[20]		✓					✓
[21]		✓					
[22]		✓	✓				
[23]		✓	✓				

Hybrid and Novel Approaches. The hybrid approach combines two distinct machine learning models or statistical and machine learning methods. This section describes some recent work that has used hybrid approaches in stock price prediction.

[24] proposed an interesting model that integrates machine learning and sentiment analysis through Twitter data for stock market predictions. The authors employ different machine learning techniques, including SVM, CNN, Wavelet CNN(W-CNN), MLP, and RF classifiers. These models are trained on the tweets dataset. The evaluation results demonstrate that the wavelet CNN algorithm performed better than other models in predicting stock prices.

[25] used a hybrid combination of Hidden Markov (HMM) and optimized LSTM networks to analyze the dynamics of Bitcoin's price movements. They

collected Bitcoin data from the Coin Base Exchange market and evaluated their model using MSE, RMSE, and MAE metrics. The authors conclude that their proposed model outperforms other models in predicting Bitcoin prices, with RMSE, MSE, and MAE values of 5.821, 33.888, and 2.510, respectively.

[26] suggest a hybrid deep learning framework, Stock GAN, for predicting stock market prices. Their proposed framework integrates a GAN with a Hybrid Prediction Algorithm (GAN-HPA). The authors demonstrate the performance of their proposed model using MAE, MSE, and correlation coefficient metrics(Corr). The findings demonstrate the effectiveness of their suggested approach, GAN-HPA, in predicting stock market prices. In [27], the authors have compared various ML algorithms, including MLP, CNN, and RNN, for predicting stock market trends using the S&P 500 data. The researchers demonstrate that CNN outperformed other models with an accuracy of 0.53 and propose a novel approach, wavelet-CNN, that outperforms CNN with an accuracy of 0.55.

In their study, [28]proposed a combination of LSTM and CNN models for predicting the next day's closing stock price, trained on the daily stock price for 7127 trading days. The proposed approach is compared with MLP, CNN, RNN, LSTM, and CNN-RNN models, using MAE, RMSE, and the R-square (R2). The results indicate that their approach surpasses other models, achieving 39.688 for RMSE, 27.564 for MAE, and 0.9646 for R2. The researchers [29] developed a stock price prediction method using the combination model of ARIMA-LS-SVM. For the evaluation process, the researchers demonstrated that their model showed higher accuracy (Acc) and lower prediction errors compared to RS-SVM (Rough Set-SVM), LS-SVM (Least Squares SVM), and traditional stock price predictions.

Table 4 presents various research methods incorporating hybrid approaches and methodologies, which have gained interest in research due to their ability to increase overall study efficiency.

Table 4. Different methods used in research using hybrid approaches.

References	Methods											
	CNN	W-CNN	MLP	SVM	RF	LSTM	HMM	GAN	RNN	RS-SVM	LS-SVM	AR-IMA
[24]	✓	✓	✓	✓	✓							
[25]						✓	✓					
[26]	✓					✓		✓				
[27]	✓		✓						✓			
[28]	✓		✓			✓			✓			
[29]										✓	✓	✓
[6]	✓					✓			✓			✓

2.3 Datasets and Evaluation Metrics

In the literature, various datasets have been processed using methods like feature selection, order reduction, and feature representation before being fed to machine learning models for stock price predictions. The two main approaches commonly used in the literature are classification and regression. Classification models are evaluated using accuracy, recall, precision, R2, and F1-score, while regression models use RMSE, MSE, and MAPE.

Table 5 summarizes the evaluation metrics, data sources, and input types used in different studies. The information contained in this table is a valuable reference for researchers and practitioners in this field.

Table 5. Summary of data sources, type of inputs, and evaluation measures most frequently used in the literature.

References	Data Source	Input	Metrics
[16]	S&P BSE-BANKEX, Yahoo	Stock prices	RMSE, MdAPE, DA
[17]	Shanghai Comp., PetroChina, ZTE	Stock prices	RMSE, MAE
[18]	NSE India	Stock market data	MAE, MSE, Acc
[19]	GOOGLE, Nike NYSE	Opening prices	MSE
[3]	MSCI UK	Closing prices	MAE, MSE, RMSE
[20]	Google stocks	Market data	RMSE
[21]	Indian stock prices	Stock prices	RMSE
[12]	Binance API	Bitcoin prices	Acc.
[15]	Yahoo Finance, Quandl	Stock prices	F1, Acc
[13]	NSE India NIFTY 50	Stock data	Acc.
[14]	Yahoo (5 cos.)	Market data	MAPE, MBE, RMSE
[27]	S&P 500	Market data	Acc.
[24]	Twitter	Sentiment analysis	RMSE, Acc, Log-Loss
[26]	NSE India	Stock prices	MAE, MSE, Corr
[9]	Apple stocks comments	Sentiment, Stock data	MSE, MAPE, MAE, R2, Acc.
[22]	S&P 500	Market data	MSE
[28]	Shanghai Comp. GEM	Closing prices	MAE, RMSE
[29]	Market segment	Market data	MSE, RMSE
[25]	Coinbase	Bitcoin prices	MSE, MAE, RMSE
[8]	S&P 500	Daily prices	MAPE, Acc.
[10]	StockTwits	Prices, Sentiment	Acc.
[11]	Twitter trends	Prices, Sentiment	Acc., Recall, F1

3 Comparison, Limitations, and Future Directions

The comparative analysis of different state-of-the-art models that have been used in stock price predictions demonstrates that there are various limitations of the existing models in the literature that need to be highlighted. Machine Learning models have shown their ability in predicting stock price, however exploring other approaches like sentiment analysis, or combining two or more approaches can significantly improve the model's performance. Also, giving more attention to the feature selection methods such as Principal Components Analysis (PCA), AutoEncoder (AE), and so on, or using hyperparameters including learning rates or regularization, etc. can achieve better prediction results. On the other hand, a large amount of historical data is available, which is also a major challenge. Traditional approaches such as the ARIMA model, which assumes a linear relationship between future, current, and past values of a given time series, are not suitable for modeling multivariate time series with non-linear residuals. Instead RNN, LSTM, and CNN approach are more suitable in the case of multivariate time series with non-linear residuals. While recurrent neural networks (RNNs), and LSTM networks can face the vanishing gradients problem during the training of deep networks. The researchers developed other architectures such as Gated Recurrent Units (GRUs) and Transformer models, which have successfully mitigated this problem.

This study focuses on traditional machine learning, deep learning, sentiment analysis, and hybrid approach, but there are other approaches that have not been presented in this review such as the Graph-Based Approach, time series forecasting models, and the combined model that used several approaches including, historical stock indicators and news, and should be examined in future work in stock price prediction.

In the field of stock market prediction using machine learning and data analysis techniques, a critical comparison of three seminal works reveals distinct methodological approaches. [30] review primarily focuses on the application of traditional ML and DL models, including Random Forest, Support Vector Machine, k-Nearest Neighbor, NB, LSTM, ARIMA, and GRU, and Artificial Neural Networks, due to their ability to handle nonlinear patterns and time series data. The paper does not extensively cover hybrid models or the integration of sentiment analysis from social media and financial news, which could enhance prediction accuracy. It also does not deeply explore novel approaches like GANs or the specific challenges related to feature extraction and model selection.

[6], deviate from other research in this field by integrating sentiment analysis with machine learning models, detailing how incorporating sentiment data from platforms like Twitter and financial news articles can significantly improve stock price predictions. It discusses a wide range of ML and DL models and their applications in this context. However, the researchers are less focused on the challenges related to large datasets, feature extraction, and the exploration of hybrid models that combine various ML techniques or integrate statistical methods with machine learning for improved stock price prediction.

In this study, we integrate both machine learning and deep learning models including LSTM, GRU, CNN, and ELM. This work not only encapsulates traditional and advanced neural network models but also uniquely integrates sentiment analysis and hybrid models. Moreover, it critically discusses the challenges and future directions in stock price prediction, offering a holistic and forward-looking view on applying these technologies in financial market analysis. This comparative review contributes to a panoramic overview that encompasses the multiple aspects of current and potential applications of machine learning in stock market forecasting.

4 Conclusion

Analyzing and predicting stock prices is challenging due to the difficulties in managing and extracting important information from data. In recent years, traditional neural network approaches have declined in effectiveness for accurately predicting prices. To address this, our review article proposes a comparative study of various ML and DL methods applied to stock price prediction. The analysis includes traditional ML algorithms (e.g., RF, KNN, SVM, Naive Bayes), as well as advanced models like CNN, ANN, and LSTM, and examines additional approaches such as sentiment analysis and hybrid models. The study also reviews several datasets and evaluation metrics used in recent literature. This comprehensive examination aims to assist researchers in choosing effective projects and improving their prediction accuracy.

References

1. Zhong, X., Enke, D.: Forecasting daily stock market return using dimensionality reduction. Expert Syst. Appl. **67**, 126–139 (2017)
2. Bhattacharjee, I., Bhattacharja, P.: Stock price prediction: a comparative study between traditional statistical approach and machine learning approach. In: 2019 4th International Conference on Electrical Information and Communication Technology (EICT), pp. 1–6 (2019)
3. Nikou, M., Mansourfar, G., Bagherzadeh, J.: Stock price prediction using DEEP learning algorithm and its comparison with machine learning algorithms. Intell. Syst. Acc., Finance Manag. **26**, 164–174 (2019)
4. Henrique, B.M., Sobreiro, V.A., Kimura, H.: Literature review: machine learning techniques applied to financial market prediction. Expert Syst. Appl. **124**, 226–251 (2019)
5. Gandhmal, D.P., Kumar, K.: Systematic analysis and review of stock market prediction techniques. Comput. Sci. Rev. **34**, 100190 (2019)
6. Shah, J., Vaidya, D., Shah, M.: A comprehensive review on multiple hybrid deep learning approaches for stock prediction. Intell. Syst. Appl. **16**, 200111 (2022)
7. Soni, P., Tewari, Y., Krishnan, D.: Machine Learning approaches in stock price prediction: a systematic review. J. Phys.: Conf. Ser. **2161**, 012065 (2022)
8. Mohan, S., Mullapudi, S., Sammeta, S., Vijayvergia, P., Anastasiu, D.C.: Stock price prediction using news sentiment analysis. In: 2019 IEEE Fifth International Conference on Big Data Computing Service and Applications (BigDataService), pp. 205–208 (2019)
9. Jin, Z., Yang, Y., Liu, Y.: Stock closing price prediction based on sentiment analysis and LSTM. Neural Comput. Appl. **32**, 9713–9729 (2020)
10. Gupta, R., Chen, M.: Sentiment analysis for stock price prediction. In: 2020 IEEE Conference on Multimedia Information Processing and Retrieval (MIPR), pp. 213–218 (2020)
11. Valencia, F., Gómez-Espinosa, A., Valdés-Aguirre, B.: Price movement prediction of cryptocurrencies using sentiment analysis and machine learning. Entropy **21**, 589 (2019)
12. Hafid, A., Hafid, A.S., Makrakis, D.: Bitcoin price prediction using machine learning and technical indicators. In: International Symposium on Distributed Computing and Artificial Intelligence, pp. 275–284 (2023)
13. Parray, I.R., Khurana, S.S., Kumar, M., Altalbe, A.A.: Time series data analysis of stock price movement using machine learning techniques. Soft. Comput. **24**(21), 16509–16517 (2020). https://doi.org/10.1007/s00500-020-04957-x

14. Vijh, M., Chandola, D., Tikkiwal, V.A., Kumar, A.: Stock closing price prediction using machine learning techniques. Procedia Comput. Sci. **167**, 599–606 (2020)
15. Kumar, I., Dogra, K., Utreja, C., Yadav, P.: A comparative study of supervised machine learning algorithms for stock market trend prediction. In: 2018 Second International Conference on Inventive Communication and Computational Technologies (ICICCT), pp. 1003–1007 (2018)
16. Balaji, A.J., Ram, D.H., Nair, B.B.: Applicability of deep learning models for stock price forecasting an empirical study on BANKEX data. Procedia Comput. Sci. **143**, 947–953 (2018)
17. Ding, G., Qin, L.: Study on the prediction of stock price based on the associated network model of LSTM. Int. J. Mach. Learn. Cybern. **11**, 1307–1317 (2020)
18. Mukherjee, S., Sadhukhan, B., Sarkar, N., Roy, D., De, S.: Stock market prediction using deep learning algorithms. CAAI Trans. Intell. Technol. **8**, 82–94 (2023)
19. Moghar, A., Hamiche, M.: Stock market prediction using LSTM recurrent neural network. Procedia Comput. Sci. **170**, 1168–1173 (2020)
20. Sunny, M.A.I., Maswood, M.M.S., Alharbi, A.G.: Deep learning-based stock price prediction using LSTM and bi-directional LSTM model. In: 2020 2nd Novel Intelligent and Leading Emerging Sciences Conference (NILES), pp. 87–92 (2020)
21. Yadav, A., Jha, C., Sharan, A.: Optimizing LSTM for time series prediction in Indian stock market. Procedia Comput. Sci. **167**, 2091–2100 (2020)
22. Asiful, M., Hossain, R.K., THulasiram, R., Bruce, N.D., Wang, Y.: Hybrid deep learning model for stock price prediction in IEEE Symposium Series on Computational Intelligence, SSCI, Bangalore, India (2018)
23. Ghosh, B.P., et al.: Deep learning in stock market forecasting: comparative analysis of neural network architectures across NSE and NYSE. J. Comput. Sci. Technol. Stud. **6**, 68–75 (2024)
24. Reddy, N.N., Naresh, E., BP, V.K.: Predicting stock price using sentimental analysis through twitter data in 2020 IEEE International Conference on Electronics, Computing and Communication Technologies (CONECCT), pp. 1–5 (2020)
25. Hashish, I.A., Forni, F., Andreotti, G., Facchinetti, T., Darjani, S.: A hybrid model for bitcoin prices prediction using hidden Markov models and optimized LSTM networks. In: 2019 24th IEEE International Conference on Emerging Technologies and Factory Automation (ETFA), pp. 721–728 (2019)
26. Polamuri, S.R., Srinivas, K., Mohan, A.K.: Multi-model generative adversarial network hybrid prediction algorithm (MMGAN-HPA) for stock market prices prediction. J. King Saud Univ. Comput. Inf. Sci. **34**, 7433–7444 (2022)
27. Di Persio, L., Honchar, O., et al.: Artificial neural networks architectures for stock price prediction: comparisons and applications. Int. J. Circ., Syst. Sig. Process. **10**, 403–413 (2016)
28. Lu, W., Li, J., Li, Y., Sun, A., Wang, J.: A CNN-LSTM-based model to forecast stock prices. Complexity **2020**, 1–10 (2020)
29. Xiao, C., Xia, W., Jiang, J.: Stock price forecast based on combined model of ARIMA-LS-SVM. Neural Comput. Appl. **32**(10), 5379–5388 (2020). https://doi.org/10.1007/s00521-019-04698-5
30. Mintarya, L.N., Halim, J.N., Angie, C., Achmad, S., Kurniawan, A.: Machine learning approaches in stock market prediction: a systematic literature review. Procedia Comput. Sci. **216**, 96–102 (2023)

Type-Theory of Algorithms with Chain-Free Memory

Roussanka Loukanova(✉)

Institute of Mathematics and Informatics, Bulgarian Academy of Sciences, Sofia, Bulgaria
rloukanova@gmail.com

Abstract. The results of the stepwise computations are saved in memory slots, for possible reuse. The extended calculus reduces unnecessary iterations of assignments, which are chains of unnecessary assignments, by copying values of terms from one memory slot to another, without any essential algorithmic changes. The primary applications of the chain-free type theory of recursion are for computational semantics of formal and natural languages, including programming languages and compilers.

Keywords: formal language · algorithms · lambda · recursion · reduction calculus · computing in memory · memorisation · chain-free computation

1 Brief Background on Moschovakis Theory of Recursion

For detailed, mathematical introduction to L^{λ}_{ar}, including extended L^{λ}_{ar} and corresponding reduction calculi, having more subtle applications and examples, see, e.g., [2,3,6–12]. The main focus of the work in these papers, including the present one, is on theoretical developments of type theory having direct applications to computational syntax-semantics interfaces of formal and natural languages, including the various areas of Artificial Intelligence (AI). Among the potential applications of the type theory of recursion and algorithms are intelligent software systems of information, especially in AI, including in robotics, that perform algorithmic procedures, which provide efficient and reliable performance. The varieties and extensions of type theory L^{λ}_{ar} are for computational syntax-semantics of natural and formal languages, and, including programming and other specification languages in Computer Science and AI. In these sequences of papers, I have already established applications of type theory of recursion and algorithms to Natural Language Processing (NLP) and computational grammars that cover computational semantics.

Moschovakis Type Theory of Acyclic Recursion (TTAR) L^{λ}_{ar} was introduced in [14], via its demonstration with examples for possible applications to computational semantics of natural language. Moschovakis [14] provides a mathematical proof that the formal theory L^{λ}_{ar} is a proper extension of TY_2 introduced

R. Chinthaginjala et al. (Eds.): DCAI 2024, LNNS 1259, pp. 65–76, 2025.
https://doi.org/10.1007/978-3-031-82073-1_7

by Gallin [1], and thus, of Montague's Intensional Logic (IL) [13]. Furthermore, Loukanova [10] extends that proof, by mathematical proofs and by applications to syntax-semantics representations of natural language. The Gallin type system of L^{λ}_{ar} in [14], especially polymorphic types [2], provides a computational facility for formalisation of states, e.g., of computer memory, and other context information.

Removing the Acyclicity Constraint (AC) on the formal language of L^{λ}_{ar}, see later Def. 1 of the L^{λ}_{ar} terms, yields the type theory of full recursion (TTFR) L^{λ}_{r}. The type theory of full recursion (TTFR) provides type-theoretic algorithmic semantics of natural and formal languages by generalizing the let-expressions of lambda-calculus, in particular, of the formal language LCF, by Scott [16] and Plotkin [15]. In [10], I provide a mathematical proof that Scott let-terms are not algorithmically equivalent with the where-recursion terms of the TTFR L^{λ}_{r}. They remain denotationally equivalent.

In this paper, in Sect. 2, I provide a brief introduction to Moschovakis Type-Theory of Acyclic Recursion, L^{λ}_{ar}, which is more in line with its recent developments and new directions of its applications.

Section 3 provides the reduction rules of L^{λ}_{ar}, introduced by [14]. I provide them also in abbreviated forms by vector notations. In the second part of the paper, Sect. 4, I extend the reduction calculus of L^{λ}_{ar}, for reducing the assignment chains that copy information from one memory location (slot) to another, without any changes in it. In Sects. 2.2–4, I demonstrate the role of the canonical forms of the L^{λ}_{ar} terms for their algorithmic and denotational semantics by examples of translations, i.e., renderings, of natural language (NL) expressions into the extended L^{λ}_{ar}, with and without chain assignments.

2 Introduction to Type-Theory of Acyclic Recursion L^{λ}_{ar}

2.1 Syntax

The set $\mathsf{Types}_{L^{\lambda}_{ar}}$ is defined recursively, in Backus-Naur Form:

$$\tau ::= \mathsf{e} \mid \mathsf{t} \mid \mathsf{s} \mid (\tau \rightarrow \tau) \qquad \text{(Types)}$$

The abbreviations in (1a)–(1c) are, correspondingly, for state-dependent types $\widetilde{\sigma}$, $\widetilde{\mathsf{t}}$, $\widetilde{\mathsf{e}}$. The type (1d) is by currying coding.

$$\widetilde{\sigma} \equiv (\mathsf{s} \rightarrow \sigma) \qquad \text{(for state-dependent types } \sigma\text{)}, \tag{1a}$$

$$\widetilde{\mathsf{e}} \equiv (\mathsf{s} \rightarrow \mathsf{e}) \qquad \text{(for state-dependent entities)}, \tag{1b}$$

$$\widetilde{\mathsf{t}} \equiv (\mathsf{s} \rightarrow \mathsf{t}) \qquad \text{(for state-dependent entities)}, \tag{1c}$$

$$(\overrightarrow{\tau} \rightarrow \sigma) \equiv (\tau_1 \rightarrow \cdots \rightarrow (\tau_n \rightarrow \sigma)) \in \mathsf{Types} \tag{1d}$$

Typed Vocabulary of L^{λ}_{ar} The formal language of L^{λ}_{ar} has, for every $\tau \in \mathsf{Types}$:

(1) denumerable sets of typed *constants*, $K_\tau = \mathsf{Consts}_\tau = \{\mathsf{c}_0^\tau, \ldots, \mathsf{c}_k^\tau, \ldots\}$
(2) two kinds of infinite, denumerable sets of typed variables:
pure variables: $\mathsf{PureVars}_\tau = \{\mathsf{v}_0^\tau, \mathsf{v}_1^\tau, \ldots\}$
recursion (memory) variables: $\mathsf{RecVars}_\tau = \{\mathsf{r}_0^\tau, \mathsf{r}_1^\tau, \ldots\}$

I call the recursion variables also *memory variables* or *memory locations (slots, cells)*, for their role in algorithmic computations. Values, which are obtained by recursive, or iterative, computations, can be saved, i.e., memorised, in typed memory locations.

Definition 1 (Terms). $\mathsf{Terms} = \mathsf{Terms}(\mathrm{L}_{\mathrm{ar}}^\lambda) = \bigcup \mathsf{Terms}_\tau$ *is the set of the terms of* $\mathrm{L}_{\mathrm{ar}}^\lambda$*, where, for each* $\tau \in \mathsf{Types}$, Terms_τ *is the set of the terms of type* τ.

$$A :\equiv \mathsf{c}^\tau : \tau \mid x^\tau : \tau \qquad \textit{(constants and variables)} \quad (2a)$$
$$\mid B^{(\sigma\to\tau)}(C^\sigma) : \tau \qquad \textit{(application terms)} \quad (2b)$$
$$\mid \lambda(v^\sigma)(B^\tau) : (\sigma \to \tau) \qquad (\lambda\textit{-abstraction terms}) \quad (2c)$$
$$\mid A_0^{\sigma_0} \text{ where } \{\, p_1^{\sigma_1} := A_1^{\sigma_1}, \ldots, p_n^{\sigma_n} := A_n^{\sigma_n} \,\} : \sigma_0 \qquad \textit{(recursion terms)} \quad (2d)$$

given that

(1) $\mathsf{c} \in K_\tau$, $x \in \mathsf{Vars}_\tau = \mathsf{PureVars}_\tau \cup \mathsf{RecVars}_\tau$, $v \in \mathsf{PureVars}_\sigma$
(2) the term (2d) *satisfies the Acyclicity Constraint (AC):*
(a) for $n \geq 0$, $i = 0, \ldots, n$, $\sigma_i \in \mathsf{Types}$, $A_i \in \mathsf{Terms}_{\sigma_i}$
(b) for $i = 1, \ldots, n$, $p_i \in \mathsf{RecVars}_{\sigma_i}$ *are pairwise different recursion variables*

Acyclicity Constraint (AC). The sequence of assignments (3a) is *acyclic* iff there is a function $\mathsf{rank} : \{p_1, \ldots, p_n\} \to \mathbb{N}$, such that:

$$\{\, p_1^{\sigma_1} := A_1^{\sigma_1}, \ldots, p_n^{\sigma_n} := A_n^{\sigma_n} \,\} \quad (n \geq 0) \qquad (3a)$$
$$\text{if } p_j \text{ occurs freely in } A_i, \text{then } \mathsf{rank}(p_i) > \mathsf{rank}(p_j) \qquad (3b)$$

Note 1 Without the Acyclicity Constraint (AC), the rules (2a)–(2d) define the language L_r^λ of full recursion.

About the Denotational Semantics of $\mathrm{L}_{\mathrm{ar}}^\lambda$. The denotation function $\mathsf{den}^{\mathfrak{A}}$ is defined by structural induction on term formations by Definition 1, see [2,3,6–12,14].

Definition 2 [Immediate and Pure Terms]. *The set of the* immediate terms *consists of all terms of the form* (4)*, for* $p \in \mathsf{RecVars}$, $u_i, v_j, \in \mathsf{PureVars}$ *(*$i = 1, \ldots, n$, $j = 1, \ldots, m$*),* $V \in \mathsf{Vars}$*:*

$$T :\equiv V \mid p(v_1)\ldots(v_m) \mid \lambda(u_1)\ldots\lambda(u_n)p(v_1)\ldots(v_m), \quad \textit{for } m, n \geq 0 \qquad (4)$$

Every term A that is non-immediate is pure.

Informally, A term (4), a variable V or of the form $\lambda(\overrightarrow{u})p(\overrightarrow{v})$ $(m, n \geq 0)$, is immediate if its denotation is obtained by the values provided via the variable valuations $g \in G$, without any algorithmic steps of computations, i.e., immediately. The denotation $\mathsf{den}(A)$ of every pure, i.e., non-immediate, term A is computed by an algorithm determined by its canonical form $\mathsf{cf}(A)$. The reduction calculi of L_r^λ reduces every term A to its $\mathsf{cf}(A)$.

2.2 Examples for Reductions to Canonical Forms

Example 1. Assume that (5a)–(5b). Thus, the denotation of the VP (5b) is designated by $\mathsf{den}(B)$. In L_{ar}^{λ} and its reduction calculi the algorithm for computing $\mathsf{den}(B)$, and thus, of the VP (5b), is determined by the canonical form (5g), which is congruent to (5h).

$$\text{is larger than} \xrightarrow{\mathsf{render}} \mathit{isLargerThan} \in \mathsf{Consts} \tag{5a}$$

$$\text{is larger than some cube} \xrightarrow{\mathsf{render}} B \tag{5b}$$

$$B \equiv \lambda(y)\Big[\big[\mathit{some}(\mathit{cube})\big](\lambda(x_o)\,[\mathit{isLargerThan}(x_o)(y)])\Big] \tag{5c}$$

$$\Rightarrow \lambda(y)\Big[\big[\mathit{some}(o) \text{ where } \{\, o := \mathit{cube} \,\}\big]\big(\lambda(x_o)\,[\mathit{isLargerThan}(x_o)(y)]\big)\Big] \tag{5d}$$

by (ap), (comp-ap / rep1), (comp-lam / rep2) from (5c)

$$\Rightarrow \lambda(y)\big[\mathit{some}(o)\big(\lambda(x_o)\,[\mathit{isLargerThan}(x_o)(y)]\big) \text{ where } \{\, o := \mathit{cube} \,\}\big] \tag{5e}$$

by (recap), (comp-lam / rep2) from (5d)

$$\begin{aligned} \Rightarrow \lambda(y)\big[\mathit{some}(o)\big(h'\big) \text{ where } \{\, & o := \mathit{cube}, \\ & h' := \lambda(x_o)\,[\mathit{isLargerThan}(x_o)(y)] \,\}\big] \end{aligned} \tag{5f}$$

by (ap), (head), (B-S), (comp-lam / rep2) from (5e)

$$\begin{aligned} \Rightarrow_{\mathsf{cf}} \mathsf{cf}(B) \equiv\ & \lambda(y)\mathit{some}(o'(y))\big(h(y)\big) \text{ where} \\ & \{\, o' := \lambda(y)\mathit{cube},\ h := \lambda(y)\lambda(x_o)\,[\mathit{isLargerThan}(x_o)(y)] \,\} \end{aligned} \tag{5g}$$

by (λ) from (5f)

$$\begin{aligned} \equiv_c\ & \lambda(y_k)\mathit{some}\big(o'(y_k)\big)\big(h(y_k)\big) \text{ where} \\ & \{\, o' := \lambda(y_k)\mathit{cube},\ h := \lambda(y_k)\lambda(x_o)\mathit{isLargerThan}(x_o)(y_k) \,\} \end{aligned} \tag{5h}$$

by renaming congruence (cong) from (5g)

$$\begin{aligned} \Rightarrow_{\gamma^*} \mathsf{cf}_{\gamma^*}(B) \equiv\ & \lambda(y)\mathit{some}(o)\big(h(y)\big) \text{ where} \\ & \{\, o := \mathit{cube},\ h := \lambda(y)\lambda(x_o)\,[\mathit{isLargerThan}(x_o)(y)] \,\} \end{aligned} \tag{5i}$$

from (5g) by γ^* [10]

Note that, after application of the by γ^* rule [7,10] to (5g) of the original reduction calculus of L_{ar}^{λ}, the result is (5i), a γ^*-canonical form, which is more efficient, without unnecessary λ-abstractions and applications.

Example 2. The canonical form $\mathsf{cf}(A)$ of A determines the algorithm for computing the denotation $\mathsf{den}(A)$ of A and (6a).

$$\text{N is larger than some cube} \xrightarrow{\text{render}} A \tag{6a}$$

$$A \equiv \big[\lambda(y)\big(\big[some(cube)\big]\big(\lambda(x_o)\,[isLargerThan(x_o)(y)]\big)\big)\big](N) \tag{6b}$$

$$\Rightarrow_{\mathsf{cf}}\ \mathsf{cf}(A) \equiv \underbrace{\big[\lambda(y_k)\big(some\big(o'(y_k)\big)\big(h(y_k)\big)\big)\big](k)}_{\text{explicit, irreducible}}\ \mathsf{where}$$

$$\{\, h := \lambda(y_k)\lambda(x_o)\,[isLargerThan(x_o)(y_k)],\ o' := \lambda(y_k)cube,\ k := N \,\}$$

from (5h), by (recap), (ap), (comp-rec / rep3), (head), (cong) (6c)

$$\Rightarrow_{\gamma^*}\ \mathsf{cf}_{\gamma^*}(A) \equiv \underbrace{\big[\lambda(y_k)some(o)\big(h(y_k)\big)\big](k)}_{\text{explicit, irreducible}}\ \mathsf{where}$$

$$\{\, h := \lambda(y_k)\lambda(x_o)[isLargerThan(x_o)(y_k)],\ o := cube,\ k := N \,\}$$

from (6c) by γ^* [10] (6d)

The γ^*-canonical form $\mathsf{cf}_{\gamma^*}(A)$ (6d) is more efficient algorithmically than (6c).

3 Reduction Calculus of $\mathrm{L}_{\mathrm{ar}}^{\lambda}$

In this section, I present the reduction rules of $\mathrm{L}_{\mathrm{ar}}^{\lambda}$, originally introduced by Moschovakis [14]. Loukanova [7,10] extends $\mathrm{L}_{\mathrm{ar}}^{\lambda}$ by adding the γ^* rule.

Congruence. The congruence relation is maintained by the reduction relation:

$$\text{If } A \equiv_c B, \text{ then } A \Rightarrow B \tag{cong}$$

Transitivity

$$\text{If } A \Rightarrow B \text{ and } B \Rightarrow C, \text{ then } A \Rightarrow C \tag{trans}$$

Compositionality. Replacement of sub-terms with correspondingly reduced ones respects the term structure by the definition of the term syntax.

$$\text{If } A \Rightarrow A' \text{ and } B \Rightarrow B', \text{ then } A(B) \Rightarrow A'(B') \tag{comp-ap / rep1}$$

$$\text{If } A \Rightarrow B, \text{ then } \lambda(u)(A) \Rightarrow \lambda(u)(B) \tag{comp-lam / rep2}$$

$$\text{If } A_i \Rightarrow B_i, \text{ for } i = 0, \ldots, n, \text{ then} \tag{comp-rec / rep3}$$

$$A_0 \ \mathsf{where}\ \{\, p_1 := A_1, \ldots, p_n := A_n \,\} \Rightarrow B_0 \ \mathsf{where}\ \{\, p_1 := B_1, \ldots, p_n := B_n \,\}$$

Head Rule. If $p_i \neq q_j$ and p_i does not occur freely in B_j, for all $i = 1, \ldots, n$, $j = 1, \ldots, m$, then

$$\begin{aligned}&(A_0 \text{ where } \{ \overrightarrow{p} := \overrightarrow{A} \}) \text{ where } \{ \overrightarrow{q} := \overrightarrow{B} \} \\ &\Rightarrow A_0 \text{ where } \{ \overrightarrow{p} := \overrightarrow{A},\ \overrightarrow{q} := \overrightarrow{B} \}\end{aligned} \tag{head}$$

Bekič-Scott Rule. If $p_i \neq q_j$ and q_j does not occur freely in A_i, for all $i = 1, \ldots, n$, $j = 1, \ldots, m$, then

$$\begin{aligned}&A_0 \text{ where } \{ p := (B_0 \text{ where } \{ \overrightarrow{q} := \overrightarrow{B} \}), \overrightarrow{p} := \overrightarrow{A} \} \\ &\Rightarrow A_0 \text{ where } \{ p := B_0, \overrightarrow{q} := \overrightarrow{B},\ \overrightarrow{p} := \overrightarrow{A} \}\end{aligned} \tag{B-S}$$

Recursion-Application Rule. If p_i does not occur freely in B, for all $i = 1, \ldots, n$, then

$$\begin{aligned}&(A_0 \text{ where } \{ \overrightarrow{p} := \overrightarrow{A} \})(B) \\ &\Rightarrow A_0(B) \text{ where } \{ \overrightarrow{p} := \overrightarrow{A} \}\end{aligned} \tag{recap}$$

Application Rule. If B is a proper term and b is a fresh recursion (memory) variable, then

$$A(B) \Rightarrow\ A(b) \text{ where } \{ b := B \} \tag{ap}$$

λ-Rule. If, for every $i = 1, \ldots, n$, p'_i is a fresh recursion (memory) variable, and A'_i is the result of the replacement of

$$A'_i \equiv A_i\{p_1 :\equiv p'_1(u), \ldots, p_n :\equiv p'_n(u)\} \equiv A_i\{\overrightarrow{p} :\equiv \overrightarrow{p'(u)}\} \tag{8a}$$

then

$$\begin{aligned}&\lambda(u)\,(A_0 \text{ where } \{ p_1 := A_1, \ldots, p_n := A_n \}) &&(\lambda)\\ &\Rightarrow \lambda(u)\,A'_0 \text{ where } \{ p'_1 := \lambda(u)\,A'_1, \ldots, p'_n := \lambda(u)\,A'_n \} \\ &\equiv \lambda(u)A_0\{\overrightarrow{p} :\equiv \overrightarrow{p'(u)}\} \text{ where } \{ \overrightarrow{p'} := \overrightarrow{\lambda(u)A\{\overrightarrow{p} :\equiv \overrightarrow{p'(u)}\}} \} &&(9)\end{aligned}$$

4 Chain Reduction Calculus

In this section, I extend the set of the reduction rules of L^{λ}_{ar} to an extended reduction calculus of L^{λ}_{ar}, by (chain) rule, for removing chain assignments. The new (chain) rule reduces repeated saving, via chain-like term assignments, $q := p,\ p := A$, of the value of a term A, which has already been computed and saved in a primary memory location p. More generally, the (chain) rule removes repeated memorization $q := \lambda(\overrightarrow{y})(p(\overrightarrow{y}))$ of the same, already computed, value of A saved in p, via a chain mode, $q := \lambda(\overrightarrow{y})(p(\overrightarrow{y})),\ p := A$. At first, I motivate the need of the extended reduction calculus with examples.

Example 3. Assume that $m, p, i, f_0, m_0 \in \mathsf{RecVars}$. The tentative denotation values of $f_0, m_0 \in \mathsf{FreeVars}(F^1)$ would be provided by variable valuations. In the term (10a), the assignments $p := i$ and $i := f_0$ save and access the tentative value of f_0 in chain-like fashion, without contribution of essential information. We would like to reduce this chain of assignments, by the (chain) rule that I introduce in this section. Note that after the application of the (chain) rule, we still have an underspecified term F, by $f_0, m_0 \in \mathsf{FreeVars}(F)$.

$$F^1 \equiv m(p) \text{ where } \{\, p := i,\ i := f_0,\ m := m_0 \,\} \tag{10a}$$
$$\Rightarrow_{\mathsf{ch}} F \equiv m(i) \text{ where } \{\, i := f_0,\ m := m_0 \,\} \tag{10b}$$

by (chain) rule

Assume now that $cube, large \in \mathsf{Consts}$.

$$F_{spec} \equiv m(i) \text{ where } \{\, i := f_0,\ m := m_0,\ f_0 := cube,\ m_0 := large \,\} \tag{11a}$$
$$\Rightarrow_{\mathsf{ch}} F'_{spec} \equiv m(f_0) \text{ where } \{\, m := m_0,\ f_0 := cube,\ m_0 := large \,\} \tag{11b}$$
$$\Rightarrow_{\mathsf{ch}} F''_{spec} \equiv m_0(f_0) \text{ where } \{\, f_0 := cube,\ m_0 := large \,\} \tag{11c}$$

by (chain)rule

Example 4 (A Less Efficient Option). Let us consider a sentence like "The cube is large" consisting of a subject noun phrase (NP), e.g., a referential definite description "the cube", and a predicative verbal phrase (VP), e.g., "is large". At first, we consider rendering of the components and of the entire sentence into terms, which are not necessarily in canonical forms. The canonical sub-forms, e.g., (13b) of (13a), and the final canonical form of the rendering term E in (14a) can be obtained afterwards, by using the reduction rules.

$$C_i \equiv the(cube) \Rightarrow_{\mathsf{cf}} the(c) \text{ where } \{\, c := cube \,\} \qquad \text{by (ap)} \tag{12}$$

The NP "the cube" can be rendered to the term C or $\mathsf{cf}(C)$ in (12), or alternatively to (13a):

$$[\text{the cube}]_{\text{NP}} \xrightarrow{\text{render}} E^0 \equiv_{\mathsf{c}} i \text{ where } \{i := \mathit{the}(\mathit{cube})\} \tag{13a}$$

$$\Rightarrow i \text{ where } \{i := \mathit{the}(c),\ c := \mathit{cube}\}$$

$$\text{by (comp-rec / rep3) from (12), (13}a\text{); and (B-S)} \tag{13b}$$

Then $T_{isLarge}(E^0)$ can be reduced to a canonical form, e.g., as in (14a)–(14h):

$$[\,[\text{The cube}]_{\text{NP}}, [\text{is large}]_{\text{VP}}\,]_{\text{S}} \xrightarrow{\text{render}} T_{isLarge}(E^0) \equiv_{\mathsf{c}} E \tag{14a}$$

$$E \equiv \big[b \text{ where } \{\, b := \mathit{isLarge}\,\}\big]\big(i \text{ where } \{i := \mathit{the}(\mathit{cube})\}\big) \tag{14b}$$

$$\Rightarrow \big[b(i \text{ where } \{i := \mathit{the}(\mathit{cube})\})\big] \tag{14c}$$

$$\text{where } \{\, b := \mathit{isLarge}\,\} \qquad \text{by (recap)}$$

$$\Rightarrow \big[b(e) \text{ where } \{\, e := [i \text{ where } \{i := \mathit{the}(\mathit{cube})\}]\,\}\big] \tag{14d}$$

$$\text{where } \{\, b := \mathit{isLarge}\,\}$$

by (ap) to the head subterm of (14c); (comp-rec / rep3) to (14c)

$$\Rightarrow b(e) \text{ where } \{\, e := \big[i \text{ where } \{i := \mathit{the}(\mathit{cube})\}\big],$$

$$b := \mathit{isLarge}\,\} \tag{14e}$$

by (head) to (14d)

$$\Rightarrow b(e) \text{ where } \{\, e := i,\ i := \mathit{the}(\mathit{cube}),\ b := \mathit{isLarge}\,\} \quad \text{by (B-S)} \tag{14f}$$

$$\Rightarrow_{\mathsf{ch}} T \equiv b(i) \text{ where } \{\, i := \mathit{the}(\mathit{cube}),\ b := \mathit{isLarge}\,\} \tag{14g}$$

by (chain) rule from (14f)

$$\Rightarrow_{\mathsf{cf}} \mathsf{cf}(T) \equiv \mathsf{cf}_{\mathsf{ch}}(T) \equiv b(i) \text{ where } \{\, i := \mathit{the}(c),\ c := \mathit{cube},$$

$$b := \mathit{isLarge}\,\} \tag{14h}$$

by (comp-rec / rep3), (B-S), from (12), (14g)

To reduce chain like assignments of recursion terms, as in (14f), I add the new reduction rule (chain) to the set of the reduction rules of $\mathrm{L}^{\lambda}_{\mathrm{ar}}$. The term (14g) doesn't have the chain assignments $e := i$, $i := \mathit{the}(\mathit{cube})$, after applying the (chain) rule to (14f). As a result, the canonical form (14h) is simplified. Thus, the (chain) rule optimises the canonical forms of the semantic representations of natural language, by providing simpler algorithmic steps.

Chain Rule. For any $A \in \mathsf{Terms}$, $p \in \mathsf{RecVars}$, $y_i \in \mathsf{PureVars}$, $y_i \notin \mathsf{FreeVars}(A)$, $i \in \{\, 1, \ldots, n\,\}$ $(n \geq 0)$:

$$A' \equiv_{\mathsf{c}} \big[A_0 \text{ where } \{\, q := \lambda(\overrightarrow{y})(p(\overrightarrow{y})),\ p := A,\, p_1 := A_1, \ldots, p_n := A_n \,\}\big] \tag{15a}$$

$$\Rightarrow_{\mathsf{ch}} A'' \equiv \big[A_0\{\, q :\equiv p \,\} \text{ where } \{\, p := A,\, p_1 := A_1\{\, q :\equiv p \,\}, \ldots, p_n := A_n\{\, q :\equiv p \,\}\,\}\big] \quad \text{(chain)}$$

Often, we shall designate the extended reductions $A \Rightarrow_{\mathsf{ch}} B$ by $A \Rightarrow B$.

Example 5 (A More Efficient Option for Rendering into Canonical Form).
An NP, e.g., "the cube", can be rendered directly into a term in a canonical form, from the components of the NP, in this example a simple common noun (N), like "cube", and the underspecified rendering (16a) of the determiner "the", as in (16b).

$$\text{the} \xrightarrow{\text{render}} d \text{ where } \{\, i := d(c),\, d := \mathit{the} \,\} \ : ((\widetilde{\mathsf{e}} \to \widetilde{\mathsf{t}}) \to \widetilde{\mathsf{e}}) \tag{16a}$$

$$[\text{the cube}]_{\text{NP}} \xrightarrow{\text{render}} T^0 \equiv i \text{ where } \{\, i := d(c),\, d := \mathit{the},\, c := \mathit{cube} \,\} \ : \widetilde{\mathsf{e}} \tag{16b}$$

$$[\text{is large}]_{\text{VP}} \xrightarrow{\text{render}} T_{isLarge} \equiv b \text{ where } \{\, b := \mathit{isLarge} \,\} \ : (\widetilde{\mathsf{e}} \to \widetilde{\mathsf{t}}) \tag{16c}$$

The composition of the components can be rendered directly into a canonical form, as in (17a)–(17b), by using the canonical forms of the components, of $T_{isLarge}(T^0) \Rightarrow T^1 \Rightarrow_{\mathsf{ch}} T^2$:

$$\{\, [\text{The cube}]_{\text{NP}}, [\text{is large}]_{\text{VP}} \,\}_{\text{S}} \xrightarrow{\text{render}} T^2 \tag{17a}$$

$$T^1 \equiv b(e) \text{ where } \{\, e := i,\, i := d(c),\, d := \mathit{the},\, c := \mathit{cube},\, b := \mathit{isLarge} \,\} \tag{17b}$$

$$T^1 \Rightarrow_{\mathsf{ch}} T^2 \equiv b(i) \text{ where } \{\, i := d(c),\, d := \mathit{the},\, c := \mathit{cube},\, b := \mathit{isLarge} \,\} \text{ by (chain) rule} \tag{17c}$$

Note that the term T^1 is derived more efficiently in (17a)–(17b), than in (14a)–(14h), directly from the terms in canonical forms $T^0 \equiv \mathsf{cf}(T^0)$ and $T_{isLarge} \equiv \mathsf{cf}(T_{isLarge})$. Then, (17c) removes the chain assignments $e := i$, $i := d(c)$, by the (chain) rule. Furthermore, the underspecified term (18) represents a wide class of predications of properties to definite descriptions. Specified, chain-free, canonical forms are obtained by specification assignments to the memory locations b and c, e.g., as for T^2 from (18):

$$T \equiv b(i) \text{ where } \{\, i := d(c),\, d := \mathit{the} \,\} \tag{18}$$

The details of algorithmic semantics of $\mathrm{L}^{\lambda}_{\mathrm{ar}}$ are given in [3,6,7,14].

Theorem 1 (Effective Reduction Calculi). *For every term* $A \in \mathsf{Terms}$, *its chain-free canonical form* $\mathsf{cf}(A)$ *is effectively computed, by the chain reduction calculus:* $A \Rightarrow \mathsf{cf}(A)$.

For each proper (i.e., non-immediate) term A, the chain-free algorithmic meaning of A consists of the computational steps provided by the assignments in the scope of the where-operator, in its chain-free canonical form $\mathsf{cf}_{\mathsf{ch}}(A)$.

5 Conclusion and Future Work

The formal languages of acyclic recursion $\mathrm{L}^{\lambda}_{\mathrm{ar}}$ and of full recursion L^{λ}_{r} provide a type-theoretic formalization of the mathematical notion of algorithm. The canonical forms of terms represent algorithms for computing values of recursive functions, which are encoded by one-argument functions via currying technique. That is, the algorithmic semantics of a proper (i.e., non-immediate) term A, is determined by its canonical form $\mathsf{cf}(A)$.

In this paper, I have extended the reduction calculus of $\mathrm{L}^{\lambda}_{\mathrm{ar}}$ introduced by Moschovakis [14] by adding the rule (chain), which eliminates assignments of the form $q := \lambda(\overrightarrow{y})(p(\overrightarrow{y}))$, $p := A$, $A \in \mathsf{Terms}$. The special case of chain assignments is $q := p$, $p := A$, i.e., a "copy" of the value of A is fetched from p and saved in q. By the introduced rule (chain), $q := p$ is removed and all the occurrences of q are replaced with p.

Thus, the chain free canonical forms of terms determine more efficient algorithms for computing their denotations.

The algorithmic semantics of *proper* (i.e., non-immediate) terms A, is determined by its canonical form $\mathsf{cf}(A)$:

$$\mathsf{alg}(A) = \mathsf{alg}(\mathsf{cf}(A)) \text{ computes } \mathsf{den}(A) = \mathsf{den}(\mathsf{cf}(A)) \tag{19}$$

$$\underbrace{\text{Syntax of } \mathrm{L}^{\lambda}_{\mathrm{ar}}\ (\mathrm{L}^{\lambda}_{r}) \Longrightarrow \text{Algorithms: } \mathsf{alg}(A) = \mathsf{alg}(\mathsf{cf}(A)) \Longrightarrow \text{Denotations } \mathsf{den}(A)}_{\text{Algorithmic Semantics of } \mathrm{L}^{\lambda}_{\mathrm{ar}}(\mathrm{L}^{\lambda}_{r})}$$

The syntax-semantic compositions of the smaller components into larger ones, by linking them, can introduce chain assignments. The introduced reduction rule (chain) removes them and results in more efficient algorithms. This is demonstrated by Examples 3, 4, 5. The type theory $\mathrm{L}^{\lambda}_{\mathrm{ar}}$ provides rich varieties of terms available for compositions depending on the choices of computational grammars of formal and natural languages, including large-scale grammars, see [4,5]. The type theory of algorithms for chain-free calculations is new. Extended studies of more properties and applications are forthcoming.

References

1. Gallin, D.: Intensional and Higher-Order Modal Logic: With Applications to Montague Semantics. North-Holland Publishing Company, Amsterdam and Oxford, and American Elsevier Publishing Company (1975). https://doi.org/10.2307/2271880
2. Loukanova, R.: Acyclic Recursion with Polymorphic Types and Underspecification. In: van den Herik, J., Filipe, J. (eds.) Proceedings of the 8th International Conference on Agents and Artificial Intelligence. vol. 2, pp. 392–399. SciTePress — Science and Technology Publications, Lda. (2016). https://doi.org/10.5220/0005749003920399
3. Loukanova, R.: Relationships between Specified and underspecified quantification by the theory of acyclic recursion. ADCAIJ: Adv. Distrib. Comput. Artif. Intell. J. **5**(4), 19–42 (2016). https://doi.org/10.14201/ADCAIJ2016541942
4. Loukanova, R.: An approach to functional formal models of constraint-based lexicalized grammar (CBLG). Fund. Inform. **152**(4), 341–372 (2017). https://doi.org/10.3233/FI-2017-1524
5. Loukanova, R.: Computational syntax-semantics interface with type-theory of acyclic recursion for underspecified semantics. In: Osswald, R., Retoré, C., Sutton, P. (eds.) IWCS 2019 Workshop on Computing Semantics with Types, Frames and Related Structures. Proceedings of the Workshop, pp. 37–48. The Association for Computational Linguistics (ACL), Gothenburg, Sweden (2019). https://www.aclweb.org/anthology/W19-1005
6. Loukanova, R.: Gamma-reduction in type theory of acyclic recursion. Fundamenta Informaticae **170**(4), 367–411 (2019). https://doi.org/10.3233/FI-2019-1867
7. Loukanova, R.: Gamma-star canonical forms in the type-theory of acyclic algorithms. In: van den Herik, J., Rocha, A.P. (eds.) Agents and Artificial Intelligence. ICAART 2018, Lecture Notes in Computer Science, book series LNAI, vol. 11352, pp. 383–407. Springer International Publishing, Cham (2019). https://doi.org/10.1007/978-3-030-05453-3_18
8. Loukanova, R.: Type-theory of acyclic algorithms for models of consecutive binding of functional neuro-receptors. In: Grabowski, A., Loukanova, R., Schwarzweller, C. (eds.) AI Aspects in Reasoning, Languages, and Computation, vol. 889, pp. 1–48. Springer International Publishing, Cham (2020). https://doi.org/10.1007/978-3-030-41425-2_1
9. Loukanova, R.: Eta-reduction in type-theory of acyclic recursion. ADCAIJ: Adv. Distrib. Comput. Artif. Intell. J. **12**(1), 1–22, e29199 (2023). https://doi.org/10.14201/adcaij.29199
10. Loukanova, R.: Logic operators and quantifiers in type-theory of algorithms. In: Bekki, D., Mineshima, K., McCready, E. (eds.) Logic and Engineering of Natural Language Semantics, Lecture Notes in Computer Science(LNCS), vol. 14213, pp. 173–198. Springer Nature Switzerland, Cham (2023). https://doi.org/10.1007/978-3-031-43977-3_11
11. Loukanova, R.: Restricted computations and parameters in type-theory of acyclic recursion. ADCAIJ: Adv. Distrib. Comput. Artif. Intell. J. **12**(1), 1–40 (2023). https://doi.org/10.14201/adcaij.29081
12. Loukanova, R.: Semantics of propositional attitudes in type-theory of algorithms. In: Bekki, D., Mineshima, K., McCready, E. (eds.) Logic and Engineering of Natural Language Semantics (LENLS 2023), Lecture Notes in Computer Science (LNCS), vol. 14569. Springer Nature Switzerland AG, Cham (2024). https://doi.org/10.1007/978-3-031-60878-0_15

13. Montague, R.: The proper treatment of quantification in ordinary English. In: Hintikka, J., Moravcsik, J., Suppes, P. (eds.) Approaches to Natural Language, vol. 49, pp. 221–242. Synthese Library. Springer, Dordrecht (1973). https://doi.org/10.1007/978-94-010-2506-5_10
14. Moschovakis, Y.N.: A logical calculus of meaning and synonymy. Linguist. Philos. **29**(1), 27–89 (2006). https://doi.org/10.1007/s10988-005-6920-7
15. Plotkin, G.D.: LCF considered as a programming language. Theor. Comput. Sci. **5**(3), 223–255 (1977). https://doi.org/10.1016/0304-3975(77)90044-5
16. Scott, D.S.: A type-theoretical alternative to ISWIM, CUCH, OWHY. Theor. Comput. Sci. **121**(1), 411–440 (1993). https://doi.org/10.1016/0304-3975(93)90095-B

Resilient Peer-to-Peer Learning Based on Adaptive Aggregation

Chandreyee Bhowmick(✉) and Xenofon Koutsoukos

Institute of Software Integrated Systems, Vanderbilt University, Nashville, TN 37209, USA
{chandreyee.bhowmick,xenofon.koutsoukos}@vanderbilt.edu

Abstract. Collaborative learning in peer-to-peer networks offers the benefits of distributed learning while mitigating the risks associated with single points of failure inherent in centralized servers. However, adversarial workers pose potential threats by attempting to inject malicious information into the network. Thus, ensuring the resilience of peer-to-peer learning emerges as a pivotal research objective. The challenge is exacerbated in the presence of non-convex loss functions and non-iid data distributions. This paper introduces a resilient aggregation technique tailored for such scenarios, aimed at fostering similarity among peers' learning processes. The aggregation weights are determined through an optimization procedure, and use the loss function computed using the neighbor's models and individual private data, thereby addressing concerns regarding data privacy in distributed machine learning. Theoretical analysis demonstrates convergence of parameters with non-convex loss functions and non-iid data distributions. Empirical evaluations across three distinct machine learning tasks support the claims. The empirical findings, which encompass a range of diverse attack models, also demonstrate improved accuracy when compared to existing methodologies.

Keywords: Distributed machine learning · Resilient learning · Peer-to-peer learning · Adaptive aggregation · Non-convex loss function · Non-iid data distribution

1 Introduction

The progression of artificial intelligence (AI), propelled by advancements in hardware acceleration and machine learning (ML) algorithms, underscores the necessity of leveraging larger datasets. However, managing such extensive data poses challenges in terms of computational resources and time-intensive computations. In response, distributed machine learning (DML) has emerged as a viable solution, involving the distribution of data across multiple machines [8]. Federated learning (FL) trains a global model across multiple workers, coordinated by a centralized parameter server [21], and is shown adept at scenarios prioritizing data security and privacy [7,22].

In contrast, a peer-to-peer (P2P) learning architecture eliminates the necessity for a central coordinator, allowing each node or worker to function as both a server and a client. This decentralized approach enables improved communication overhead,

R. Chinthaginjala et al. (Eds.): DCAI 2024, LNNS 1259, pp. 77–87, 2025.
https://doi.org/10.1007/978-3-031-82073-1_8

enhanced resilience, and offers advantages in scalability, resource utilization, and eliminates single point of failure [11,24]. P2P learning provides superior data privacy compared to FL, as the central server in FL is susceptible to data privacy breaches [17]. The workers in a P2P paradigm learn their own model using their private dataset, and periodically exchange model information with neighbors to improve learning. Decentralized Parallel Stochastic Gradient Descent (D-PSGD) [20] represents the foundational P2P algorithm, combining stochastic gradient descent (SGD) with a gossip averaging algorithm [26]. Building upon this, Decentralized Momentum Stochastic Gradient Descent (DMSGD) introduced momentum to D-PSGD [3]. Further research endeavors have sought to enhance and broaden the applicability of D-PSGD across various scenarios [1,16].

However, P2P distributed learning faces vulnerabilities to potential attacks and threats, akin to other ML algorithms. The malfunctioning or malicious workers within the network can compromise the collective learning process [23], especially in algorithms lacking robust design, such as D-PSGD [23]. Various attack models have been devised that compromise data stored on workers or shared information between them [5]. While data poisoning attacks tamper with samples in local datasets, parameter poisoning represents a more potent attack, often involving compromising communication between workers. Research efforts have focused on developing resilient algorithms in decentralized systems. However, most algorithms are based on the assumption that the data is independent and identically distributed (IID) across workers, and the loss function is convex. Yet, in many real-world applications, data distributions among workers vary significantly (non-IID) based on the user pool [14], and convex loss is overly restrictive for most ML tasks. The challenge of resilient P2P learning becomes even more daunting when it accommodates non-uniform data distribution and non-convex loss.

There are research efforts that have focused on resilient P2P learning, but there is a lack of papers that attempt to address all three challenges mentioned above. For instance, some resilient algorithms only allow convex loss functions, such as ByRDiE algorithm [28] that employs a decentralized coordinate-descent-based learning approach, and MOZI [12] that incorporates a two-step filtering operation to limit the tolerable number of Byzantine workers. The work in [23] devised Byzantine-resilient distributed learning in the presence of non-uniformly distributed data and time-varying networks. On the other hand, [10] developed resilient P2P algorithms for both convex and non-convex loss functions, but requires the data distribution to be iid.

There are growing concerns, and thereby, a pressing need to develop resilient P2P learning algorithms capable of accommodating non-uniform data distributions and non-convex loss functions. Among the few initiatives addressing this challenge, [13] considers a broader class of loss functions and non-iid data distributions. However, it operates under restrictive assumptions such as the knowledge of attacked workers' identities by normal workers and the impractical insignificance of weights associated with attacked workers. In another work [9], the resilient algorithm ICwTM necessitates the exchange of both local gradients and local models, thus increasing the communication overhead, and is also restrictive as it requires all the workers to have the same initialization. Both these works have restrictive bound of the tolerable adversarial workers, improving which is the main motivation for this work.

In this paper, we focus on developing a distributed P2P learning algorithm that tolerates more Byzantine agents than any state-of-the-art methods, and works well in non-iid distribution of data and non-convex loss functions. Our contribution is a novel loss-based adaptive aggregation method that is resilient in the presence of an arbitrary number of adversarial neighbors. The aggregation uses loss computed using each worker's own data and the parameters received from its neighbor. A smaller loss indicates that their objectives are related, so a higher aggregation weight is assigned. This idea has been proven effective in context of multi-task learning [19] and multi-agent reinforcement learning [6].

It should be noted that the proposed approach considers a broad class of model poisoning attacks but it does not use any information about specific attacks targeting the workers. By ensuring that the parameters of the normal workers converge to a common model, the approach achieves better accuracy independent of the specific attacks. The accuracy improvement is a result of assigning larger aggregation weights to workers corresponding to smaller loss indicating that their objectives are aligned. Theoretical analysis prove that the parameters of the normal workers aggregated using the proposed method converge, when the loss is non-convex and the data distribution is non-iid. A detailed empirical evaluation that considers three different classification tasks and a wide range of attack models demonstrate that the proposed algorithm achieves better accuracy compared to other resilient methods.

2 Preliminaries of Distributed Peer-to-Peer Learning

2.1 Background on Peer-to-Peer Learning

Consider a connected network of N workers, denoted by a static and undirected graph $\mathcal{G} = \{\mathcal{V}, \mathcal{E}\}$, where $\mathcal{V} = \{1, 2, \cdots, N\}$ is the set of N nodes that represent the workers, and the set of edges $\mathcal{E}$ represents the communication between the workers. Specifically, $(i, j) \in \mathcal{E}$ implies that node i and node j share information with each other. The neighborhood of worker k is the set $\mathcal{N}_k = k \cup \{l \in \mathcal{V} | (l, k) \in \mathcal{E}\}$. In P2P learning, multiple workers with private data collaborate with each other to solve a supervised ML problem in a distributed manner. Each worker k has access only to a local training set $S_k = \{\xi_k^i\}_{i=1}^{|S_k|}$, where $\xi_k^i = (x_k^i, y_k^i)$ is a data sample of worker k, with $x_k^i \in \mathbb{R}^{d_x}, y_k^i \in \mathbb{R}^{d_y}$. The distribution of data among the workers can be independent and identically distributed (iid) or non-iid. We use $\ell(w_k; \xi_k)$ to denote a non-negative and possibly regularized loss function that maps (w, ξ) to corresponding loss $\ell(w, \xi)$. The regularized loss function is often referred to as the risk function $r(\cdot)$. We assume the model w in this paper to be parametric, i.e., $w \in \mathbb{R}^d$.

In P2P learning, each worker aims to learn a model by minimizing the local risk function over its local data. The objective of P2P learning is thus formulated as follows:

$$w^* = \min_{\{w_1, w_2, \cdots, w_N\}} \frac{1}{N} \sum_{j=1}^{N} r_j(w_j), \tag{1}$$

where $r_j(w_j) := \frac{1}{|S_j|} \sum_{i=1}^{|S_j|} \ell(w_j; \xi_j^i)$ is the local empirical risk function of worker j.

During each training epoch t, worker k minimizes its local risk using stochastic gradient descent (SGD) on local dataset. The local update step is given by:

$$\hat{w}_k^t = w_k^{t-1} - \mu_k^t \cdot \nabla l(w_k^t; \xi_k^t), \tag{2}$$

where $\hat{w}_k^t$ is the intermediate updated parameters of worker k. Subsequently, every worker transmits these parameters to their neighbors, which is followed by the aggregation step, where each worker combines the parameters received from its neighbors to compute the updated parameters. The aggregation step is given as follows

$$\begin{aligned} w_k^t &= \sum_{l \in \mathcal{N}_k} c_t(l,k) \cdot \hat{w}_l^t \\ \text{subject to} &\sum_{l \in \mathcal{N}_k} c_t(l,k) = 1, \quad c_t(l,k) \geq 0, \quad c_t(l,k) = 0 \text{ for } l \notin \mathcal{N}_k, \end{aligned} \tag{3}$$

where $c_t(l,k)$ is the aggregation weight assigned by worker k to worker l, computed based on the aggregation rule being utilized.

We consider that F out of N workers can be adversarial, meaning that the parameters shared by them are malicious. The sets of normal and adversarial workers are denoted as $\mathcal{V}_c \subseteq \mathcal{V}$ and $\mathcal{V}_f \subseteq \mathcal{V}$, respectively. Despite the fact that the exchange of models between peers is intended to improve learning performance, peers must exercise caution to ensure that the information from malicious neighbors does not affect their learning. The resilient aggregation technique is employed to ensure this, where the aggregation weights are so designed that the learning of normal peers remains unaffected when the number of malicious neighbors is below a specific threshold.

2.2 Attacks on Peer-to-Peer Learning

Model poisoning attacks alter the information shared by the workers with their neighbors [2]. The following attack models have been explored in the literature in context of P2P learning. The workers under *arbitrary Byzantine attack* [25] transmit arbitrary values of matching dimensions to their neighbors. *Sign-flipping (SF) attack* manipulates the model information sent by worker i as follows: $-\sigma_{SF} \cdot \hat{w}_k^t$, where $\sigma_{SF} \in \mathbb{R}^+$ is the attack strength. A class of state-of-the-art attacks, called the time-coupled attacks, operate by introducing slight perturbations to the model, resulting in the aggregated model deviating significantly from the correct model in course of time. *A-little-is-enough (ALIE)* [4] is a non-omniscient time-coupled attack reliant on minor perturbations in the attack vector. Another example is the *Fall-of-empire (FoE) attack* [27], which manipulates the inner-product between the true and the aggregated parameters. The proposed approach considers these broad classes of model poisoning attacks but it is attack agnostic and does not use any information about specific attacks targeting the workers.

3 Peer-to-Peer Learning Using Loss-Based Adaptive Aggregation

The proposed resilient P2P learning algorithm is based on the following assumptions:

Assumption 1. *The risk function $r_k(\cdot)$ is non-convex and a.s. twice differentiable in the first argument. Let $\mathcal{W}_s^*$ denote the set of all first-order stationary points of the statistical risk $R_k(w) = \mathbb{E}_\xi\left[r_k(w)\right]$, i.e., $\mathcal{W}_s^* = \{w \in \mathbb{R}^d : \nabla R_k(w) = 0\}$. Then for any $w_s^* \in \mathcal{W}_s^*$, the statistical risk is locally m-strongly convex in a sufficiently large neighborhood of w_s^*, i.e., there exist two positive constants m and β such that $\forall w \in \mathbb{B}(w_s^*, \beta), \nabla^2 R_k(w) \geq mI$, where $\nabla^2 R_k(w)$ is the Hessian of $R_k(w)$.*

Assumption 2. *The risk function $r_k(\cdot)$ has $\mathcal{L}$-Lipschitz continuous gradient; i.e., $\forall w_1, w_2, \left\|\nabla r(w_1; (x,y)) - \nabla r(w_2; (x,y))\right\| \leq L\|w_1 - w_2\|$, where $L > 0$ is the Lipschitz constant.*

Assumption 3. *For every normal worker $k \in \mathcal{V}_c$, the stochastic gradient $\nabla \ell(w_k^t; \xi_k^t)$ is an unbiased estimate of $\nabla r_k(w_k^t)$, i.e., $\mathbb{E}\left[\nabla \ell(w_k^t; \xi_k^t)\right] = \nabla r_k(w_k^t)$, for all $t \in \mathbb{N}$.*

Assumption 4. *For every normal worker $k \in \mathcal{V}_c$, there exists $a_k \geq 1$, such that for all $t \in \mathbb{N}$, $\mathbb{E}\left[\left\|\nabla \ell(w_k^t; \xi_k^t)\right\|_2^2\right] \leq M + a_k\left\|\nabla r_k(w_k^t)\right\|_2^2$.*

Assumption 1 guarantees local strong convexity of the risk function, but it does not imply global convexity. This assumption is widely used in the literature for analysis of non-convex optimization in ML [15]. It helps in proving local convergence in the sense that they hold as long as the parameters of the workers are initialized within a sufficiently small neighborhood of a stationary point. Assumption 2 implies that the risk function itself is also $\mathcal{L}$-Lipschitz continuous, i.e., $\forall w_1, w_2$, $\left\|r(w_1; (x,y)) - r(w_2; (x,y))\right\| \leq L'\|w_1 - w_2\|$, where $L' > 0$ is the Lipschitz constant.

We propose a distributed P2P learning algorithm with a novel model aggregation method that promotes similarity between the workers. Each normal worker aggregates the model of its neighbors using an adaptive sum. The aggregation weights are so designed that the normal workers are resilient against an arbitrary number of adversarial neighbors. The proposed approach revolves around learning the optimal aggregation weights for each normal worker k, with the objective of minimizing the quadratic distance between the aggregated model parameters w_k^t and the optimal parameters w_k^*, formulated as $\min_{C_k}\left\|w_k^t - w_k^*\right\|^2$. Using (3), we find an equivalent optimization problem

$$\min_{C_k}\left\|\sum_{l \in \mathcal{N}_k} c_t(l,k)\hat{w}_k^t - w_k^*\right\|^2 \quad (4)$$
$$\text{subject to} \sum_{l \in \mathcal{N}_k} c_t(l,k) = 1, \quad c_t(l,k) \geq 0, \quad c_t(l,k) = 0 \text{ for } l \notin \mathcal{N}_k.$$

We assume that there exists a positive constant Γ such that for all $j \in \mathcal{V}_c$ and $t \in \mathbb{N}$, $\left\|w_j^t - w_s^*\right\| \leq \Gamma$, where $w_s^* \in \mathcal{W}_s^*$. Assuming that the w_j^0 for all $j \in \mathcal{V}_c$ are initialized within $\mathbb{B}(w_s^*, \Gamma)$, the solution to (4) are found by following the standard optimization [18] technique[1]:

$$c_t(l,k) = \begin{cases} \frac{r_k(\hat{w}_l^t)^{-1}}{\sum_{p \in \mathcal{N}_k} r_k(\hat{w}_p^t)^{-1}} & \text{for } l \in \mathcal{N}_k, \\ 0 & \text{for } l \notin \mathcal{N}_k. \end{cases} \tag{5}$$

We can improve the performance of the learning by making the workers collaborate with a selected set of neighbors. In this, each worker aggregates the parameters of those neighbors that have a smaller risk than its own, while keeping the aggregation weights same as what was derived earlier. The updated aggregation weights are computed as

$$c_t(l,k) = \begin{cases} \frac{r_k(\hat{w}_l^t)^{-1}}{\sum_{p \in \mathcal{N}_k^+} r_k(\hat{w}_p^t)^{-1}} & \text{for } l \in \mathcal{N}_k^+, \\ 0 & \text{otherwise}, \end{cases} \tag{6}$$

where $\mathcal{N}_k^+$ is the set of neighbors of normal worker k for which $r_k(\hat{w}_l^t) \leq r_k(\hat{w}_k^t)$.

Remark 1. The aggregation rule (6) does not rely on any assumption on the data distribution among the workers. Therefore, this algorithm can be applied to iid as well as non-iid data distributions among workers. While the empirical evaluation presented in this paper specifically targets non-iid data distributions, it is worth noting that the proposed method is equally effective for iid data distributions.

The overall distributed learning algorithm with the proposed aggregation method is outlined in Algorithm 1.

Algorithm 1. Peer-to-peer distributed learning with adaptive model aggregation.

Input: number of epochs T, communication graph $\mathcal{G} = \{\mathcal{V}, \mathcal{E}\}$, step sizes $\{\mu_i^t\}_{t \geq 0}$ and initial parameters $\{w_i^0\}$ for all $i \in [1, 2, \cdots, N]$.
for all epoch $t \in [1, 2, \cdots, T]$ **do**
  for all agent $i \in [1, 2, \cdots, N]$ **do** ▷ Local learning step
    update $\hat{w}_i^t$ using (2).
    Send $\hat{w}_i^t$ to the neighbors $\mathcal{N}_i$.
  for all agent $i \in [1, 2, \cdots, N]$ **do** ▷ Model aggregation step
    Compute risk function $r_i(\hat{w}_l^t)$ for each neighbor $l \in \mathcal{N}_i$.
    Determine the set of better performing neighbors $\mathcal{N}_i^+$.
    Compute aggregation weights using (6).
    Aggregate model parameters using (3) to find w_i^t.

We proceed to analyze the time complexity of the adaptive aggregation step. Determining the weights in (6) necessitates computing batch loss for each neighbor using sampled data. With a constant batch size, the computational time scales linearly with the dimension of the model parameters and the size of the neighborhood. Hence, the time complexity of aggregation step for agent k is $O(d \cdot |\mathcal{N}_k|)$.

[1] The details of this derivation are given in the supplementary material.

4 Theoretical Analysis

Here we present the analytical findings that the aggregated parameters converge in presence of an arbitrary number of adversarial neighbors. Under the assumption of local strong convexity of the risk function, the aggregated parameters of each normal worker $k \in \mathcal{V}_c$ converge towards optimality, maintaining a bounded gap.

Lemma 1. [2] *Suppose every worker be initialized with parameters within $\mathbb{B}(w_s^*, \Gamma)$. For every normal worker $k \in \mathcal{V}_c$ that uses aggregation step* (3) *with aggregation weights given by* (5)*, the following inequality holds*

$$\mathbb{E}\left[r_k(w_k^t) - r_k(w_k^*)\right] \leq \frac{1}{|\mathcal{N}_k|} \sum_{l \in \mathcal{N}_k} \mathbb{E}\left[r_k(\hat{w}_l^t) - r_k(w_k^*)\right] \tag{7}$$

Theorem 1 *(Convergence of parameters). Suppose Assumptions 1–4 hold and the parameters of each worker be initialized within $\mathbb{B}(w_s^*, \Gamma)$. Then for every normal worker $k \in \mathcal{V}_c$ that runs the local learning step* (2)*, followed by an aggregation step* (3) *with weights computed by* (6)*, their parameters converge towards optimality w_k^* as*

$$\lim_{t\to\infty} \mathbb{E}\left[r_k(w_k^t) - r_k(w_k^*)\right] \leq \frac{\mu_k LM}{2m}, \tag{8}$$

for fixed stepsize $\mu_k \in \left(0, \frac{1}{La_k}\right]$, in presence of an arbitrary number of adversarial neighbors.

The optimality gap is bounded and a function of the learning rate. The P2P algorithm where the workers use diminishing learning rate also achieves similar results, with the optimality gap being slightly different than what is given in (8).

5 Evaluation

The algorithm is evaluated on three ML tasks featuring non-iid data distributions[3]. Using neural networks as the workers' model validates the premise of non-convex loss.

Tasks and Datasets. The first task is human activity recognition on UCI HAR dataset[4]. We use 30 workers for this task, where each worker corresponds to one individual performing these activities. This means each subset has unique characteristics and distributions, making the overall data distribution naturally non-iid. Next is handwritten digit classification on MNIST[5] dataset using 10 workers. We generate pathological non-iid data distribution among workers by making only two labels available at each worker [21]. The final one is binary classification problem of spam detection. For this task, Spambase dataset[6] is distributed non-uniformly among 10 workers. We use fully

[2] All proofs are included in the supplementary material.
[3] https://github.com/cbhowmic/ResilientP2P.
[4] https://archive.ics.uci.edu/ml/datasets/human+activity+recognition+using+smartphones.
[5] https://www.kaggle.com/datasets/hojjatk/mnist-dataset.
[6] https://archive.ics.uci.edu/ml/datasets/spambase.

connected neural networks for activity recognition and spam detection tasks, and a convolutional neural network for digit classification. More details regarding each experiment are given in the supplementary material(see footnote 3). The performance is measured by the worst accuracy of the normal workers on the testing set.

Attack Models and Baselines. Four general classes of model poisoning attacks are considered: sign-flipping (SF), arbitrary byzantine, fall-of-empires (FoE) and a-little-is-enough (ALIE). For SF attack, the attack strength σ is chosen as 1. In case of arbitrary Byzantine attack, random signals from the range $(-0.5, 0.5)$ are multiplied with the actual parameters. We consider three variants of the BRIDGE algorithm [10], using trimmed mean (TM), coordinate-wise mean (CM) and Krum. We also simulate average and medoid as baselines. For TM, the same number of parameters as the number of adversarial workers is trimmed. In case of no attack, 10% of the parameters received are trimmed.

Results and Discussion. For MNIST and Spambase datasets, we consider 3 adversarial workers, which is the bound for Krum aggregation (i.e., $N \geq 2F + 3$). In case of HAR task, 13 adversarial workers are considered for the same reason. The test accuracy achieved using various baselines on HAR, MNIST and Spambase are given in Fig. 1, Fig. 2, and Fig. 3, respectively. The performance depends on the task and the attack, but in all cases, the adaptive aggregation method performs better than all other baselines. Specifically, in case of activity recognition under ALIE attack, all methods except the adaptive aggregation fail to learn. For digit classification task, CM and medoid baselines perform comparable with the proposed method. Additional empirical results can be referred to in the supplementary material.

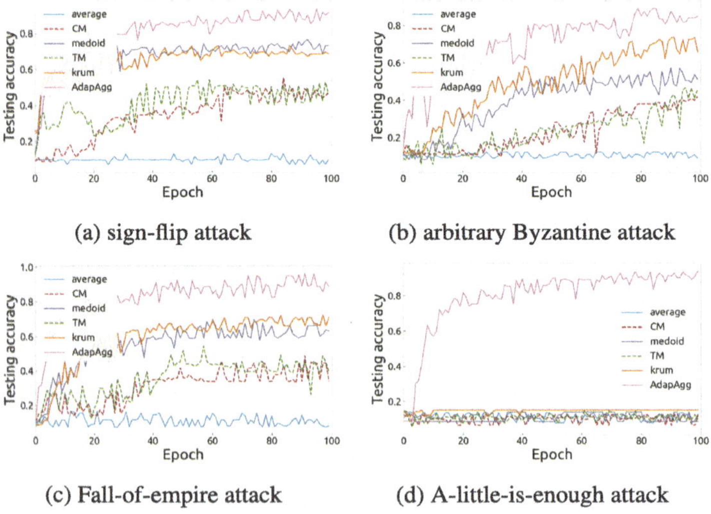

(a) sign-flip attack (b) arbitrary Byzantine attack

(c) Fall-of-empire attack (d) A-little-is-enough attack

Fig. 1. Test accuracy for activity recognition task with 13 adversarial workers.

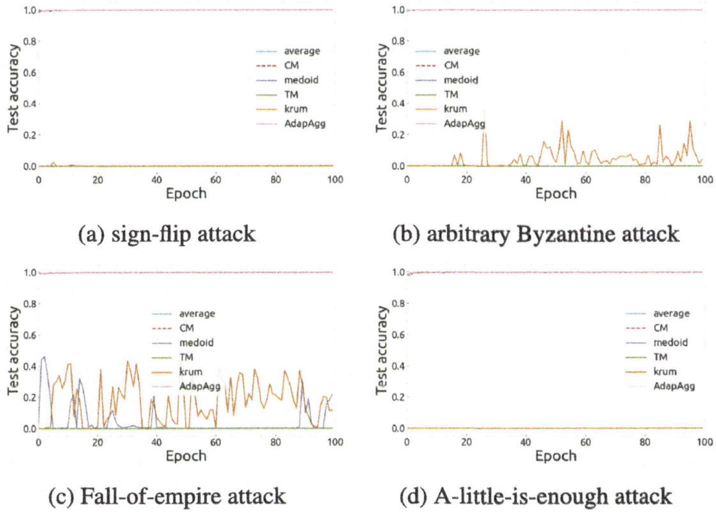

(a) sign-flip attack (b) arbitrary Byzantine attack

(c) Fall-of-empire attack (d) A-little-is-enough attack

Fig. 2. Test accuracy for digit classification task with 3 adversarial workers.

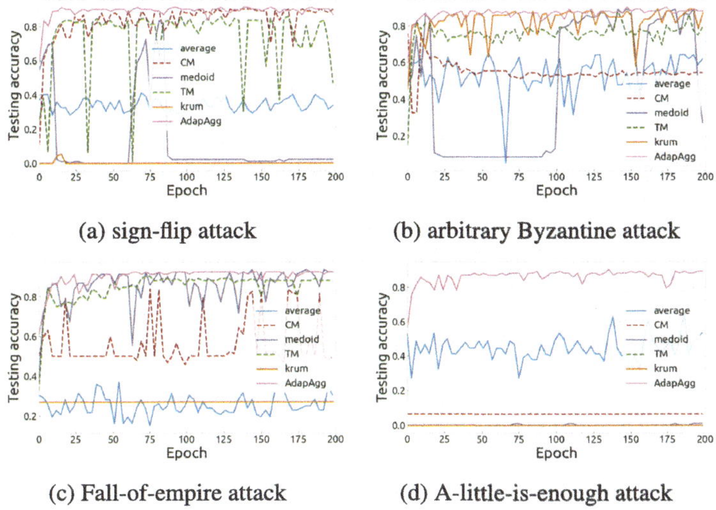

(a) sign-flip attack (b) arbitrary Byzantine attack

(c) Fall-of-empire attack (d) A-little-is-enough attack

Fig. 3. Test accuracy for spam detection task with 3 adversarial workers.

6 Conclusion

This paper presented a novel aggregation technique for peer-to-peer distributed learning that is resilient under non-convex loss function and non-iid data distribution. The aggregation rule relies on capturing the similarity between the workers' learning, and forms an adaptive sum of the neighbor's parameters based on that. We analytically show convergence of the proposed method with assumption of local strong convexity. The empirical results demonstrate the superior performance of this algorithm compared to baseline methods when different ML tasks were solved in presence of a variety of attacks. Being an intuitive aggregation based on the similarities in learning performance, this idea can also be used in other learning setups where promoting similarities is preferred.

References

1. Assran, M., Loizou, N., Ballas, N., Rabbat, M.: Stochastic gradient push for distributed deep learning. In: International Conference on Machine Learning, pp. 344–353. PMLR (2019)
2. Bagdasaryan, E., Veit, A., Hua, Y., Estrin, D., Shmatikov, V.: How to backdoor federated learning. In: International Conference on Artificial Intelligence and Statistics, pp. 2938–2948. PMLR (2020)
3. Balu, A., Jiang, Z., Tan, S.Y., Hedge, C., Lee, Y.M., Sarkar, S.: Decentralized deep learning using momentum-accelerated consensus. In: ICASSP 2021-2021 IEEE International Conference on Acoustics, Speech and Signal Processing (ICASSP), pp. 3675–3679. IEEE (2021)
4. Baruch, G., Baruch, M., Goldberg, Y.: A little is enough: circumventing defenses for distributed learning. Adv. Neural Inf. Process. Syst. **32** (2019)
5. Bhagoji, A.N., Chakraborty, S., Mittal, P., Calo, S.: Model poisoning attacks in federated learning. In: Proc. Workshop Secur. Mach. Learn.(SecML) 32nd Conf. Neural Inf. Process. Syst.(NeurIPS), pp. 1–23 (2018)
6. Bhowmick, C., Li, J., Koutsoukos, X.: Adaptive learning from peers for distributed actor-critic algorithms. In: International Symposium on Distributed Computing and Artificial Intelligence. pp. 54–64. Springer (2023)
7. Bonawitz, K., et al.: Practical secure aggregation for privacy-preserving machine learning. In: proceedings of the 2017 ACM SIGSAC Conference on Computer and Communications Security, pp. 1175–1191 (2017)
8. Dehghani, M., Yazdanparast, Z.: From distributed machine to distributed deep learning: a comprehensive survey. J. Big Data **10**(1), 158 (2023)
9. El-Mhamdi, E.M., Farhadkhani, S., Guerraoui, R., Guirguis, A., Hoang, L.N., Rouault, S.: Collaborative learning in the jungle (decentralized, byzantine, heterogeneous, asynchronous and nonconvex learning). Adv. Neural. Inf. Process. Syst. **34**, 25044–25057 (2021)
10. Fang, C., Yang, Z., Bajwa, W.U.: Bridge: byzantine-resilient decentralized gradient descent. IEEE Trans. Signal Inf. Process. over Netw. **8**, 610–626 (2022)
11. Foster, I., Iamnitchi, A.: On death, taxes, and the convergence of peer-to-peer and grid computing. In: Peer-to-Peer Systems II: Second International Workshop, IPTPS 2003, Berkeley, CA, USA, February 21-22, 2003. Revised Papers 2, pp. 118–128. Springer (2003)
12. Guo, S., et al.: Byzantine-resilient decentralized stochastic gradient descent. IEEE Trans. Circuits Syst. Video Technol. **32**(6), 4096–4106 (2021)
13. He, L., Karimireddy, S.P., Jaggi, M.: Byzantine-robust decentralized learning via self-centered clipping. arXiv preprint arXiv:2202.01545 (2022)
14. Hsieh, K., Phanishayee, A., Mutlu, O., Gibbons, P.: The non-IID data quagmire of decentralized machine learning. In: International Conference on Machine Learning, pp. 4387–4398. PMLR (2020)
15. Jain, P., Kar, P., et al.: Non-convex optimization for machine learning. Foundations and Trends® Mach. Learn. **10**(3-4), 142–363 (2017)
16. Koloskova, A., Loizou, N., Boreiri, S., Jaggi, M., Stich, S.: A unified theory of decentralized SGD with changing topology and local updates. In: International Conference on Machine Learning, pp. 5381–5393. PMLR (2020)

17. Koloskova, A., Stich, S., Jaggi, M.: Decentralized stochastic optimization and gossip algorithms with compressed communication. In: International Conference on Machine Learning, pp. 3478–3487. PMLR (2019)
18. Lewis, F.L., Vrabie, D., Syrmos, V.L.: Optimal Control. John Wiley & Sons (2012)
19. Li, J., Abbas, W., Koutsoukos, X.: Byzantine resilient distributed multi-task learning. Adv. Neural. Inf. Process. Syst. **33**, 18215–18225 (2020)
20. Lian, X., Zhang, C., Zhang, H., Hsieh, C.J., Zhang, W., Liu, J.: Can decentralized algorithms outperform centralized algorithms? A case study for decentralized parallel stochastic gradient descent. Adv. Neural Inf. Process. Syst. **30** (2017)
21. McMahan, B., Moore, E., Ramage, D., Hampson, S., y Arcas, B.A.: Communication-efficient learning of deep networks from decentralized data. In: Artificial Intelligence and Statistics, pp. 1273–1282. PMLR (2017)
22. McMahan, H.B., Ramage, D., Talwar, K., Zhang, L.: Learning differentially private recurrent language models. In: International Conference on Learning Representations (2018)
23. Peng, J., Li, W., Ling, Q.: Byzantine-robust decentralized stochastic optimization over static and time-varying networks. Signal Process. **183**, 108020 (2021)
24. Verbraeken, J., Wolting, M., Katzy, J., Kloppenburg, J., Verbelen, T., Rellermeyer, J.S.: A survey on distributed machine learning. Acm comput. surv. (csur) **53**(2), 1–33 (2020)
25. Wu, Z., Ling, Q., Chen, T., Giannakis, G.B.: Federated variance-reduced stochastic gradient descent with robustness to byzantine attacks. IEEE Trans. Signal Process. **68**, 4583–4596 (2020)
26. Xiao, L., Boyd, S.: Fast linear iterations for distributed averaging. Syst. Control Lett. **53**(1), 65–78 (2004)
27. Xie, C., Koyejo, O., Gupta, I.: Fall of empires: Breaking byzantine-tolerant SGD by inner product manipulation. In: Uncertainty in Artificial Intelligence, pp. 261–270. PMLR (2020)
28. Yang, Z., Bajwa, W.U.: ByRDiE: byzantine-resilient distributed coordinate descent for decentralized learning. IEEE Trans. Signal Inf. Process. over Netw. **5**(4), 611–627 (2019)

An Explainable AI-Based Demand Response Optimization Framework for Smart Buildings

Muhammad Ibrar[1], Hayla Nahom[1], Abegaz Mohammed[1], Sergio Márquez-Sánchez[2], Javier Hernandez Fernandez[3], Juan Manuel Corchado[2], and Aiman Erbad[4(✉)]

[1] Hamad Bin Khalifa University, Ar-Rayyan, Qatar
{mibrar,nhayla,mabegaz}@hbku.edu.qa

[2] BISITE Research Group, University of Salamanca, Calle Espejo s/n. Edificio Multiusos I+D+i, 37007 Salamanca, Spain
{smarquez,sergio.alro}@usal.es

[3] Iberdrola Innovation Middle East, 210177 Doha, Qatar
j.hernandezf@iberdrola.com

[4] College of Engineering, Qatar University, 2713 Doha, Qatar
aerbad@qu.edu.qa

Abstract. Smart building energy demand response (DR) plays a paramount role in the decarbonization of energy use by leveraging different techniques that adjust energy consumption in response to fluctuating prices and demands. In this paper, we present an explainable AI-integrated DR optimization framework for smart buildings. This adaptive XAI-based DR approach intelligently forecasts energy demands and schedules the energy consumption of the buildings based on the forecasted demand patterns, grid conditions, and energy prices. This approach allows the stakeholders to make informed and optimal DR decisions based on the forecasted demand to maximize energy efficiency and user comfort while minimizing electricity costs. The simulation results demonstrate that the proposed approach promotes energy efficiency and user comfort.

Keywords: Demand response optimization · Energy management · Explainable artificial intelligence · K-means · Random forest regressor

1 Introduction

Demand response (DR) in smart buildings is a technical approach designed to manage energy systems by balancing electricity supply and demand in response to external signals, such as changes in electricity prices, grid conditions, or environmental factors. The design of adaptive and optimal DR strategies helps to efficiently manage energy demand, optimize consumption, and ensure grid stability by shifting non-essential energy consumption (EC) to off-peak hours, temporarily reducing or interrupting non-critical energy loads during peak demand,

R. Chinthaginjala et al. (Eds.): DCAI 2024, LNNS 1259, pp. 88–98, 2025.
https://doi.org/10.1007/978-3-031-82073-1_9

dynamically adjusting EC of specific building systems or equipment, and on-site power generation for local consumption or grid support [1].

The DR optimization strategies in smart building systems may involve using Artificial Intelligence (AI) techniques to manage EC based on changing demand patterns, grid conditions, and energy prices. These strategies can be price-based or incentive-based DR strategies. The price-based DR strategies allow consumers to adapt their EC demand based on the change in electricity prices, including time-of-use pricing, critical peak pricing, and real-time pricing. The incentive-based DR strategies include direct load control and interruptible load programs that offer financial incentives to consumers who agree to reduce their energy usage during peak hours. To optimize these DR strategies, various optimization techniques have been used like binary particle swarm optimization, grasshopper optimization, and artificial neural networks that enhance DR programs by predicting power consumption. In [2], machine learning (ML) models, particularly LSTM and SVM, are utilized to predict electricity demand in buildings. These models analyze historical data, climate conditions, and occupancy patterns. Moreover, several studies have explored advanced ML techniques, such as reinforcement learning (RL) and deep RL (DRL), to present optimal DR algorithms and these algorithms have shown promising results by outperforming traditional time-of-use pricing strategies and delivering substantial cost savings in smart buildings [3]. These works have shown that model-free RL may require significant data to outperform conventional approaches. To address these issues, the above-mentioned works have attempted to mix planning and real-world data. However, this attempt has not consistently improved the energy system performance [4]. Further, most ML-based DR research studies may not be trained to provide explanations to energy consumers. Thus, energy consumers may not be aware of the reasons behind ML-driven decisions and energy cost fluctuation, and they cannot understand EC patterns or identify anomalies in EC data[5].

In this regard, explainable AI (XAI) promises to make the DR decision-making process of AI algorithms transparent and understandable to humans[6]. With XAI, model transparency is ensured since XAI employs inherently interpretable ML models like decision trees, linear regression, and XGBoost, allowing short-term electric power consumption forecasting to minimize peak demand and stabilize the power system [7]. Furthermore, XAI techniques can be used to determine what factors or features contribute most to EC and demand and explain why certain rules are triggered under specific conditions, enhancing transparency and trust in the system. These models provide insights into how stakeholders make decisions, allowing them to comprehend the reasons behind EC forecasts and optimization strategies; these enable them to make informed decisions about energy management strategies. Compared to conventional AI systems, XAI offers significant advantages. One critical benefit is its ability to provide a clear, step-by-step explanation of AI decision-making processes, ensuring interpretability and transparency[8]. Dobrovolskis *et al.* [9] examined the increasing significance of XAI within smart systems and introduced an agent-based approach for crafting XAI solutions within an IoT context alongside an experiment conducted to show the viability and efficacy of the proposed method in developing an understandable smart system for managing various smart devices, for example, lighting, ventilation, and heating. The existing literature presented foundational investigations on XAI applications in DR optimization, but to the best of our knowledge, its potential is not fully realized.

Motivated by the limitations of conventional AI systems and the potential of XAI in building energy management systems, we present a novel XAI-based DR optimization framework combined with a K-means clustering algorithm for smart buildings that accurately forecast EC to help end users make optimal energy reduction decisions, minimizing energy costs and maximizing user satisfaction. We group the home appliances into three clusters based on their EC patterns and operational characteristics to improve the accuracy of the EC forecasting model and easily identify important factors influencing consumption patterns and stakeholders' energy management decisions. The main objective of this paper is to maximize user comfort and minimize costs by exploring XAI in smart buildings for DR Optimization. The main contributions of this work are:

- We propose an XAI-based adaptive DR optimization framework for energy management in smart buildings. This approach forecasts EC demand and energy pricing while providing the reasoning behind the ML process to help stakeholders make optimal DR decisions based on the forecasted demand.
- We utilize an adaptive K-means clustering-based electrical appliance clustering algorithm for smart building applications. This algorithm considers EC patterns and device characteristics to accurately cluster the devices.
- We adopt SHapley Additive eXplanations (SHAP) to explain the predictions of complex models used in DR optimization, providing insights into the factors that influence the model's output and identifying potential biases.

The remaining sections of this paper are structured as follows: The proposed XAI-Based DR Framework for smart buildings is presented in Sect. 2. Section 3 presents the numerical results. Finally, Sect. 4 concludes this article.

2 The Proposed XAI-Based DR Optimization Framework

In this section, we discuss the proposed XAI-based DR optimization in smart buildings, which consists of various smart devices or sensors, such as smart lights, smart thermostats, and smart security cameras connected to the network and controlled remotely, data analytics, centralized energy management server (EMS), and interconnected systems that allow for intelligent management and control of different aspects of the smart building, as shown in Fig. 1. The smart devices aim to optimize energy usage, streamline operations, and enhance user comfort, and the EMS fetches periodic electricity price updates from the energy market and manages EC across a range of smart devices to minimize energy costs while ensuring users' satisfaction levels are met. The energy consumption demand data could be trained to the XAI model to add an explanation to the stakeholders of the reasoning about the factors in the energy consumption demand fluctuations. The stakeholders then observe the real-time energy consumption forecast and environmental conditions, to make optimal DR decisions.

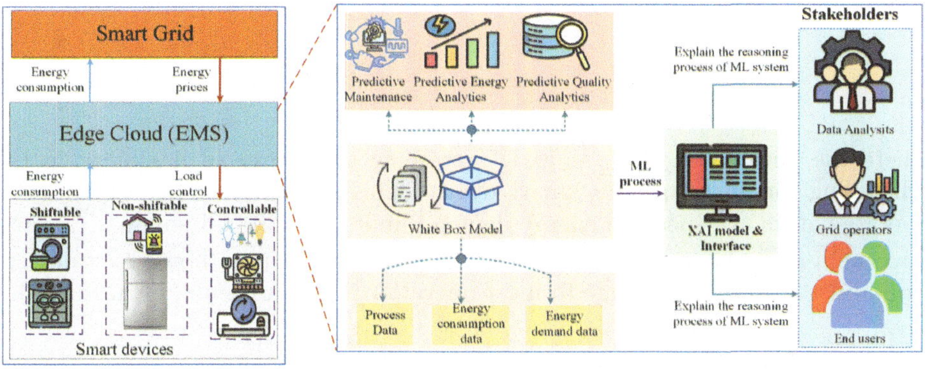

Fig. 1. System Model

2.1 Energy Consumption Model

In this proposed model, we classified the smart devices in smart buildings into three logical clusters based on characteristics and priorities using the K-means algorithm: non-shiftable load smart devices (i.e., refrigerator, alarm system), adaptable or controllable load smart devices (i.e., lighting, AC, heating), and time-adjustable or shiftable load smart devices (i.e., washing machine, dishwasher). In the proposed XAI-based framework, each smart IoT device, denoted as $d \in \mathbf{D}$, has an associated EC for task execution, which is defined as $\mathbf{E}_d^t$. We define a scheduling window $\mathbf{T}_d = \left[t_d^{ini}, t_d^{end}\right]$ for task execution, where t_d^{ini} denoted the task execution starting time and t_d^{end} denote the task execution end time. Moreover, we define a binary control variable $\partial_d^t \in \{0, 1\}$ to determine if smart IoT device d is scheduled to operate in time slot t ($\partial_d^t = 1$) or not ($\partial_d^t = 0$).

Non-shiftable Load Smart Devices: The devices in this category are characterized by their essential demand for a persistent energy supply without any interruptions. Hence, the control variable $\partial_{d,nsl}^t$ for a non-shiftable load must satisfy the below expression to ensure uninterrupted operation.

$$\partial_{d,nsl}^t = \begin{cases} 1, \, t \in \left[t_{d,nsl}^{ini}, t_{d,nsl}^{end}\right] \\ 0, \quad otherwise. \end{cases} \tag{1}$$

This means the EC of load non-shiftable IoT devices must be associated with their energy demand, which is mathematically expressed as: $\mathbf{E}_{d,nsl}^t = \partial_{d,nsl}^t \cdot e_{d,nsl}$, where $e_{d,nsl}$ denotes the energy demand of the load non-shiftable IoT device d.

Controllable Smart Devices: For the devices in this category, we can control and dynamically modify their EC loads due to some external factors, such as weather conditions, DR programs, and hourly dynamic electricity prices. We define the control variable for such controllable IoT devices by $\partial^t_{d,con}$ and mathematically represent it in Eq. (2).

$$\partial^t_{d,con} = \begin{cases} 1, \chi^t_{d,con}, & t \in \left[t^{ini}_{d,con}, t^{end}_{d,con}\right] \\ 0, & otherwise, \end{cases} \tag{2}$$

where the parameter $\chi^t_{d,con} \in [0, 1]$ represents the degree of flexibility in EC scheduling for an adaptable, smart device. This enables the smart devices to operate within the specified energy demand limits, i.e., the minimum $e^{t^{min}}_{d,con}$ and maximum $e^{t^{max}}_{d,con}$ energy demands. The decimal value denotes how smart devices can adjust their energy usage to meet operational requirements, as in Eq. (3).

$$\mathbf{E}^t_{d,con} = e^{t^{min}}_{d,con} + \partial^t_{d,nsl} \cdot \left(e^{t^{max}}_{d,con} - e^{t^{min}}_{d,con}\right). \tag{3}$$

Time-Adjustable or Shiftable Smart Devices: The smart devices in this category can be programmed to operate at a time when there are low EC demands. Suppose the variables $\mathbf{T}^{t_{st}}_{d,sft}$ and $\mathbf{T}^{t_{rq}}_{d,sft}$ show the start time and duration required for the shiftable loads of smart device to complete its task, respectively. So, we define the control variable $\partial^t_{d,sft}$ for load shiftable smart device d, which governs the EC load subject to certain constraints to optimize its operation as mathematically expressed in Eq. (4).

$$\partial^t_{d,sft} = \begin{cases} 1, \text{ if } \partial^{\mathbf{T}^{st}_{d,sft}}_{d,sft} = 1 \text{ and } \sum^t_{\Gamma=t^{ini}_{d,sft}} \partial^t_{d,sft} \leq \mathbf{T}^{rq}_{d,sft} \\ 1 \text{ if } \sum^t_{\Gamma=t^{ini}_{d,sft}} \partial^t_{d,sft} = 0, \text{ and } t^{end}_{d,sft} - t = \mathbf{T}^{rq}_{d,sft} \\ 0, \quad \text{otherwise.} \end{cases} \tag{4}$$

In Eq. (4), the first condition ensures that the operation of shiftable loads of smart devices remains undisturbed to prevent any interruptions in the process, while the second condition shows the specific deadline, the facility must complete the task within the prescribed time-frame successfully. Additionally, in Eq. (4), Γ represents the time variable or time parameter in the given piecewise function, which takes different values or conditions. We calculate the EC of the load shiftable smart devices at time slot t as: $\mathbf{E}^t_{d,sft} = \partial^t_{d,sft} \cdot e_{d,nsl}$.

2.2 Energy Consumption Prediction and Demand Forecasting

We cluster the smart devices into three categories based on their operational characteristics using the k-means algorithm: namely, load non-shiftable, controllable load, and shiftable loads smart devices. In this work, we utilize the PecanStreet dataset with large values to group the devices into three clusters. We use a random forest regressor (RFR) to predict the EC of smart devices and forecast the consumption demand of end users to make optimal DR decisions.

RFR is indeed a powerful algorithm for EC prediction and demand forecasting due to its ability to handle complex relationships and interactions in the data[10]. We use historical data on EC from the PecanStreet dataset to train the RFR model, which includes relevant features such as date/time, weather conditions, economic indicators, and any other factors that could impact EC. In the first step, we clean the data by removing missing values, outliers, and inconsistencies. In the next step, we convert categorical variables to numerical representations and split data into training and testing sets. We then add new features and extract relevant information from the existing features, such as the day of the week, hour of the day, season, etc. After training the RFR model with the training data, we evaluate the trained model with the testing data using the commonly used evaluation metric known as Root Mean Squared Error (RMSE) which is given as:

$$RMSE = \sqrt{\frac{1}{n}\sum_{i=1}^{n}(y_i - \hat{y_i})^2}, \tag{5}$$

where y_i and $\hat{y}_i$ are the actual and expected values of the ith observation, respectively, and n is the number of observations.

2.3 Explainable AI-Based Energy Consumption Prediction

SHAP explains individual predictions made by ML models to provide a way to understand the impact of different features on the model's predictions[5]. This means the SHAP values can be useful in understanding which features contribute the most to the errors measured by MRSE. For example, if some features often have high SHAP values for large error predictions, this may indicate that those features are more influential when the model struggles to make accurate predictions. **Algorithm 1** presents the procedure for the SHAP-based EC prediction.

3 Performance Evaluation

3.1 Simulation Settings

We consider evaluating the EC of smart devices in three clusters, including 5 non-shiftable loads (i.e., refrigerator), 17 controllable (i.e., lighting, AC, heating), and 8 load shiftable (i.e., washing machine, dishwasher, and clothes dryer) using the PecanStreet dataset ($New\_York\_15\_min$[1])[11].

3.2 Analyzing SHAP Summary Plots

Figure 2a, Fig. 2b, and Fig. 2c present an explicit visualization of the impact of individual IoT device on a predictive model's output within the context of EC. The features in cluster 1 have a narrow range of SHAP values, indicating

[1] https://dataport.pecanstreet.org/academic.

Algorithm 1. Shapley Value-Based Energy Consumption Prediction Algorithm

Input: Dataset with features and clusters (clusters1, clusters2, clusters3)
Output: Shapley values for EC, model training, and predictions
Procedure: Initialize
  Load dataset and cluster devices using K-means algorithm
  Calculate the total EC for the dataset
End procedure
Procedure: Calculate Shapley Values
  For each feature f **do**
    Initialize $Shapley(f) = 0$
    For each subset S of features excluding f **do**
      Calculate marginal contribution $MC(f, S)$
      $Shapley(f) \leftarrow Shapley(f) + \frac{|S|!(n-|S|-1)!}{n!}(MC(f, S))$
    end for
  end for
  Normalize Shapley values based on total EC
End procedure
Procedure: Train Model
  Split dataset into training and test sets
  Use SHAP values as feature weights or to select key features
  Train machine learning model on the training set
End procedure
Procedure: Predict
  Apply trained model to test set or new data
  Output predictions for analysis or decision-making
End procedure

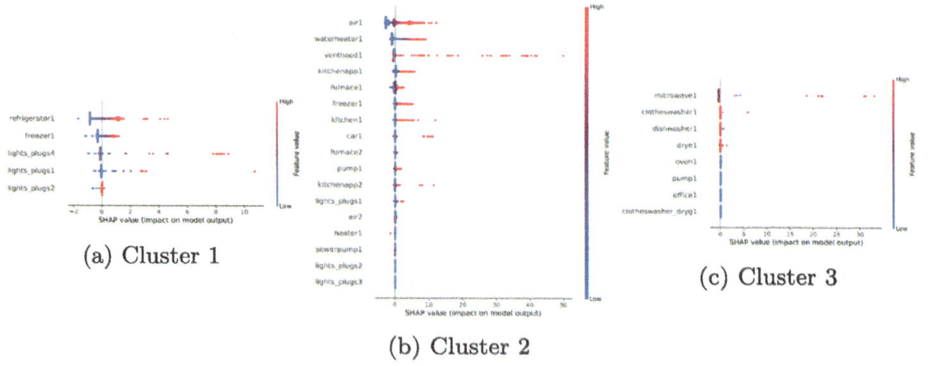

(a) Cluster 1 (b) Cluster 2 (c) Cluster 3

Fig. 2. SHAP Summary Plot for the three clusters.

that their impact on the model's output is relatively uniform. This is because the non-shiftable loads, such as $'lights\_plugs1'$ and $'refrigerator1'$, cannot be easily changed or rescheduled.

Figure 2b reveals that cluster 2 has a wider range of SHAP values, indicating greater variability in how each factor affects model output because loads can be adjusted in real-time to manage EC. Figure 2c shows that cluster 3 has lower SHAP values, with exceptions such as '*washingmachine*1' and '*dishwasher*1' having greater values. This is because shiftable loads can be moved off-peak to balance energy demand.

3.3 Energy Consumption(KW)

Figure 3a illustrates the daily EC in kilowatts (KW) of three distinct clusters over a period from May to November of the specified year. The figure illustrates the high levels of EC in cluster 1, with a fairly consistent pattern over time. This is because non-shiftable loads typically operate continuously at a consistent rate, reflecting the necessity and constancy of their services. The EC pattern in cluster 2 has a more varied pattern with significant peaks and troughs, as shown in Fig. 3b. The reason is the controllable nature of these loads likely contributes to the variability. The same figure shows that cluster 3 has the lowest and most erratic EC profile. This inconsistency aligns with the shiftable nature of these devices, which may be used intensively at certain times and remain idle at others.

We evaluate the EC (KW) in peak and off-peak times across three identified

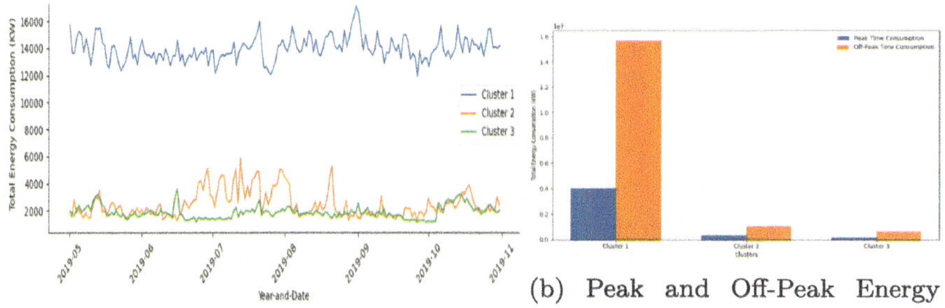

(a) Daily Energy Consumption

(b) Peak and Off-Peak Energy Consumption

Fig. 3. Energy Consumption by each cluster.

clusters of smart devices, as shown in Fig. 3b. These devices consume less energy during peak times, possibly indicating that the cluster contains devices that are either used less or have energy-saving features that reduce their load during high-demand periods. Cluster 2 has the highest EC during off-peak times. This may reflect a strategic shift of device operation to off-peak times to lower energy costs or alleviate grid stress through effective load control management. on the other hand, cluster 3 has balanced EC between peak and off-peak periods, because shiftable loads are flexible in operation timing/can be moved to off-peak times.

Figure 4a displays a comparison between actual EC at time T and predicted EC at time $T+1$ for Cluster 1. The data points primarily lie within the lower half of the EC range for this cluster and RMSE is 5.797.

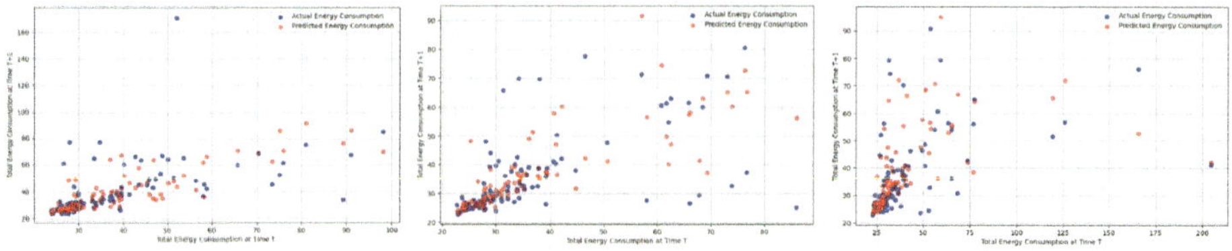

(a) Cluster 1 (b) Cluster 2 (c) Cluster 3

Fig. 4. Actual vs. Predicted Energy Consumption for Each Cluster.

Figure 4b shows an even distribution of actual EC values across a broader range, with most data points again congregating at the lower end of the scale, and RMSE is 6.243. Its predictions tend to be close to the actual values for lower and mid-range EC but show variability as the consumption level rises. Figure 4c shows a distinct EC pattern in clusters, with actual EC values spread over a much wider range and RMSE is 5.404, which seems the prediction is less accurate overall for this cluster.

3.4 Demand Response Adjustment

Figure 5 presents the EC patterns of smart devices in three clusters before and after the implementation of DR adjustments over a 24-hour. For Cluster 1, the 'before' and 'after' lines appear to overlap, indicating that the non-shiftable loads exhibit little to no change due to DR strategies. The DR impact on Cluster 2 is evident, with a marked reduction in EC during peak hours post-adjustment. This suggests that controllable loads offer flexibility, allowing for the strategic curtailment of energy use without compromising the end-user experience. The 'after' line dips below the 'before' line during these peak periods.

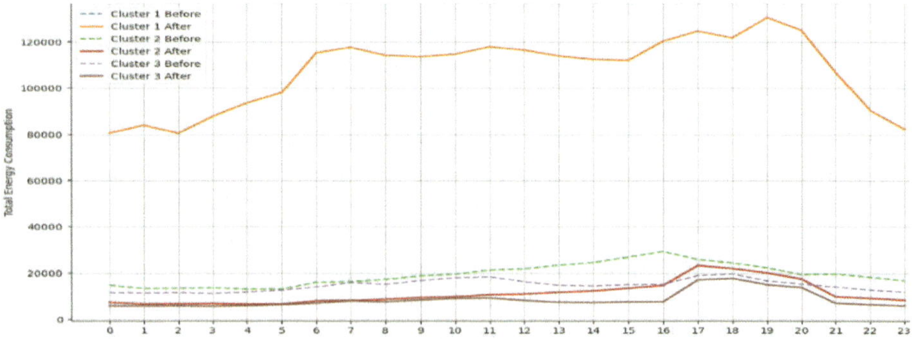

Fig. 5. Demand Response-Adjustment of Energy Consumption

Cluster 3 demonstrates the most significant shift, with a pronounced reduction and flattening of the consumption curve post-DR implementation. The 'after' curve is lower in EC during peak hours and the daily load is uniform.

4 Conclusions

In this paper, we introduced an innovative approach to smart buildings by integrating XAI into DR optimization and adopted the K-means algorithm to cluster smart devices based on their characteristics. The proposed XAI-based DR technique effectively forecasts energy demands and enhances the transparency and interpretability of the decision-making process within EMS, allowing stakeholders to understand and trust the system and make informed DR decisions based on the forecasted demand and pricing signal to maximize energy efficiency and user comfort while minimizing energy costs.

Acknowledgments. This publication was made possible by NPRP grant NPRP13S-0128-200187 from the Qatar National Research Fund (a member of Qatar Foundation). The findings achieved herein are solely the authors' responsibility.

Disclosure of Interests. The authors have no competing interests to declare that are relevant to the content of this article.

References

1. Jurjevic, R., Zakula, T.: Demand response in buildings: a comprehensive overview of current trends, approaches, and strategies. Buildings **13**(10), 2663 (2023)
2. Pallonetto, F., Jin, C., Mangina, E.: Forecast electricity demand in commercial building with machine learning models to enable demand response programs. Energy AI **7**, 100121 (2022)
3. Jang, D., Spangher, L., Nadarajah, S., Spanos, C.: Deep reinforcement learning with planning guardrails for building energy demand response. Energy AI **11**, 100204 (2023)
4. Renzhi, L., et al.: Deep reinforcement learning-based demand response for smart facilities energy management. IEEE Trans. Industr. Electron. **69**(8), 8554–8565 (2021)
5. Wang, X., Wang, H., Bhandari, B., Cheng, L.: Ai-empowered methods for smart energy consumption: a review of load forecasting, anomaly detection and demand response. Int. J. Precis. Eng. Manuf.-Green Technol. 1–31 (2023)
6. Javed, A.R., Ahmed, W., Pandya, S., Maddikunta, P.K.R., Alazab, M., Gadekallu, T.R.: A survey of explainable artificial intelligence for smart cities. Electronics **12**(4), 1020 (2023)
7. Balakumar, P., Ramu, S.K., Vinopraba, T.: Dynamic pricing for load shifting: reducing electric vehicle charging impacts on the grid through machine learning-based demand response. Sustain. Cities Soc. 105256 (2024)
8. Machlev, R., et al.: Explainable artificial intelligence (XAI) techniques for energy and power systems: review, challenges and opportunities. Energy AI **9**, 100169 (2022)

9. Dobrovolskis, A., Kazanavičius, E., Kižauskienė, L.: Building XAI-based agents for IoT systems. Appl. Sci. **13**(6), 4040 (2023)
10. Pang, X., Luan, C., Liu, L., Liu, W., Zhu, Y.: Data-driven random forest forecasting method of monthly electricity consumption. Electr. Eng. 1–15 (2022)
11. van Tilburg, J, Siebert, L.C., Cremer, J.L.: MARL-iDR: multi-agent reinforcement learning for incentive-based residential demand response. In: 2023 IEEE Belgrade PowerTech, pp. 1–8 (2023)

Healthchain - Bidimensional Blockchain for Collaborative and Secure Deep Learning in Healthcare

Rafael Marin Machado de Souza[1,2(✉)], Andrew Holm[3], Marcio Biczyk[1], and Leandro Nunes de Castro[1,2,3]

[1] In.lab, Clinics Hospital, Medicine Faculty, São Paulo University (HCFMUSP), São Paulo, SP, Brazil
rafael.marin@hc.fm.usp.br
[2] Graduate Program in Technology, School of Technology, State University of Campinas (Unicamp), Limeira, São Paulo, SP, Brazil
[3] Florida Gulf Coast University, Fort Myers, FL 33965, USA
ldecastrosilva@fgcu.edu

Abstract. The growing interest in ensuring patient privacy has led hospitals to increasingly restrict access to medical records and exams data. On the one hand, it has increasingly guaranteed security, on the other hand, it has led them to become silos of private data, making it difficult or even preventing the training of machine learning models to aid diagnoses. Seeking to reconcile the best of these two scenarios, *swarm learning* came as a framework that has enabled model training without the need for data sharing, unifying the distributed training of Federated Learning with the security of Blockchain. This paper introduces the structure and functioning of the *Healthchain project*, a swarm network that allows the training of multiple machine learning models and the sharing of results between different hospitals through a bidimensional architecture based on blockchain. A Proof of Concept (PoC) was developed to validate its operation. In the experiments performed, the network was used for *transfer learning*, sharing models and training, and for carrying out FedAVG filtered by nodes, a federated learning aggregation method that allows selecting the training that will be used for the calculation, avoiding the use of node training with low data quality. The paper also shows the evaluation of growth rates of the size of training blocks according to the data used.

Keywords: Patient data privacy · Swarm Learning · Federated Learning · Blockchain

1 Introduction

The application of artificial intelligence (AI) in the medical field has revolutionized the way of diagnosing and treating diseases. However, several challenges

R. Chinthaginjala et al. (Eds.): DCAI 2024, LNNS 1259, pp. 99–108, 2025.
https://doi.org/10.1007/978-3-031-82073-1_10

need to be overcome, among them, the incompatibility between systems (there are cases in which in the same hospital there may be several different systems), the lack of data for training, the sensitivity of patient data making hospitals private data silos increasingly more isolated from one another. These can generate models with low training quality, which in turn can provide inaccurate information to the physicians, making diagnosis difficult or even putting the patient at risk [1,2].

Seeking to solve these challenges, two technologies stand out: *blockchain* and *federated learning*. Blockchain, a distributed ledger technology, has gained prominence with the rise of cryptocurrencies. However, its potential goes far beyond that. Blockchain's ability to provide transparency, security, and immutability makes it an important tool in various applications. Federated learning, by contrast, is an emerging machine learning paradigm that emphasizes collaboration between multiple agents or nodes to achieve a common goal without exposing sensitive data and allowing shared training. Developed by Google [3,4] to typing prediction on mobile devices, it is now used in several applications. The combination of these two technologies is called *swarm learning*, a distributed form of collaborative machine learning [5,6].

This paper proposes an open architecture for swarm learning, the *healthchain project*, an approach that integrates hospitals for multiple machine learning models training without the need for data sharing. This architecture is designed to address the challenges of data privacy and security in healthcare, while enabling collaborative development.

The paper is organized as follows. Section 2 provides an overview of the swarm learning and blockchain technologies to enlighten the used principles, their applications, and some related works. Section 3 presents the healthchain blocks, a blockchain-based bidimensional structure, and the network arrangement. Section 4 assesses the developed proof-of-concept with transfer and federated learning for a deep learning model, and shows the growth rates of blocks with a large language model (LLM). Section 5 discusses the results and further works.

2 Swarm Learning

Swarm Learning is a decentralized machine learning architecture [5,6] whose main objective is to preserve the data's privacy for training, testing, and application, using distributed processing. Model training takes place on edge devices, directly at the location that holds and owns the data used, thus eliminating the need for data sharing and increasing security and privacy.

As can be seen in Fig. 1, the *swarm network* is made up of components known as *nodes*. The Swarm Edge Nodes are responsible for training machine learning models locally on your own data and the Swarm Leader Nodes are responsible for merging and aggregating the training and making the trained data available to other nodes via blockchain and it is elected for each new round of training [5,7].

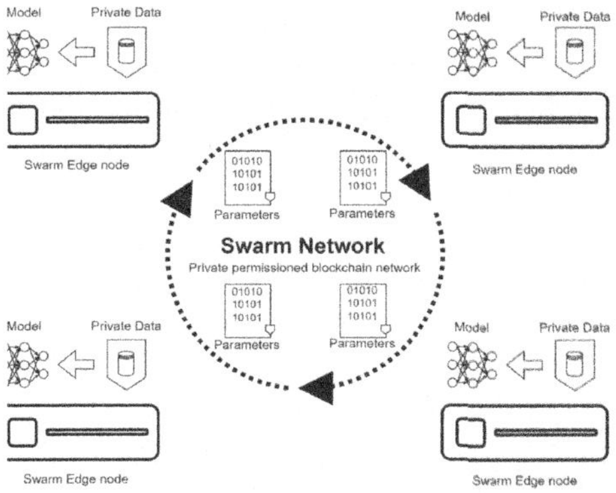

Fig. 1. Swarm network architecture [5,7].

2.1 Blockchain Architecture

Blockchain architecture forms the fundamental framework that underlies the decentralized and secure nature of blockchain technology. This section explores the key components that together constitute the structure and functionality of a blockchain network, as originally introduced by Satoshi Nakamoto [8] (Fig. 2).

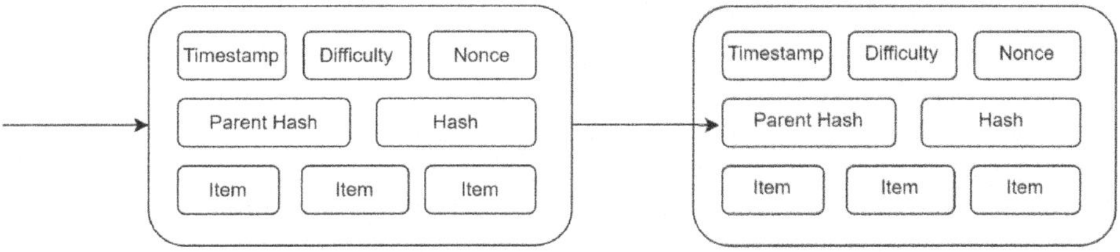

Fig. 2. Blockchain architecture as proposed by Satoshi Nakamoto [8].

A *block* stands as a foundational element within the blockchain architecture, constituting a list of data authenticated by a hash signature. It is formed by control structures to ensure the security of information and the continuity of the blockchain. A block is composed of a *timestamp* (the date and time of the creation of the block), a *difficulty* (reflects the number of leading zero digits required in the calculated hash), a *nonce* (a number is used once to solve the challenge proposed by the difficulty parameter), and a *parent hash* (the reference to the hash of the preceding block, except for the genesis block, the first block of the chain).

In a blockchain network, each node is provided with a complete chain to facilitate the creation of new blocks, forming a foundational element of the network's operation. As the network contemplates a blockchain update, the size of the chain emerges as a critical consideration, particularly concerning larger, more established chains which often present scalability challenges impacting transaction speed and overall efficiency. Within this dynamic environment, the *Longest Chain rule* serves as the guiding principle for determining the legitimacy of the blockchain's state amidst the decentralized consensus process. When multiple nodes propose competing blocks, this rule stipulates that the chain exhibiting the greatest cumulative Proof-of-Work (PoW, a consensus algorithm used to ensure that all nodes agree on the validity of the blocks) effort is recognized as the valid chain. Through collective acknowledgment of the chain showcasing the most significant computational investment, nodes actively contribute to strengthening the integrity and immutability of the blockchain.

Swarm Learning allows data to be applied to deep learning models closer to the data source [6,9]. Blockchain technology allows multiple edge devices, collectively called a swarm, to share learning with each other reliably and prevents malicious actors from gaining unauthorized access to the swarm.

2.2 Swarm Learning Applications and Related Works

Swarm learning applications can be found in several areas. In medicine, for example, it is used to share image analyses for classification and diagnosis models of various diseases, such as some types of cancer, COVID-19, tuberculosis, leukemia and lung pathologies [5,10].

It also assists in training computer vision models of autonomous cars, enabling the collaborative use of algorithms in autonomous driving systems while preserving data privacy. By sharing model training instead of raw data, it is possible to create effective and privacy-respecting self-driving systems [11,12].

Swarm learning has also been used in industrial robotics, assisting in the dynamic layout optimization of reconfigurable robotic assembly cells using multiple Swarm Edge nodes as software agents for distributed training [13]. In finance, it is used to perform credit analysis and scoring [9], improve fraud detection [14], allowing systems to learn from a variety of data sources to make more accurate predictions. In the energy area, it is used to predict demand for electrical energy consumption through distributed analysis of renewable energy grids [15] or even for trading and sharing between IoT devices [16].

These are just some sample applications of swarm learning. It has the potential to impact any field that benefits from machine learning, distributed data analysis, and privacy preservation. In the next section we propose an architecture for this technology, a project called Healthchain, where we seek to demonstrate its use for distributing both models and training gradients.

3 The Healthchain Project

The objective of this work is to develop an architecture that allows the sharing of multiple models that can be trained with private data, so that the training can also be shared in a distributed way between hospitals without compromising the patients' privacy and without dependence on proprietary standards. To this end, a blockchain-based format is proposed to be used for storing and distributing models and training, as can be seen in Fig. 3.

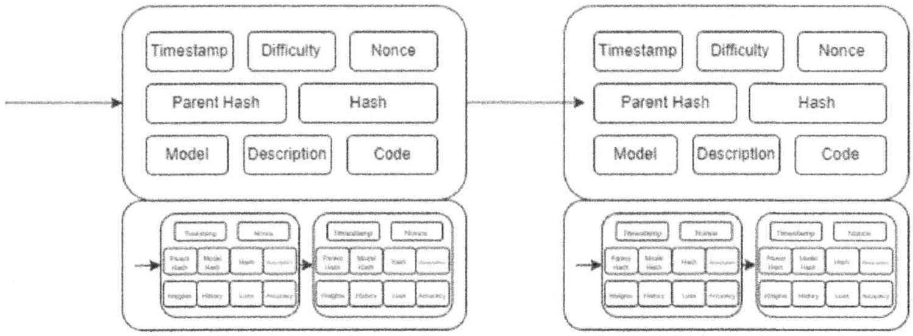

Fig. 3. Healthchain blocks architecture.

The healthchain stores the serialized data of the models in the main blocks and the training blocks stores the serialized training. This approach seeks to allow models and training to be sent to edge devices separately. Along with the models, usage descriptions are also made available, with instructions for their correct use and the model code, if the developer intends to make it available.

A training block is composed of *description* (a complete description of the training's content), *weights* (the gradients resulting from the training carried out by the device), *history* (the evolution of training in relation to epochs), *loss* (a measure of efficiency defined by cost functions, which measures the errors of the training. By reducing this index indicates a better performance of the models' training), and *accuracy* (another measure of training efficiency, that indicates how accurate the model is, according to the comparisons to the labels).

3.1 Healthchain Architecture

In the Healthchain architecture, a swarm network is formed by hospitals that cooperate to update the chain. These must be pre-authorized and properly inserted into the network, receiving a complete copy of the blockchain, as illustrated in Fig. 4.

The Healthchain is distributed among the Swarm Leader nodes via peer-to-peer communication. Every time a hospital's chain is updated, it is sent to all other hospitals, which in turn update their own chains using the propagation principle of the largest current, previously mentioned.

As it is a two-dimensional chain, both dimensions are tested during the update to validate hashes of model blocks and training blocks to guarantee the security and integrity of the information. If there is competition in the update, model blocks are prioritized.

Communication between swarm edge nodes and leader nodes is carried out via Rest API, and edge nodes have more autonomy in communication, being able to add new models for the network or new training for existing models.

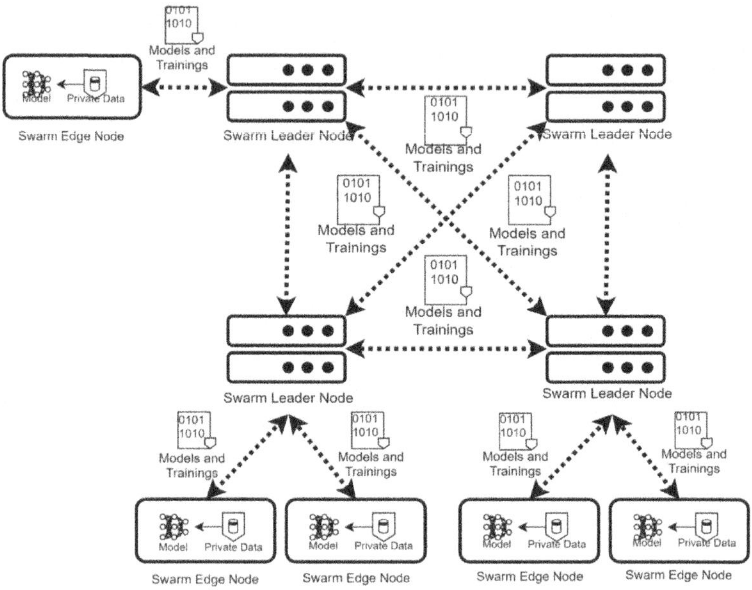

Fig. 4. Healthchain network architecture.

3.2 Proof-of-Concept

The Healthchain network aims to store deep learning models and allow their sharing and training between nodes, being composed of two main dimensions: 1) the *model dimension*, where each node can upload its deep learning models to the network. Models are stored immutably in blockchain blocks and the model metadata (architecture and hyper-parameters) are recorded serialized; and, 2) the *training dimension*, where nodes can access the training carried out by others, as well as send the training they carry out with their own data to the network.

A proof of concept was developed in Node.js. During testing, the edge nodes share each change with the network, which is validated by the leaders and then distributed to the other leaders, using the following operational flow:

- *Model Registration*: A node uploads its deep learning model and requests the creation of a new block in the first dimension of the healthchain, recording the model details.
- *Training*: The node that uploaded the model can perform training with its data and request the inclusion of a block in the model's training dimension.
- *Resource Sharing*: Once validated and created by the leader node, both the model and training are made available to the network. The other leader nodes receive the new chain, perform continuity validation, and update their own chain using the longest chain rule.
- *Access to Trained Models*: Models and training are available for download to other nodes, which can use them to carry out new training with their own

data and request the inclusion of a node in the training dimension of the model used.

- *Training and Validation*: In this way, no data is edited during the process, guaranteeing immutability and the validations carried out by the network guarantee the authenticity and integrity of the entire chain.

4 Performance Assessment

To test the proposed framework, Healthchain was used by adding to the model block a neural network with two hidden layers of 32 neurons each, and 10 neurons in the output layer, developed in Python using Tensorflow [23]. It was applied to the Mnist dataset [21], containing 60,000 images, with an initial training of 5 epochs applied to a random 30% of this dataset.

In the first scenario (Task A), the previous pre-trained model was used to continue the training applied to another random section of 30% of the same dataset, applying another 5 training epochs to this new data. In a second scenario (Task B), the training result from Task A was used as a starting point. The dataset was changed to contain 40% of false classifications in order to generate noise in the training and reduce the model's accuracy. In the last scenario (Task C), the average of training with accuracies greater than 0.8 was applied to another random section of 30% of the dataset. All three scenarios were run 20 times to evaluate the best results per task. The test results can be seen in Fig. 5.

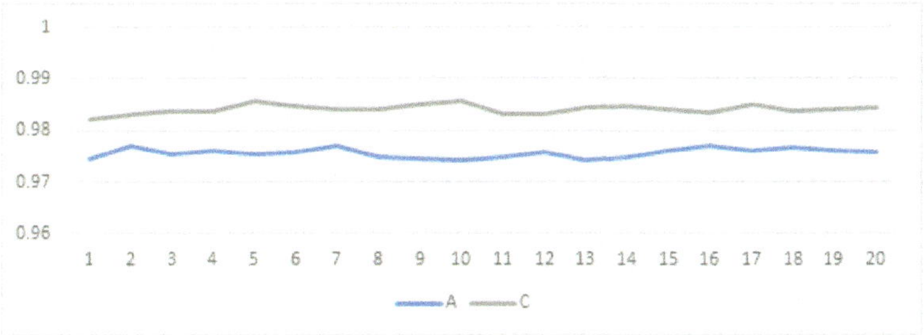

Fig. 5. Model accuracy for each task.

As Task B is noisy, it was not presented in the figure, and all the accuracies of this task were below 0.62. In all rounds, Task C, which had the FedAVG filtered by nodes, showed better performance when compared to task A, which presented transfer learning from the main model as a starting point for training new data. This shows to have avoided the noise generated by nodes with low data quality.

To evaluate the growth rate per Healthchain training block, the nanoGPT project [20], developed in Python using Pytorch [22], was used as a model. The model was chosen because it is an implementation of a Large Language Model (LLM) that allows the use of databases of different sizes, with training in token format. To carry out the training, it were used free e-books provided by the Guttenberg project [19]. As can be seen in Table 1, ten cumulative tests were carried out with 500 training epochs per set, where in each test a new book was added to the model's training data. In the last test ten books were used.

Table 1. Growth rate of training block

Books	Characters	Tokens	Growth rate (%)	Training Block Size (MB)	Growth rate (%)
1	565618	509056	0.00	32.64	0.0000
2	806474	725826	42.58	32.66	0.0478
3	1126344	1013709	99.14	32.67	0.0919
4	1359792	1223812	140.41	32.70	0.1838
5	1878817	1690935	232.17	32.71	0.2145
6	2490745	2241670	340.36	32.74	0.3064
7	2807037	2526333	396.28	32.74	0.3064
8	3231076	2907968	471.25	32.75	0.3370
9	3683614	3315252	551.25	32.77	0.3983
10	4009101	3608190	608.80	32,77	0.3983

Table 1 also shows that, although the size of the training block is large (32 Mb) compared to the Mnist test model (approx. 1.5 Mb), its growth rate does not keep up in proportion to the data, which demonstrates that training sizes are related to the models' architecture.

5 Discussion and Future Trends

The Healthchain architecture was used in the context of deep learning, offering an open solution for immutability, integrity, decentralization and collaboration, where the Healthchain network allows multiple users to share computing resources to train models. It does not create dependency on any proprietary standard, can be developed in any modern language, and allows models and training to be shared through serialization. It can be used for Transfer Learning and Federated Learning, enabling the implementation of a FedAVG filtered by nodes, preventing the result from being influenced by low data quality nodes, and allowing the average to be carried out with all training that is above a custom indicator, as shown in the proof-of-concept section.

To evaluate the Healthchain growth pattern, an LLM model requiring large processing and storage capacities for training was used. Free e-books were used for model training, showing that increasing data growth does not influence the size of the stored training. This allows large models also to be used in the architecture, such that nodes with less computational capacity use large models that have already been trained.

Some further works are planned for the project, including the implementation of more security layers, usage logs, and training data compression. Security will be enhanced through advanced protocols and encryption to ensure data integrity and confidentiality. Training and usage logs will provide transparency and traceability, enabling efficient audits and accountability. Training Data compression will allow Healthchain to handle larger data volumes, improving network efficiency.

Acknowledgments. This study was funded by FAPESP (grant number 2021/11.905-0 and process number 2023/13355-3).

Disclosure of Interests. The authors have no competing interests to declare that are relevant to the content of this article.

References

1. Bajwa, J., Munir, U., Nori, A., Williams, B.: Artificial intelligence in healthcare: transforming the practice of medicine. Future Healthc J **8**(2), e188–e194 (2021)
2. Jiang, L., et al.: Opportunities and challenges of artificial intelligence in the medical field: current application, emerging problems, and problem-solving strategies. J. Int. Med. Res. **49**(3), 03000605211000157 (2021)
3. McMahan, H.B., Moore, E., Ramage, D., Agüera y Arcas, B.: Federated learning of deep networks using model averaging. ArXiv. (2016). https://doi.org/10.48550/arXiv.1602.05629
4. McMahan, H.B., Moore, E., Ramage, D., Hampson, S., Aguera y Arcas, B.: Communication-efficient learning of deep networks from decentralized data. In: Proceedings of the 20th International Conference on Artificial Intelligence and Statistics (AISTATS) 2017, vol. 54 (2017)
5. Warnat-Herresthal, S., Schultze, H., Shastry, K.L., et al.: Swarm learning for decentralized and confidential clinical machine learning. Nature **594**, 265–270 (2021). https://doi.org/10.1038/s41586-021-03583-3
6. HPE - Hewlett Packard Enterprise Swarm Learning Framework. https://developer.hpe.com/platform/swarm-learning/home/. Accessed 02 Mar 2024
7. Han, J., Ma, Y.F., Han, Y., Zhang, Y., Huang, G.: Demystifying swarm learning: a new paradigm of blockchain-based decentralized federated learning. ArXiv preprint arxiv: arXiv:2201.05286 (2022)
8. Nakamoto, S.: Bitcoin: a Peer-to-Peer Electronic Cash System (2008). https://bitcoin.org/bitcoin.pdf
9. John, A.P., Devaraj, J., Gandhimaruthian, L.: Swarm learning based credit scoring for P2P lending in block chain. Peer-to-Peer Netw. Appl. **16**, 2113–2130 (2023). https://doi.org/10.1007/s12083-023-01526-5
10. Truhn, D., et al.: Encrypted federated learning for secure decentralized collaboration in cancer image analysis. Med. Image Anal. (2024). https://www.sciencedirect.com/science/article/pii/S1361841523003195, https://doi.org/10.1016/j.media.2023.103059

11. Mishra, A., Jefferson, O.P.J., Kapur, P., Kannur, K., Agarwal, P., Arya, A.: Swarm learning in autonomous driving: a privacy preserving approach. In: Proceedings of the 2023 15th International Conference on Computer Modeling and Simulation (ICCMS '23), pp. 271–277.Association for Computing Machinery, New York, NY, USA (2023). https://doi.org/10.1145/3608251.3608271
12. Huang, X., Yin, H., Chen, Q., Zeng, Y., Yao, J.: DAG-based swarm learning: a secure asynchronous learning framework for Internet of Vehicles. Digital Commun. Netw. (2023). https://doi.org/10.1016/j.dcan.2023.10.004
13. Wang, L., Wang, Z., Gumma, K.: Multi-agent cooperative swarm learning for dynamic layout optimisation of reconfigurable robotic assembly cells based on digital twin. J. Intell. Manuf. (2024). https://doi.org/10.1007/s10845-023-02229-7
14. Asrori, S.S., Wang, L., Ozawa, S.: Permissioned blockchain-based XGBoost for multi banks fraud detection. In: Neural Information Processing 29th International Conference (ICONIP 2022). Virtual Event, November 22–26. Springer-Verlag, Berlin, Heidelberg (2022). https://doi.org/10.1007/978-3-031-30111-7
15. Antal, M., Mihailescu, V., Cioara, T., Anghel, I.: Blockchain-based distributed federated learning in smart grid. Mathematics (2022). https://doi.org/10.3390/math10234499
16. Zekiye, A., Özkasap, Ö.: Blockchain-based federated learning for decentralized energy management systems. In: 2023 Fifth International Conference on Blockchain Computing and Applications (BCCA), Kuwait, Kuwait, pp. 186–193 (2023). https://doi.org/10.1109/BCCA58897.2023.10338852.
17. Wang, H., Mao, D., Chen, Z., Rao, H., Li, Z.: Blockchain-based decentralized Federated learning model. In: 2023 4th International Conference on Information Science, Parallel and Distributed Systems (ISPDS) (2023). https://doi.org/10.1109/ispds58840.2023.10235493
18. Gaži, P., Ren, L., Russell, A.: Practical settlement bounds for longest-chain consensus. In: Handschuh, H., Lysyanskaya, A. (eds.) Advances in Cryptology – CRYPTO 2023. CRYPTO 2023. Lecture Notes in Computer Science, vol. 14081. Springer, Cham (2023). https://doi.org/10.1007/978-3-031-38557-5_4
19. Guttenberg Project – Online library of Free eBooks. https://www.gutenberg.org/. Accessed 8 Mar 2024
20. Karpathy, A.: NanoGPT Project. https://github.com/karpathy/nanoGPT. Accessed 08 Mar 2024
21. LeCun, Y., Cortes, C., Burges, C.: MNIST handwritten digit database. http://yann.lecun.com/exdb/mnist. Accessed 20 Feb 2024
22. PyTorch Project. www.pytorch.org/. Accessed 08 Mar 2024
23. TensorFlow – an end-to-end platform for machine learning. https://www.tensorflow.org/. Accessed 20 Feb 2024

An Ant Colony Optimization Approach to Optimize Biblical Sermons

Bruno C. S. Lima[1(✉)], Ismar S. Frango[1], and Leandro N. de Castro[1,2,3]

[1] Mackenzie University, São Paulo, SP, Brazil
bcmusic2@hotmail.com
[2] Graduate Program in Technology, School of Technology, State University of Campinas (Unicamp), Limeira, São Paulo, SP, Brazil
[3] Human-Centered Artificial Intelligence and Natural Computing Research Group, Florida Gulf Coast University, 33965 Fort Myers, FL, USA
ldecastrosilva@fgcu.edu

Abstract. Religious literature is among the most widely read ones. The Bible is the normative corpus for Christians. However, it is not a trivial text to read, due to its high degree of literary and linguistic variability, making sermon construction a laborious activity. This work proposes a method, called SwarmaBle, for automating the construction of Biblical sermons through a combination of natural language processing and swarm intelligence. The study combines quantitative and qualitative assessments of SwarmaBle, exploring its sensitivity to two hyperparameters and presenting a discussion about expert-evaluated recommended passages. The results show a consistent performance of SwarmaBle, achieving an average performance above 70% in the quantitative analysis. Domain experts reported positive evaluations regarding SwarmaBle's ability to recommend meaningful Biblical passages for the given topic.

Keywords: Biblical Sermon Construction · Swarm Intelligence · Ant Colony Optimization · Natural Language Processing · Holy Bible

1 Introduction

Religion plays a pivotal role in shaping human culture and morality, with the Bible serving as a significant influence for over 72% of the global population [1–3]. This impact extends beyond spirituality, making the Bible a subject of interest for both religious individuals and researchers in diverse fields.

A *sermon* is a discourse delivered by a religious leader, such as a Priest or Pastor, during religious services [4]. Found in various religious traditions, sermons primarily aim to provide religious instruction, guidance, and spiritual insight. Religious leaders craft sermons through a process involving prayer, reflection, theme selection, Biblical passage choice, and exegesis [5,6]. From a *Natural Language Processing* (NLP) perspective, analyzing the Bible poses challenges due to its complex nature, encompassing diverse languages, genres, figurative language, translation differences, historical contexts, and theological depth [7–10].

R. Chinthaginjala et al. (Eds.): DCAI 2024, LNNS 1259, pp. 109–118, 2025.
https://doi.org/10.1007/978-3-031-82073-1_11

The *Ant Colony Optimization* (ACO) algorithm, inspired by ant foraging behavior, is a *Swarm Intelligence* technique [11]. ACO, efficient in solving combinatorial optimization problems, is proposed in this paper, integrated with NLP techniques, to automate the selection of Biblical passages for sermons. ACO ants navigate a graph of Biblical passages, selecting contextually relevant routes.

Given the popularity of the Holy Bible in society, and the social benefits that religious literature has provided [12,13], this article may represent a significant step towards the automation of pastoral homiletic and, consequently, the expansion of the benefits historically derived from biblical literature.

2 SwarmaBle: An ACO Approach to Select Passages for Biblical Sermons

SwarmaBle is the name given to the Swarm Intelligence Approach based on the Ant Colony Optimization algorithm developed to select Biblical passages.

2.1 SwarmaBle Process Flow

The SwarmaBle flowchart can be summarized as follows: **1) Load the Biblical Corpus** to be used to find the passages that will compose the sermon. The user can choose to use the whole Bible or specific Books to build the sermon; **2) Input Theme or Biblical Passage** either by simply choosing a theme (e.g., spiritual growth, happiness at work, harmonious family, etc.), or by selecting a specific Biblical passage that contains the main message to be delivered in the sermon; **3) Parameterize the Output** by choosing the number (N) of passages in the sermon. Also, define the search amplitude, that is, which Book is going to be included in the search and sermon; **4) Build the Bible Graph** containing all the Bible verses as its nodes and their connections to the target theme or passage. The edge weights are proportional to the similarity degree between the passages (edges). This graph is named *Bible Graph* and is the main structure to be used in the ACO search; and **5) Search for the Optimal Sermon** by running the ant colony optimization algorithm over the graph until a pre-defined stopping criterion is met.

The first three steps of the SwarmaBle flowchart are basically initialization and parametrization steps. Building the Bible graph and performing the ACO search require more complex stages detailed in the following subsections.

2.2 Building the Bible Graph

The main steps for text preparation used in SwarmaBle are: tokenization, stop-words removal, and text representation. In the specific case of text representation, SwarmaBle uses FastText, a method for text representation that generates *word embeddings* [14,15] that capture semantic information by considering sub-word information. The result of using FastText embeddings is a vector representation of words in the text, where each word is associated with a high-dimensional vector in a continuous space.

The Bible graph is built from the results obtained in the NLP phase, in which the similarities between Biblical verses are calculated for the whole Bible verses or the specific Book chosen by the user. In the Bible graph, the nodes or vertices are Bible verses, and the edges are the connections among verses. The graph is fully connected and the weights of the edges are their similarities, calculated by the Cosine similarity measure.

Formally, the Bible graph, Bg, is a complete graph $Bg = (V, E)$, where V is the set of vertices or nodes (Biblical verses), and E is the set of edges that connect all verses. The set of vertices V is defined as $V = v_1, v_2, v_3, ..., v_n$, where v_i represents the i-th Biblical verse. The set of edges E is defined so that each pair of verses is connected by an edge, therefore, for a complete graph $E = (v_i, v_j)|v_i, v_j \in V, i \neq j$. This indicates that there is an edge connecting each verse v_i to every other verse v_j, where $i \neq j$.

Therefore, the Bible graph assembled for use by SwarmaBle is fully connected and weighted. It is important to highlight, however, that if the sermon theme is a new text instead of a Bible verse, then it is necessary to include this text in the graph and calculate its similarity to all other vertices in the graph. Else, if the sermon theme is a Bible verse, then, it is labelled as the initial vertex for the search and no new similarity has to be calculated.

The length of the path between nodes is given by the inverse of the sum of similarities. This means that the larger the sum of similarities, the shorter the length of the path. Equation 1 illustrates this concept.

$$L_k = \frac{1}{\sum_{E=1}^{n} \eta_{kE}} \tag{1}$$

where, L_k is the length of the path, η is the similarity, and E is the verse.

2.3 ACO to Search the Bible Graph

The search in SwarmaBle has the following steps:

1. **Initialization**: In SwarmaBle, the number N of ants is equal to the number of verses that will compose the sermon. Hyperparameters α and β will be initialized with different values to assess the sensitivity of the algorithm to them. Parameters ρ and Q will be set using standard values from the literature. After initializing the hyperparameters, the N ants are placed in N randomly chosen nodes (verses) of the graph and the iterative process starts and continues until the maximal number of iterations, $maxit$, is achieved.
2. **Search Strategy**: Each ant has to select a next node to move to until it has passed through N nodes in the graph. Differently from the standard ACO in which all nodes in the graph have to be traversed, in SwarmaBle only a fraction of the graph's nodes is traversed. The selection of the next node to move to follows the probabilistic rule:

$$p_{ij} = \frac{(\tau_{ij})^{\alpha}(\eta_{ij})^{\beta}}{\sum_{k \in N_i} (\tau_{ik})^{\alpha}(\eta_{ik})^{\beta}} \tag{2}$$

where τ_{ij} is the amount of pheromone on edge (i, j), η_{ij} is the desirability of edge (i, j), N_i is the feasible neighborhood of node i, α and β control the relative importance of the pheromone versus the desirability, respectively.
The last step in this procedure is the deposition of pheromone in the traversed edges following Eq. 3:

$$\Delta\tau_{ij} = \frac{Q}{L_k} \tag{3}$$

where $\Delta\tau_{ij}$ is the amount of pheromone laid on edge (i, j) by the k-th ant, Q is a constant pheromone amount, and L_k is the cost of the k-th ant's tour.

3. **Pheromone Update**: The last procedure in the ACO algorithm is responsible for the pheromone evaporation (Eq. 4) and the application of elitism, that is, the reinforcement of one or more best routes (ants) in the swarm:

$$\tau_{ij} = (1 - \rho)\tau_{ij} + \Delta\tau_{ij} \tag{4}$$

where ρ is the evaporation rate, and $\Delta\tau_{ij}$ is the amount of pheromone ants deposit on edge (i, j): $\Delta\tau_{ij} = \sum_k \Delta\tau_{ij}^k$, and k is the index of ants.

4. **Fitness Function and Objective**: The fitness is the sum of all edge weights (cosine similarity) in each route (sermon) constructed by the ants:

$$f(k) = \sum_{i=1, i \in S}^{N} w_i(k) \tag{5}$$

where $f(k)$ is the fitness of ant k, N is the number of verses in the sermon, and S is the set of verses selected by ant k.
The objective of the SwarmaBle search is to maximize function $f(k)$:

$$max_k(f(k)). \tag{6}$$

In other words, the problem consists of finding the path with the highest fitness, which means that the passages chosen to compose the sermon have the highest possible similarity among themselves when brought together in a sermon.

3 Performance Evaluation and Discussion

This paper performs two types of assessments with SwarmaBle: a quantitative and a qualitative analysis. In the quantitative analysis some different combinations of α and β are tested and the performance of the algorithm is measured using Eq. 5. In the qualitative analysis a panel of experts is created and given a number of questions whose answers are used to qualify the proposed passages for the sermons.

3.1 Experimental Methodology

It were performed a sequence of SwarmaBle executions varying α and β, which control the relative importance of pheromone and the heuristic desirability, respectively: $\{0.5, 1.0, 2.0, 5.0\}$. The other parameters were fixed, as follows: $maxit = 3,000$, $\rho = 0.001$, and $Q = 0.01$. For each possible combination of the parameters, 5 executions will be run to quantify the mean and standard deviation of the algorithm's performance, measured by Eq. 5.

Two distinct approaches to input the themes for the search algorithm (Step 2) will be tested:

1. **Using a Biblical passage** as an input sentence, i.e., choosing a verse that synthesizes the theme for the sermon. The following passages were tested:
   - **Gn 1.1**: "In the beginning God created the heavens and the earth."
   - **Jn 3.16-17**: "For God so loved the world that he gave his one and only Son, that whoever believes in him shall not perish but have eternal life. For God did not send his Son into the world to condemn the world, but to save the world through him."
   - **Mk 9.1**: "And he said to them, 'Truly I tell you, there are some standing here who will not taste death until they see that the kingdom of God has come with power."'
2. **Constructing a theme** by typing a sentence or question containing the theme for the sermon development. The following questions were constructed:
   - $Q1$: "How to achieve happiness and faith in the midst of suffering?"
   - $Q2$: "How should we pray to God?"
   - $Q3$: "What do you need to do to live a life of sanctification and repentance?"

3.2 Quantitative Analysis

Table 1 summarizes the results. For each search sentence, we tested the four different values for α, which were shown in four sequential rows in the table, and β, shown in the table columns. The values presented are the mean and standard deviation over five runs for each combination.

From the results, in most cases a moderate value of α (around 1 or 2) yields the best performances. This suggests that while the collective knowledge of the ant colony is important, relying too heavily on it (as in the case of $\alpha = 5$) can lead to suboptimal solutions, possibly because it makes the algorithm less capable of exploring new solutions. The results also show that higher values of β (around 2 or 5) tend to yield better performances, suggesting that the individual knowledge of the ants is crucial for finding good solutions.

In summary, an increase in β usually produced an increase in the fitness function value, whilst an increase in α either did not significantly change the performance, or promoted its decrease, mainly for larger values. These results suggest that privileging the similarity between pairs of Biblical passages had a greater positive influence in the sermon quality when compared with the pheromone influence in the search process.

3.3 Qualitative Analysis

The qualitative analysis of the recommended passages was performed by domain experts, e.g., a Priest or a Pastor. A *Panel of Experts* (PE) was established and the criteria for composing the PE included the requirement of pastoral experience and theological training, ensuring that all evaluators had an academic background associated with the creation and presentation of sermons.

Table 1. Summary of results obtained by SwarmaBle.

Beta					
Sentence	**Alpha**	0.5	1	2	5
Gn 1.1	0.5	0.49 ± 0.29	0.50 ± 0.08	0.51 ± 0.05	0.52 ± 0.05
Gn 1.1	1	0.51 ± 0.01	0.52 ± 0.07	0.52 ± 0.08	0.52 ± 0.05
Gn 1.1	2	0.50 ± 0.08	0.50 ± 0.01	0.50 ± 0.02	0.51 ± 0.05
Gn 1.1	5	0.47 ± 0.08	0.47 ± 0.08	0.48 ± 0.03	0.51 ± 0.07
Jn 3.16-17	0.5	0.58 ± 0.01	0.60 ± 0.04	0.62 ± 0.05	0.63 ± 0.04
Jn 3.16-17	1	0.61 ± 0.05	0.63 ± 0.07	0.63 ± 0.05	0.63 ± 0.04
Jn 3.16-17	2	0.61 ± 0.01	0.61 ± 0.01	0.63 ± 0.01	0.63 ± 0.04
Jn 3.16-17	5	0.52 ± 0.01	0.54 ± 0.57	0.58 ± 0.01	0.63 ± 0.05
Mk 9.1	0.5	0.67 ± 0.01	0.67 ± 0.01	0.69 ± 0.08	0.71 ± 0.0
Mk 9.1	1	0.69 ± 0.01	0.70 ± 0.05	0.70 ± 0.05	0.70 ± 0.04
Mk 9.1	2	0.66 ± 0.01	0.67 ± 0.02	0.68 ± 0.01	0.71 ± 0.0
Mk 9.1	5	0.61 ± 0.02	0.65 ± 0.01	0.65 ± 0.01	0.69 ± 0.08
Phrase 1	0.5	0.88 ± 0.01	0.89 ± 0.01	0.90 ± 0.05	0.92 ± 0.0
Phrase 1	1	0.90 ± 0.01	0.92 ± 0.01	0.92 ± 0.01	0.92 ± 0.04
Phrase 1	2	0.85 ± 0.08	0.86 ± 0.01	0.90 ± 0.01	0.92 ± 0.05
Phrase 1	5	0.83 ± 0.02	0.82 ± 0.02	0.85 ± 0.01	0.90 ± 0.01
Phrase 2	0.5	0.76 ± 0.01	0.78 ± 0.01	0.80 ± 0.01	0.81 ± 0.0
Phrase 2	1	0.79 ± 0.08	0.80 ± 0.01	0.80 ± 0.01	0.81 ± 0.0
Phrase 2	2	0.75 ± 0.01	0.77 ± 0.02	0.80 ± 0.05	0.80 ± 0.08
Phrase 2	5	0.73 ± 0.01	0.73 ± 0.03	0.76 ± 0.01	0.79 ± 0.01
Phrase 3	0.5	0.80 ± 0.08	0.83 ± 0.01	0.86 ± 0.07	0.86 ± 0.04
Phrase 3	1	0.84 ± 0.08	0.85 ± 0.08	0.86 ± 0.08	0.87 ± 0.0
Phrase 3	2	0.81 ± 0.03	0.80 ± 0.02	0.84 ± 0.01	0.86 ± 0.04
Phrase 3	5	0.74 ± 0.03	0.75 ± 0.03	0.80 ± 0.02	0.85 ± 0.08

The methodology to evaluate the process involved conducting six SwarmaBle executions, each one providing a set of ten biblical passages for each input sentence or sermon theme. The PE had a general presentation about the Swarmable goal, and eight questions were formulated for them to score using a Likert scale ranging from 0 (wort) to 5 (best):

1. Did the selected passages for the sermon met your expectations?
2. Do you believe the selected passages exhibit sufficient diversity?
3. Do you believe the selected passages are immediately related with the theme?
4. Do you believe the selected passages are remotely related with the theme?
5. Do you believe the selected passages have a theological context related to the theme or the entry sentence?
6. Is it possible to create a sermon using the recommended passages?
7. Would you use the proposed algorithm to select passages for your sermons?
8. What grade would you attribute to the overall performance of the SwarmaBle algorithm as a recommender of Biblical passages based on the input?

The list below shows an example of the passages (verses) selected by SwarmaBle for each input phrase or verse:

1. Search Sentence: Gn 1.1 ($\alpha = 5$ and $\beta = 5$)
   Sermon Verses: 1 Ch 16.14; Jn 1.2; Gn 1.1; Ps 68.8; Hb 11.16; 1 Kgs 8.27; Is 37.16; Act 17.24; 2 Kgs 19.15; Rev 14.19
   Best Solution Fitness: 53%
2. Search Sentence: How to achieve happiness and faith in the midst of suffering ($\alpha = 5$ and $\beta = 5$)
   Sermon Verses: Pv 9.10; Gal 5.22; 1 Tim 2.15; Phil 2.1; Eph 4.19; 1 Thess 1.3; 1 Pe 5.9; Col 1.11; Jn 3.15; Rom 2.7
   Best Solution Fitness: 86%
3. Search Sentence: Mar 9.1 ($\alpha = 5$ and $\beta = 2$)
   Sermon Verses: 1 Kgs 22.8; Matt 16.28; 2 Kgs 3.14; Num 9.8; 1 Sam 16.5; Ezek 13.6; Exod 3.7; Mar 9.1; Lk 9.27; Is 35.4
   Best Solution Fitness: 69%
4. Search Sentence: How should we pray to God? ($\alpha = 5$ and $\beta = 1$)
   Sermon Verses: Jon 2.1; 1 Cor 3.17; 2 Cor 6.16; Ps 43.4; 2 Kgs 19.15; Jer 23.23; Act 7.32; Mal 1.9; 1 Jn 4.15; Jn 9.31
   Best Solution Fitness: 84%
5. Search Sentence: Jn 3.16-17 ($\alpha = 5$ and $\beta = 5$)
   Sermon Verses: Mar 10.45; Ezr 10.2; Jn 3.16; Gal 1.4; Matt 27.54; 1 Jn 4.9; Hb 11.7; Rom 5.10; Jas 1.27; Tit 1.2
   Best Solution Fitness: 63%
6. Search Sentence: What do you need to do to live a life of sanctification and repentance? ($\alpha = 5$ and $\beta = 5$)
   Sermon Verses: Matt 10.39; Dt 30.19; 1 Tim 4.8; Rom 2.7; Lk 18.30; Hb 7.25; Jn 12.25; 2 Thess 1.11; 2 Tim 3.7; 1 Cor 3.22
   Best Solution Fitness: 80%

To illustrate the output, let us take the case of **Gn 1.1**: "In the beginning God created the heavens and the earth". The recommended verses were:

- **1 Chronicles 16:14**: "He is the Lord our God; his judgments are in all the earth."
- **John 1:2**: "He was with God in the beginning."

- **Genesis 1:1**: "In the beginning, God created the heavens and the earth."
- **Psalm 68:8**: "the earth shook, the heavens poured down rain, before God, the One of Sinai, before God, the God of Israel."
- **Hebrews 11:16**: "Instead, they were longing for a better country-a heavenly one. Therefore God is not ashamed to be called their God, for he has prepared a city for them."
- **1 Kings 8:27**: "But will God really dwell on earth? The heavens, even the highest heaven, cannot contain you. How much less this temple I have built!"
- **Isaiah 37:16**: "Lord Almighty, the God of Israel, enthroned between the cherubim, you alone are God over all the kingdoms of the earth. You have made heaven and earth."
- **Acts 17:24**: "The God who made the world and everything in it is the Lord of heaven and earth and does not live in temples built by human hands."
- **2 Kings 19:15**: "And Hezekiah prayed to the Lord: 'Lord, the God of Israel, enthroned between the cherubim, you alone are God over all the kingdoms of the earth. You have made heaven and earth."'
- **Revelation 14:19**: "The angel swung his sickle on the earth, gathered its grapes and threw them into the great winepress of God's wrath."

It can be observed that the passages selected by SwarmaBle for the input phrase "In the beginning God created the heavens and the earth" (Gn 1.1) cover a range of themes and contexts, which can be seen in the following analysis:

- **1 Chronicles 16:14** and **John 1:2** reiterate the concept of God as the creator and ruler of all the earth, aligning well with the input phrase.
- **Genesis 1:1** is the exact input verse, indicating that the algorithm can identify and return the input if it is appropriate.
- **Psalm 68:8** and **Isaiah 37:16** provide a more detailed depiction of God's power over the natural world, expanding on the theme of creation.
- **Hebrews 11:16** and **1 Kings 8:27** introduce the concept of a heavenly country or dwelling, which could be linked to the idea of 'heavens' in the input phrase.
- **Acts 17:24** and **2 Kings 19:15** emphasize God's sovereignty and the insignificance of human-made temples, potentially providing a contrast or extension to the theme of God's creation.
- **Revelation 14:19**, while less directly related to the input phrase, introduces the theme of judgement, which could be used to explore the consequences of God's sovereignty over the earth.

In summary, the selected passages offer a diverse range of verses that touch on the themes presented, providing a rich set of resources for sermon creation. Figure 1 presents the scores attributed by each of the three PE members.

According to the evaluators, SwarmaBle demonstrated satisfactory performance in optimizing Bible passages for sermon preparation. Specifically, the PEs' responses to Q1 indicated that the passages returned met their expectations, resulting in above-average performance. The results to Questions 6 to 8, which focused on the evaluators' willingness to use the technology, indicated

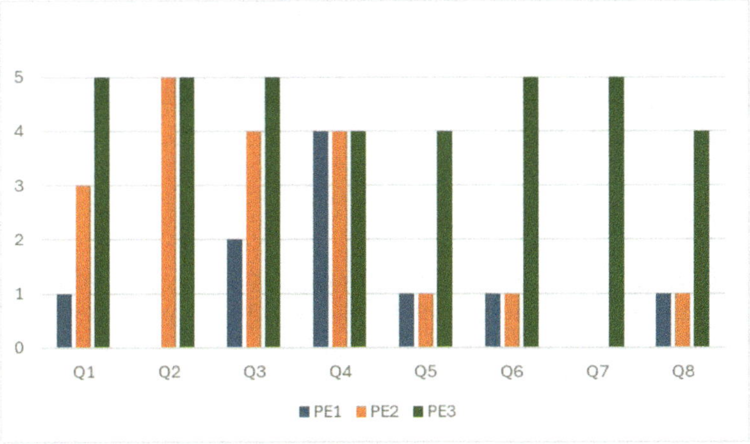

Fig. 1. Grouped bar chart for the panel of experts.

some resistance to the adoption of AI techniques, suggesting a need to address taboos and promote the benefits of these technologies within the community.

4 Conclusions and Future Works

This paper explored the feasibility of applying the ACO to optimize Bible passages for sermon preparation. Despite the unique characteristics of the biblical corpus, which make it a challenging text set to manipulate and interpret, ACO proved to be effective. The results showed consistent behavior of ACO across various experiments, providing confidence in the algorithm's reliability. The algorithm exhibited relatively fast convergence, typically within approximately 1,000 iterations. Among the hyperparameters, the heuristic information (beta) had the most significant positive influence on improving the fitness function.

Domain experts unanimously agreed that SwarmaBle performed well in generating Bible passages with contextual diversity relevant to the theme or input sentence, indicating its utility. However, the validation process also revealed resistance from theological field experts towards adopting new technologies. This underscores the importance of promoting information and awareness campaigns to highlight the benefits of technology and reduce this resistance.

Unlike technologies such as ChatGPT, which aim for textual creativity from a prompt, the proposed model facilitates the sermon creation process for Priests and Pastors rather than replacing them. Future research will explore other natural computing techniques, such as evolutionary and immune algorithms, to optimize the biblical corpus further. Additionally, there are plans to investigate other linguistic phenomena beyond semantics, such as syntax and lexicography. Gathering more expert feedback, especially understanding why certain experts gave low scores, will provide valuable insights for improvement. The authors also aim to explore whether this algorithmic structure can be adapted to index other literary works.

Acknowledgments. This research is funded by Grant 2021/11905-0, São Paulo Research Foundation (FAPESP).

References

1. Maoz, Z., Henderson, E.A.: The World Religion Dataset, 1945-2010: Logic, Estimates, and Trends. Int. Interact. **39**(3), 265–291 (2013). Routledge (2013). https://doi.org/10.1080/03050629.2013.782306
2. Amore, R.C.: Religion and Politics: New Developments Worldwide. MDPI, Basel, Switzerland (2019)
3. Beyers, J.: Religion and culture: revisiting a close relative. HTS Theological Stud. **73**(1), 1–9, Scieloza (2017)
4. Eugene, L.: Lowry. The Sermon as Narrative Art Form, Westminster John Knox Press, The Homiletical Plot (2001)
5. van Eijnatten, J.: Preaching, Sermon and Cultural Change in the Long Eighteenth Century, Brill, Leiden, The Netherlands (2008). https://doi.org/10.1163/ej.9789004171558.i-414
6. Klein, W.W., Blomberg, C.L.: Robert L. Hubbard, Introduction to Biblical Interpretation, Thomas Nelson (2004)
7. Piotrowski, M.: Natural Language Processing for Historical Texts. Morgan & Claypool Publishers (2012)
8. Murai, H.: Introducing scientific methods for the interpretation of the Bible: quantitative analysis of Christian documents. In: 2012 13th ACIS International Conference on Software Engineering, Artificial Intelligence, Networking and Parallel/Distributed Computing, IEEE (2012)
9. Murai, H.: Exegetical science for the interpretation of the bible: algorithms and software for quantitative analysis of Christian documents. In: Software Engineering. Artificial Intelligence, Networking and Parallel/Distributed Computing, Springer (2013)
10. Popa, R.C., Goga, N., Goga, M.: Extracting knowledge from the Bible: a comparison between the old and the new testament. In: 2019 International Conference on Automation, Computational and Technology Management (ICACTM), IEEE (2019)
11. Dorigo, M.: Thomas Stützle. MIT Press, Ant Colony Optimization (2004)
12. Sen, H., Colucci, L., Browne, D.: Keeping the faith: religion, positive coping, and mental health of caregivers during COVID-19. Front. Psychol. **12**, 67–86 (2022)
13. Aten, J., et al.: The psychological study of religion and spirituality in a disaster context: a systematic review. Psychol. Trauma **11**, 597–613 (2019)
14. Joulin, A., Grave, E., Bojanowski, P., Douze, M., Jégou, H., Mikolov, T.: FastText.zip: compressing text classification models (2016). https://arxiv.org/abs/1612.03651
15. Bojanowski, P., Grave, E., Joulin, A., Mikolov, T.: Enriching Word Vectors with Subword Information (2017). https://arxiv.org/abs/1607.04606

AI Folk: Sharing Machine Learning Models in a Multi-Agent Community

Andrei Olaru(✉), Alexandru Sorici, Mihai Nan, and David-Traian Iancu

National University of Science and Technology POLITEHNICA Bucharest, 313 Splaiul Independentei, Bucharest, Romania
{andrei.olaru,alexandru.sorici,mihai.nan,david_traian.iancu}@upb.ro

Abstract. With the rapid growth in the number of available pre-trained machine learning (ML) models for common tasks, with different performance, focus, and capabilities, complex problems can increasingly be solved through adequate choice of model, more than through training or tuning new models. In this paper we introduce the AI Folk methodology to address the challenge of autonomously managing ML models in a community of agents which can use and exchange semantic information about the models that they are using. We present a proof-of-concept implementation in an autonomous driving setting tackling various practical challenges which arise when dealing with this goal.

Keywords: machine learning · multi-agent systems · semantic description

1 Introduction

Machine learning (ML) models are now used all around us, in various sizes, on various devices, handling tasks of various level of specialization. Libraries and stores of ML models are available online, such as Hugging Face[1] and the recently launched Qualcomm AI Hub[2], allowing the quick download of pre-trained models for a variety of tasks.

However, in complex scenarios, such as, for instance, autonomous driving, in which the activity of an agent lasts for a longer period of time, it may be impossible to say from the start what is the complete set of models that is going to be needed by an agent controlling an autonomous car, or what is the complete set of situations that the agent is going to have to address. Given the ability to use multiple models for the same task over its lifetime, an agent may benefit from information other agents have regarding those models.

The goal of this paper is to illustrate a practical means for agents in a multi-agent system (MAS) to use pre-trained machine learning models as a community,

[1] Hugging Face https://huggingface.co.
[2] Qualcomm AI Hub https://aihub.qualcomm.com/.

R. Chinthaginjala et al. (Eds.): DCAI 2024, LNNS 1259, pp. 119–128, 2025.
https://doi.org/10.1007/978-3-031-82073-1_12

being able to search for and exchange experience information regarding ML models.

This goal involves challenges related to formalizing information about the ML models, as well as practical issues regarding loading, using, and switching ML models during execution.

We place examples and a proof-of-concept scenario in the domain of autonomous driving,in cases where unexpected conditions arise, for which a specialized ML model is be more appropriate than what the agent currently uses:

> *An autonomous car travelling through a city enters a university campus, where the streets are shared between cars and pedestrians. The car encounters a large number of pedestrians walking in the streets and the current model for controlling the car stops the car abruptly, very often. Other autonomous cars in the campus use specialized models for navigating the campus. The agent contacts other agents in the local community and transfer a campus-specific model which it is able to use so that it will run slower, closer to pedestrians, and more smoothly.*

The issues in this scenario generalize to being able to describe ML models in order to be able to look for more appropriate models in a given situation; to identify models which match a given set of properties; and to integrate, in a running framework, new ML models and switch to them as appropriate. This is, in a nutshell, the goal of the AI Folk project[3].

We address the issue of dynamically selecting the model to use in situation when the distribution of the input data becomes very different from the training data distribution. This can happen in the case of lighter ML models or when the current input is very niche. Unlike federated learning, where a larger model is trained distributively, we target selecting the most appropriate pre-trained model.

Given our research goal, we have established the following objectives:

O1. Give agents the ability to use pre-trained machine learning models. Models are abstracted as black-box entities which have an input and an output;
O2. Give agents the ability to detect when the ML models currently in use become inappropriate for their current situation. This is the moment when they should attempt to identify a more appropriate model;
O3. Give agents the ability to create descriptions of their current situation and, conversely, to identify models that match a given description. This way, the agent can send and respond to queries regarding ML models;
O4. Give agents the ability to integrate new ML models in their model library and to switch between models as needed by the current situation;
O5. Create a methodology for achieving objectives O1-O4 in any scenario, identifying which aspects are scenario-invariant and which must be developed specifically for each scenario, and how that should be done.

[3] The AI Folk project https://aifolk.upb.ro.

We have instantiated the *AI Folk Methodology* – Objective O5 – on a proof-of-concept scenario similar to the one described above, implementing all the objectives for agents and separating generic and scenario-specific implementations, thus demonstrating how the methodology can be applied.

After discussing related work, we present the AI Folk architectural elements and the methodology in Sect. 3, followed by practical challenges involved by the implementation in Sect. 4. The design of the experimentation protocol 5 is followed by the conclusions.

2 Related Work

The focus of AI Folk is on leveraging crowd-based reasoning to increase the efficacy and efficiency of individual agent decision making in the face of dynamic changes in the *characteristics* of the environment in which they operate. To do so, it is necessary for agents to be able to *describe*: (i) task-specific characteristics of the data they perceive, and (ii) characteristics of the *models* they use in prediction and decision making, including the data on which those models have been trained. When there is a mismatch between observed data characteristics and used model characteristics (a domain shift), an agent will seek to update its decision making model using *experiences* of other agents.

ML Model sharing is an increasingly common practice under the paradigm of Federated Learning (FL). However, the aspects and challenges that are most frequently addressed in the FL literature refer to federation architectures, optimization of the communication mechanisms or security and privacy [2]. In [1], for example, the focus is on using ideas from MAS coalition formation, to address challenges in FL based on non-IID data, where individual agents that train models are assumed to be self-interested. The authors of [6] use similar mechanisms of auctioning and coalition formation to enable communication-efficient FL under a collaborative Internet of Vehicles setup.

However, in such works the *learning task* is already known before hand and the focus is on improving ML models, which are already known to be compatible, based on distributed local learning and a periodic synchronization with a global ML model. In contrast, the focus of the AI Folk approach is on being able to *identify* the most relevant ML models among peer agents, based on a description centred on the *data and model characteristics* of the learning task. It therefore becomes important to look at structured methods that enable such descriptions.

Describing ML Models. In AI Folk, the ML models we focus on are based on deep neural networks (DNN). The vast architectural space of DNNs makes the selection of an appropriate architecture for a task challenging. Ontologies emerge as a promising solution to describe DNNs in a structured manner, enabling support for querying, analysis and comparisons of various DNN architectures, as well as their configuration, training methodologies and evaluation procedures.

FAIRnets [7], for example, is an RDF dataset that stores information about existing neural networks. Using RDF and OWL, the system enables SPARQL querying for desired characteristics of DNNs. Additionally, detailed information and further links about network architectures are provided.

In AI Folk we choose to extend the ANNETT-O ontology [3], which offers a structured and computationally accessible vocabulary, enabling systematic documentation and analysis of DNN design and experimentation, with emphasis on aspects of evaluation, topology, and training.

Semantic Data Description. The other need in AI Folk is to use knowledge graphs (KG) to *semantically describe the data* on which a model is trained and operates. KG descriptions allow inclusion of existing *domain knowledge* into the desired functionality of data-driven prediction models. Consequently, ML models get better at predicting under setups of high data variability (e.g. HAR in videos on the Kinetics dataset [5]), and knowledge-based *explainability* can be achieved in multiple domains (e.g. image recognition, recommender systems, natural language applications) [11]. Closest to the AI Folk interest is the use of KG to index both general and task-specific *characteristics* of the data on which a model operates. OntoDM is a suite of ontologies which is extended to provide dataset-, task- and algorithm-level semantic descriptions for applications in Data Mining [4], with use cases ranging from neurodegenerative disease to earth observation datasets, to autonomous driving and scene explanation . The work highlights the nuScenes-QA dataset [9] as an enabler for improved driving scene understanding, since DNN models can setup learning tasks that answer to queries regarding the *type*, *number* and *status* of encountered traffic participants. Such works show that there is a need and a benefit in terms of understanding, performance and explainability in semantically describing data. In AI Folk we extended ANNET-O in a modular fashion, creating a *core*[4] and a *domain specific* vocabulary that facilitates describing learning tasks and their data characteristics. The resulting ontology allows agents to identify *changes* in data characteristics which are relevant for a specific learning task.

3 Architecture and Methodology

The main goal of the project is to create a generic architecture for a variety of scenarios in which a number of agents have to make decisions using machine learning models and agents have the possibility to exchange information about the models they are using. Differently from a distributed system in which entities collaborate on the same task, in a multi-agent system agents may have different goals, but cooperation my lead to better results for the agents involved.

During this research, we have addressed two application scenarios – autonomous driving and disaster response. In both of them, the global outcome can be improved when agents are able to switch the ML models that they are using. While this process could be useful for a greater variety of scenarios, we have defined several invariant elements and the AI Folk methodology specifies how they can be applied and adjusted for other types of scenarios.

Achieving the objectives stated in the Introduction requires several elements:

[4] The core ontology can be found on the project website.

- an agent deployment framework allowing communication between agents (O1 and O4);
- a means to work with ontologies (written in OWL-DL, for instance) (O2 and O3);
- a means to dynamically load and run machine learning models (O1 and O4);

In addition, development is substantially aided by means to validate designed functionality. Therefore, we have developed a method to run scenarios, that is, to provide agents with specific inputs at specific times and to get the output for verification.

For our proof-of-concept implementation, we have relied on a multi-agent system deployed using the FLASH-MAS framework[5], a Java-based modular platform built around the generic concept of *entities*, which can be agents, but also other persistent first-class entities in a MAS, such as nodes and communication infrastructures [8]. Although not mandatory, agents can be implemented as collections of *agent shards*, each shard implementing a specific functionality, with shards of an agent interacting by means of an event queue. We have chosen FLASH-MAS for its significant modularity and flexibility, not least because of the ability to implement non-agent components as first-class entities of the framework (as opposed to Jade, for instance).

Machine learning models are generally implemented in Python, so we have integrated in FLASH-MAS an interface to a local RESTful web service which can load and run ML models (see more in Sect. 4).

3.1 General Methodology Principles

The AI Folk approach applies to agent systems in which participants have to execute their predictive tasks in environments in which the degree of input variability is too large to be covered by a single prediction model. Such situations require agents to form *experience sharing communities*, in which agents collaborate to exchange appropriate prediction models with each other. The mentioned operating conditions necessitate the support of a framework that is based on the following design principles and requirements.

1. *Formation of experience sharing communities:* refers to giving support to agents to dynamically identify and communicate with the set of peers that have a *relevant experience*, in terms of similarity of task and data descriptions on which a prediction model has been successfully employed (where success is measured based upon standard or mutually agreed upon task-specific metrics). Thus, the agent community building principle is driven foremost by an information retrieval concern, as opposed to one related to logical / physical vicinity or speed of communication.
2. *Identification of salient prediction model, task and data characteristics:* refers to enabling to agents to describe *what* is changing in the task for which they have to use prediction models. AI Folk enlists the use of semantic web

[5] Fast and Lightweight Agent Shell https://github.com/andreiolaru-ro/FLASH-MAS.

technologies, such as ontologies, to describe task types, general and task-specific data attributes, DNN model architectures, DNN model evaluation datasets and evaluation performance. Note that we currently focus on the case where data characteristics that agents should pay attention to (because they can cause distribution-shifts) are defined based on human domain knowledge. The methodology can be extended in future work to include approaches where a data-driven model itself is responsible to highlight and semantically map the data characteristics which it considers are relevant in the prediction and which have changed.

3. Transfer, loading and running of prediction models: refers to giving support to agents to reference shared models, transfer model parameters, dynamically configure a model deployment environment and run the inferences required by their prediction task. AI Folk envisions a system that supports both (i) cloud-based setups, where agents run predictions in an already configured cloud environment for which a web-based entry point and access credentials are provided when sharing it with other agents, and (ii) an edge-based setup that enables local running of prediction models and which requires methods to efficiently serialize and transfer model parameters and deployment configurations (in a manner similar to Federated Learning setups). Note that in AI Folk we currently focus on building support for cloud-based setups.

3.2 Scenario Specific Design

Our reference scenario presents the case of autonomous driving, where agents controlling the vehicles can encounter frequently changing *driving scenes* in an extended urban setting (involving both main street, as well as residential area locations). The specific prediction task we focus on is that of semantic segmentation of the drivable path based on video input. The community of experience sharing agents is that of the extended urban environment and agents use the cloud-based setup to run and exchange segmentation models.

The most important concern to address from the principles outlined in Sect. 3.1 is the definition of the semantic segmentation task specific vocabulary to describe the conditions under which the task is being performed. Informed by domain expertise, we adopt a competency question based procedure [10] to extend the general AI Folk ontology vocabulary with one that describes the specifics of data involved in a semantic segmentation task for the autonomous driving domain. By answering questions meant to qualify which aspects in a driving scene should be accounted for when analyzing the performance of a semantic segmentation model, the resulting vocabulary addresses descriptions of: (i) the type of driving scene (e.g. parking lot, residential area, main city street, rural area), (ii) the weather and illumination conditions (e.g. in daylight, at dusk or at night, in sunny, overcast or rainy conditions), (iii) properties giving statistics of the type of roads (e.g. with cross or T-shaped intersections) or the min/avg/max counts of different types of traffic participants (e.g. pedestrians, other vehicles). The extension of the core AI Folk ontology[6] to the needs of the

[6] The ontology for the auto driving scenario can be found on the project website.

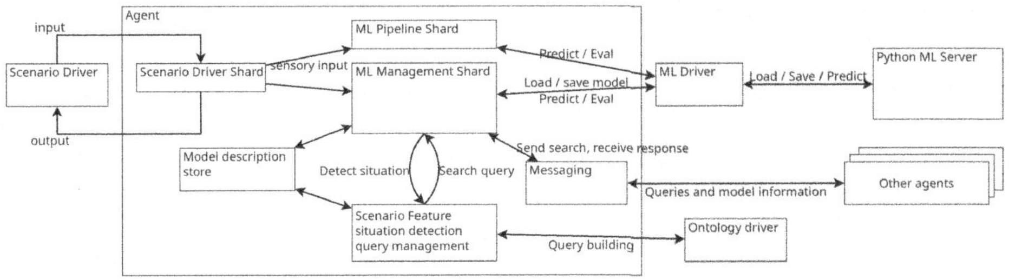

Fig. 1. The AI Folk architecture, showing the interaction between components inside and outside the agent.

reference scenario is used by agents to both describe their current task, as well as the datasets on which semantic segmentation models have been trained and evaluated on. The common description vocabulary is what enables search and fit of existing segmentation models to the current data characteristics of the task. The showcasing of agent communication, task description, model search and model use, as enabled by the AI Folk approach for the reference scenario, is given in Sect. 5.

4 Practical Challenges

Implementing the AI Folk methodology raised several relevant challenges. In terms of agentification, the challenge was to separate scenario-independent and scenario-specific components, and to model non-agent and sub-agent entities supporting the implementation of the methodology.

For support functionality, the open and modular nature of FLASH-MAS allowed us to create a new type of entity in – the *Driver*, which is local to a node (a machine) and is available to all agents on that node. We have implemented 3 drivers: the `ML Driver` interfaces with the Python server to relay operations related to ML models; the `Ontology Driver` interfaces with an OWL-DL ontology that integrates the description of models and driving situations; and the `Scenario Driver` is used in validation to feed input to agents and to verify their output, according to a scenario file.

Scenario-specific functionality – such as the logic for recognizing situations and choosing appropriate model – has been isolated in instances of a *Feature* sub-agent entity which is used by each agent in a scenario.

We have encapsulated AI Folk functionality inside the agent in three types of shards; the `ML Management Shard` keeps a list of the models currently usable by the agent, as well as the models currently in the pipeline, and it evaluates, via the scenario specific *feature*, whether there is a need to switch to a different model; the `ML Pipeline Shard` picks the input from the event queue and sends it to the ML model which is currently in use, posting the output back into the agent's event queue; the `Scenario Driver Shard` supports validation by feeding input data from the `Scenario Driver` to the agent's event queue, and by picking the corresponding output to send back to the driver for validation.

A challenge in deploying the AI Folk scenarios was the FLASH-MAS framework is implemented in Java, but most machine learning models are implemented in Python, and using dedicated ML libraries such as PyTorch and Tensorflow . We have solved this challenge by interfacing the `ML Driver` to a local Flash web server started automatically by the driver, offering a RESTful API supporting several operations: *listing* configured models, *adding* a new model, adding information about a *new dataset*, *exporting* a model, and using a model to *predict* an output based on the input.

ML models are not standardized. Each model comes with is own set of prerequisites, instantiation procedure and configuration parameters, input format and processing, and output processing. Therefor, to each model we have attached (1) a YAML file which contains configuration options, and (2) a Python file containing the methods to process the input and the output of the model according to its specific requirements. These transformations can be reused for multiple models, if possible. Hence, a model is fully defined by the model file (e.g. a `.pth` file), a YAML configuration, and a Python file with transformation code.

5 Design of AI Folk Evaluation Protocol

To apply the methodology presented in Sect. 3.1 we design an experimentation protocol, describing the application conditions, the experiment steps and the evaluation method.

The AI Folk approach is highly relevant in situations where in the environment of an agent a *data distribution shift* is expected to occur sufficiently often. Our reference scenario of semantic segmentation in autonomous driving, implying a large mobility of the self-driving agent and the possibility to experience significant changes of driving scenes, is a good example of this application condition. To simulate such conditions from existing datasets, the following steps are followed.

First, a domain of application (e.g. autonomous driving) and a main task (e.g. drivable road segmentation) are specified. For the main task, attributes which can influence it are identified (e.g. min/average/max number of pedestrians encountered over a recent time window). This leads to an instance of the core AI Folk ontology expressing the data characteristics which are deemed to influence an ML algorithm by existing human domain knowledge (see also Sect. 3.2).

Second, for the identified data attributes, a set of domain-knowledge informed *change conditions* are defined, whose objective it is to determine whether there is a *mismatch* between the characteristics of the data on which the currently employed ML model (which is solving the main task) was trained and the characteristics of the data from the currently experienced scenario (e.g. the minimum number of pedestrians detected over frames collected for the past minute is higher than the maximum number of pedestrians detected on any frame of the training set for the currently used segmentation model).

A detected mismatch signals the need to search for a replacement model, obtained from the *experience* of other agents in the community (see Sect. 3.1). Step three involves setting filtering criteria for searched models. New models need to have been trained on data that has *a suitable similarity* with respect to the characteristics of the attribute that caused the original mismatch. The filter is set up as a domain-knowledge informed reasoning rule that looks at the suitability of the data characteristics (e.g. the difference in average number of pedestrians is below a threshold), as well as the *performance* of the model on the main task (e.g. the pixel-wise accuracy for the segmentation of the drivable road class is above 95%).

The next step involves simulating the condition mismatch using one or more existing dataset(s) (e.g. Cityscapes or KITTI-360 for our segmentation task). A ML model (e.g. a DeepLab-v3 model) is finetuned on the main task on splits of data from a first selected dataset, which correspond to mismatching conditions on the identified attribute (e.g. number of pedestrians). A second ML model (e.g. YOLOv8) is finetuned on the same main task on a second dataset, on a split of data respecting the characteristics of the identified data attribute. Variation can be created in this case, by creating dataset splits that correspond to different characteristics of the selected data attribute (e.g. splits of data corresponding to different average pedestrian detection per 100 frames). Each finetuned model is given to an agent in the modeled *experience sharing* communities. Each agent has an evaluation holdout dataset.

To evaluate the experiment, agents run the mismatch identification rules and the corresponding search for a replacement model from the *experience* of other agents. The agent receives the most appropriate replacement model, based on the infrastructure and interactions described in Sect. 4. The current model and the replacement model are evaluated on the evaluation holdout dataset and if the percentage difference in the evaluation metric for the performance on the main task (e.g. pixel-wise accuracy / Dice-score for drivable road segmentation) shows, for all agents, a net increase in performance on the main task, then this constitutes an indication that reasoning on *domain knowledge informed mismatch conditions for key data characteristics* is a key proxy for anticipating performance improvement if a ML model were to be retrained on the current distribution shift. The main advantage of the AI Folk methodology is that, under the assumption of high data distribution shifts for high mobility of agents in their environment, reasoning over similar experiences in terms of domain-knowledge informed data characteristics is faster and more convenient than retraining or updating of the local ML model (as is typical in a federated learning setup).

6 Conclusion and Future Work

We have approached the issue of managing pre-trained ML models in a multi-agent community via semantic descriptions of the model and the ability to dynamically, seamlessly switch between models at runtime.

As future work, the AI Folk methodology will be applied to more complex scenarios and more effort will go into evaluating the performance of models such that it can be factored into the decision process.

Acknowledgments. This work was supported by a grant of the Ministry of Research, Innovation and Digitization, CNCS - UEFISCDI, project number PN-III-P1-1.1-TE-2021-1422, within PNCDI III.

Disclosure of Interests. The authors have no competing interests to declare that are relevant to the content of this article.

References

1. Arisdakessian, S., Wahab, O.A., Mourad, A., Otrok, H.: Coalitional federated learning: improving communication and training on non-IID data with selfish clients. IEEE Trans. Serv. Comput. **16**(4), 2462–2476 (2023)
2. Beltrán, E.T.M., et al.: Decentralized federated learning: fundamentals, state of the art, frameworks, trends, and challenges. IEEE Comm. Surv. Tutorials (2023)
3. Klampanos, I.A., Davvetas, A., Koukourikos, A., Karkaletsis, V.: ANNETT-O: an ontology for describing artificial neural network evaluation, topology and training. Intl J. Metadata Semant. Ontol. **13**(3), 179–190 (2019)
4. Kostovska, A., Džeroski, S., Panov, P.: Semantic description of data mining datasets: an ontology-based annotation schema. In: International Conference on Discovery Science, pp. 140–155. Springer (2020)
5. Ma, Y., et al.: Visual knowledge graph for human action reasoning in videos. In: Proceedings of the 30th ACM International Conference on Multimedia, pp. 4132–4141 (2022)
6. Ng, J.S., et al.: Joint auction-coalition formation framework for communication-efficient federated learning in UAV-enabled internet of vehicles. IEEE Trans. Intell. Transp. Syst. **22**(4), 2326–2344 (2020)
7. Nguyen, A., Weller, T.: FAIRNets search-a prototype search service to find neural networks. In: SEMANTiCS (Posters & Demos) (2019)
8. Olaru, A., Sorici, A., Florea, A.M.: A flexible and lightweight agent deployment architecture. In: 22nd Intl Conf on Control Systems and Computer Science (CSCS), Bucharest, Romania, 28-30 May 201, pp. 251–258. IEEE (2019)
9. Qian, T., Chen, J., Zhuo, L., Jiao, Y., Jiang, Y.G.: nuScenes-QA: a multi-modal visual question answering benchmark for autonomous driving scenario. In: AAAI Conference on Artificial Intelligence, vol. 38, pp. 4542–4550 (2024)
10. Ren, Y., Parvizi, A., Mellish, C., Pan, J.Z., van Deemter, K., Stevens, R.: Towards competency question-driven ontology authoring. In: Presutti, V., d'Amato, C., Gandon, F., d'Aquin, M., Staab, S., Tordai, A. (eds.) ESWC 2014. LNCS, vol. 8465, pp. 752–767. Springer, Cham (2014). https://doi.org/10.1007/978-3-319-07443-6_50
11. Tiddi, I., Schlobach, S.: Knowledge graphs as tools for explainable machine learning: a survey. Artif. Intell. **302**, 103627 (2022)

Android Malware Detection: Leveraging Deep Learning with Process Control Block Information

Heba Alawneh[1], Hamza Alkofahi[1], and David Umphress[2(✉)]

[1] Jordan University of Science and Technology, Irbid, Jordan
{hzalawneh,hoalkofahi}@just.edu.jo
[2] Auburn University, Auburn, AL, USA
david.umphress@auburn.edu

Abstract. The security and privacy of Android device users are seriously threatened by the sophistication and endurance of malware. The risks associated with malware assaults on Android are larger than ever since mobile devices are being utilized for sensitive transactions and data storage more and more. This paper proposes a dynamic malware detection system that mines data from Process Control Blocks (PCBs) during process execution to identify malicious activity using deep learning techniques. Our method provides robust detection of various threats by examining processes at this fundamental level. It precisely monitors changes in PCB parameters for all application threads functioning at the kernel level, in contrast to user-level malware detection. Our PCB mining approach outperforms existing methods in accuracy and scalability of captured PCB records. Our testing findings revealed a high success rate, capturing over 99 percent of context transitions. Utilizing a manageable sequence size of 12 PCBs, our deep learning detection model successfully identified malicious activity at various points in the process execution, achieving an accuracy of 95.6%.

Keywords: Dynamic Malware Detection · Deep Learning · Process Control Block mining · kernel-level monitoring

1 Introduction

Malware attacks have become increasingly common and sophisticated in recent years, posing a significant threat to individuals, businesses, and organizations worldwide. According to AV-Test Statistics [1], the number of newly developed malware has exceeded 90 million, encompassing viruses, worms, Trojan horses, rootkits, and ransomware. These malicious programs can cause diverse harm, ranging from credential theft to system damage and unauthorized remote access. Conventional malware detection methods are often inefficient, slow, or have limited code coverage. Real-time malware detection is challenging, especially on mobile devices, due to hardware and power limitations. While many research

R. Chinthaginjala et al. (Eds.): DCAI 2024, LNNS 1259, pp. 129–138, 2025.
https://doi.org/10.1007/978-3-031-82073-1_13

studies have focused on malware detection[2–8], dynamic malware detection on smartphones remains a challenge.

Our study utilizes Process Control Block Information (PCB) to detect Android malware. The PCB, also known as *task_struct* in the Linux OS, is a data structure that holds vital information about running processes in an operating system. It includes, but is not limited to, CPU scheduling, memory management, and signal information. Memory management information tracks memory usage, page faults, registers, and resource utilization, while signal information includes resource utilization, time, and thread counters. The *task_struct* also tracks the process's current state, relationships with other processes, and group threads.

In terms of process behavior, benign processes strictly adhere to access control policies. In contrast, malicious processes exploit system vulnerabilities to engage in unauthorized activities, such as stealing or spying. Notably, certain information stored in the PCB can indicate the presence of malicious behavior. This approach was first introduced by [8], and subsequent research studies have investigated leveraging PCB information for efficient malware detection [9–11].

The contribution of this research is as follows. We propose a new method for PCB mining in kernel space, synchronized with CPU utilization. Our approach effectively monitors the changes in PCB parameters throughout process execution, enabling comprehensive tracking of the process behavior and associated threads. Therefore, it facilitates dropper malware identification. We utilized the collected PCB dataset to propose a deep-learning-based detection model for Android malware. Moreover, we present a malware analysis framework that enables concurrent testing of various Android apps on multiple devices. To validate our approach, we collected the PCB data from 2,610 malware-free and 2,500 malware-infected Android apps.

The structure of the paper is outlined as follows. Section 2 offers an overview of relevant studies. Section 3 outlines our method for addressing the research problem. In Sect. 4, we discuss our experiments and results. Section 5 offers a comparative analysis of our study against previous comparable studies. Lastly, Sect. 6 summarizes our findings and outlines our future work.

2 Related Work

Based on the characteristics employed in the analysis, malware detection approaches may be classified into dynamic and static categories. Static detection, such as signature-based methods [2], compares program files against a database of known malware signatures, searching for matching code sequences. Other static methods [3,4] analyze the code and program structure without executing it. While static methods are typically fast, they fall short in identifying complex and zero-day threats.

In contrast, dynamic analysis involves monitoring programs behavior to detect malicious activities. Dynamic detection is more comprehensive and competent in identifying new malware, making it better suited for real-time detection. However, it is typically computationally and time-intensive. Existing dynamic detection methods examine malware through hardware usage, system activity, and information flow [5–7]. Several research efforts have explored the utilization of PCB information for detecting malicious behavior in programs running on Linux-based operating systems [8–11].

In [8], the author used customized system calls to extract PCB information from the target process at 1 ms intervals, identified by matching its name with the list of running processes. In contrast, previous studies [9–11] also employed kernel modules for PCB data transfer but initiated the process every 1 ms after identifying the target process. These prior methods, however, faced issues with userspace processes being preemptive, leading to delays and missed PCB records. Additionally, the frequent data collection introduced the risk of duplicate records. Furthermore, these studies primarily tracked the main thread of the application, potentially missing malicious behavior in other threads.

In conclusion, existing PCB-based approaches suffer from PCB record misses and long PCB data capturing and malware detection times. Our approach improves the efficiency and minimizes the processing overhead of PCB data capturing by mining the PCB data within the kernel space and reducing the required PCB records for detection.

3 Methodology

The main goals of this research were to develop an Android malware detection system capable of capturing Process Control Block (PCB) information and training a deep learning-based classifier for near real-time identification of malicious behavior. Our methodology involved modifying the Android kernel to collect a sequence of PCB parameters (119 kernel parameters) during process execution. We utilized over 11 million PCB records from 1,610 benign and 2,620 malware-infested applications to evaluate our approach for detecting Android malware.

3.1 PCB Mining

The PCB data collection process involves four key components: the central management unit (CMU) running on Linux, the application testing unit (ATU) running on Android in userspace, and the data collector and custom CPU scheduler running on Android at kernel level. The CMU directs the required actions for application testing while the ATU executes the tests and reports the results. The data collector creates communication channels between userspace and kernel space to transfer the PCB data gathered by the custom CPU scheduler.

The Central Management Unit (CMU) assigns and monitors tasks for connected test devices concurrently via multi-threading. It also monitors safe battery levels for application testing and user requests to suspend or terminate ongoing operations. The task manager (within the CMU) extracts the target apk's unique package name, handles application installation, permission granting, and installs data-gathering modules on the test device. The package name is then transmitted to the ATU to commence application testing and PCB mining.

Afterward, the task manager monitors the test's progress and reports success or failure. Upon success, it summarizes program context switches and their miss ratio (if any), storing the PCB file in a predetermined location. Failed tests are retried up to a maximum of r times before being skipped. Following each test, regardless of outcome, the device is reset to a fresh state for the next test.

The Application Testing Unit (ATU): Upon task assignment by the CMU, the ATU initiates the PCB buffer reader to transfer collected PCB records from kernel to userspace. It also sends the target package name (pname) to the data collector and activates the application exerciser (*monkeyrunner*) for n seconds, launching the target application. Upon completion of *monkeyrunner*'s execution, the ATU terminates the PCB buffer and signals the data collector to end. Refer to Fig. 1 for the Android application testing and PCB collection flowchart.

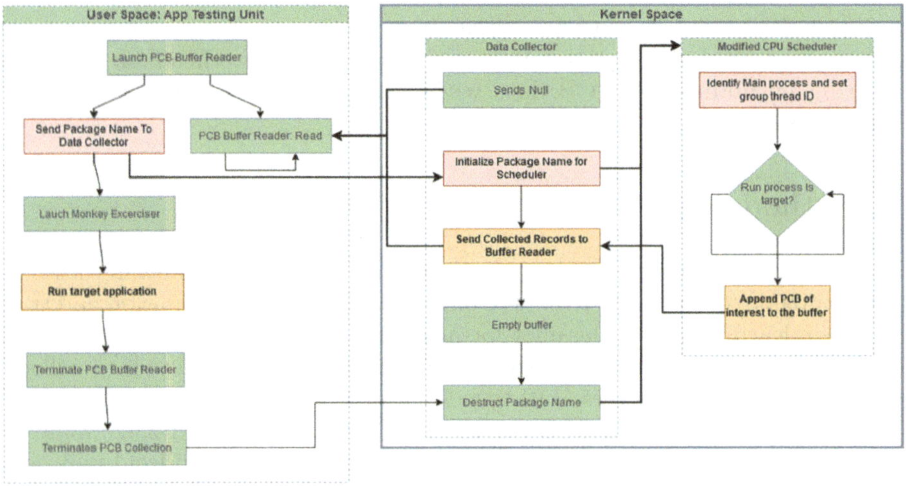

Fig. 1. The Control Flow Diagram of PCB Data Gathering During Application Testing.

The Data Collector, a runtime kernel module, enhances kernel functionality by facilitating communication channels between the kernel and the ATU for PCB gathering. It controls the start and termination of PCB gathering via signals to the CPU scheduler, ensuring reliability and preventing buffer overwriting through non-preemptive operations.

The CPU Scheduler manages CPU utilization for all threads, utilizing context switching to transition between active jobs. We customized the context-switching procedure to identify the main process and other threads executing the target application, storing their associated PIDs for identification. During context switches, the scheduler preserves outgoing target process PCB records, encapsulating behavior during the last CPU-time slice. Although context switch misses may occur before PID identification, this approach provides insights into the process's behavior throughout runtime.

3.2 Identifying Android Malware Attacks with Deep Learning and PCB Data

Our deep-learning classifier targets identifying malicious behavior through a sequence of n PCB records for an application. We collected PCB information for 2620 malware and 1610 benign samples. Each sample featured a variable thread count (ranging from 10 to 250), with the number of PCB records per thread dependent on its lifespan. The dataset was constructed by capturing and storing PCB data for all threads running each application over 25 s, resulting in 11 million PCB records. Given the dataset's scale and the objective of identifying malicious behavior, we employed a deep learning-based classifier. The classifier's input comprises a sequence of n PCB records, each containing 119 features. Preprocessing readied the data for model implementation.

Model Architecture: We adopted an RNN architecture to model sequence data, resembling the approach proposed in [12]. Illustrated in Fig. 2, our model integrates a sequence of Convolutional layers (CNNs) for knowledge extraction, a variant of recurrent layers (LSTM) for sequencing data management, and fully connected layers for conclusive decisions. Each layer's output shape is detailed above it. Our model partitions the time sequence into four segments, processing them individually through CNN, dropout, and pooling layers for feature extraction and reduction. Subsequently, the extracted data is flattened and treated as a sequence of four blocks for LSTM processing, facilitating the learning of generated sequences. We employed various techniques for generalization and downsampling, including dropout and pooling layers. The utilization of pooling layers aids in preventing exact sequence memorization, particularly beneficial given the variability in the onset of malicious behavior across samples. The architecture is concluded with one or more fully connected layers and a final output layer. Both CNN and LSTM layers employ ReLU activation.

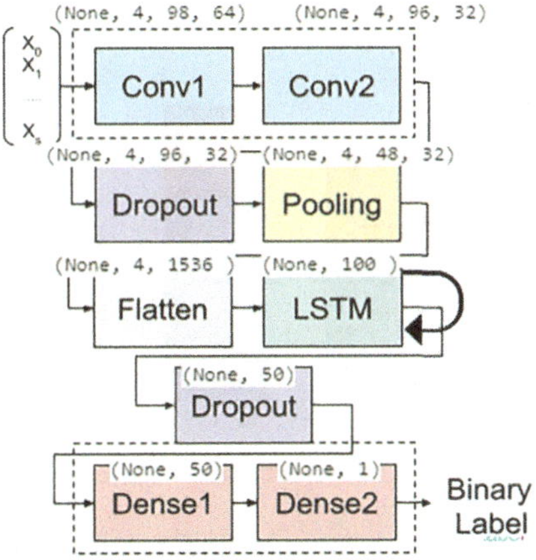

Fig. 2. The Neural Network Architecture for Detecting Android Malware.

4 Evaluation and Results

Various tools were used in this study. Raccoon [13] facilitated the collection of benign samples by directly downloading apps from the Google Play Store onto the PC. Malware samples were obtained from the Virustotal website through an academic research access license. Testing was conducted on OnePlus 5t smartphones running Android version 8.1. Initially, attempts to root the official OnePlus OS resulted in Wi-Fi functionality loss, leading to the adoption of LineageOS [14] and a custom kernel (v4.4.78) to meet research objectives.

The testing environment was automated using monkeyrunner [15], simulating human interaction through a sequence of clicks and touches. INetSim [16] further emulated diverse internet services, providing responses to network requests and stimulating applications to unveil additional functionalities.

A secured testing environment was set up using an ASUS RT-AC66U router running DD-WRT firmware, test Android devices, and a Raspberry Pi 3 hosting INetSim. The router was configured to use the Raspberry Pi as the DNS server, with AP-isolation enabled to prevent direct communication between connected devices and mitigate the risk of malware spreading across the network.

4.1 The PCB Data

PCB data was gathered from 1610 benign and 2620 malware-infected apps, forming the dataset available in [17]. The number of PCB sequences collected per app aligns with the runtime thread count. Figure 3 depicts the thread count distribution in benign and malware-infected apps, indicating that malware apps generally employ fewer threads than benign ones.

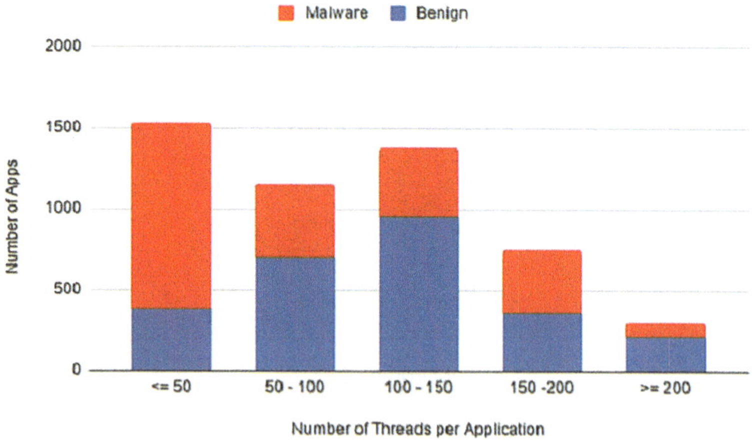

Fig. 3. The Distribution of the Number of Threads in Both Benign and Malware-infested Apps.

The automated test for each application involved 15,000 interactions over an average duration of 25 s, resulting in an average of 18,500 context switches. Figure 4 depicts the distribution of application tests based on the PCB miss ratio, reflecting the proportion of uncaptured PCB records compared to total context switches among all running threads within an application. Notably, our approach successfully recorded over 99% of context switches in 90% of tested applications, underscoring its effectiveness in accurately collecting PCB data from Android applications for precise behavior analysis.

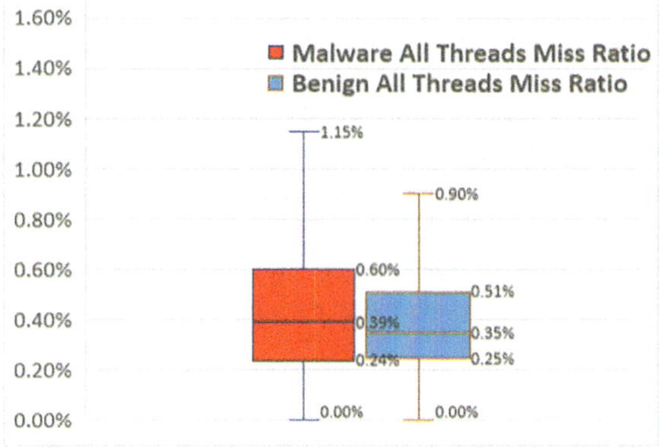

Fig. 4. Comparative Analysis: PCB Miss Ratios in Benign and Malware Applications Across All Threads.

4.2 Identifying Android Malware Using Deep Learning

We shuffled and partitioned the dataset into 60–40% training-testing. The PCB sequences were segmented into sub-sequences of size n, padded with zeros as needed while retaining original labels. Both training and testing datasets were reshuffled before use. Three classifier models were trained with sequence sizes n (12, 100, and 300 PCBs), evaluating accuracy on the test set. These sequence sizes were chosen to gauge how input sequence lengths impact model performance, spanning short, medium, and long sequences.

The accuracy achieved by the trained classifier on the test dataset indicated a high detection rate for collected malware samples across different execution stages, with the best classifier achieving an averageaccuracy of 95.6% using a 12-PCB sequence length, underscoring the effectiveness of our approach in identifying malicious activities within a concise PCB sequence. The models trained using sequence sizes 100 and 300 achieved a similar score of 95.2%.

5 Discussion

Methods proposed in [8–11] extract PCB records at 1 ms intervals using system calls from userspace. Yet, this strategy faces delays and interruptions as userspace processes have lower priority than kernel processes, risking PCB misses and duplicates. Moreover, timed PCB inquiries heighten the likelihood of missing or duplicating entries, and transferring one PCB record at a time incurs substantial overhead from kernel-to-userspace process switching.

Our approach retrieves PCB data in kernel space triggered by process context switching, significantly reducing the occurrence of misses and duplicates. Collecting and transmitting PCB records through a buffer also minimizes kernel process switching overhead. Another advantage is our ability to retrieve PCB data from all application threads, enabling more comprehensive analysis.

Utilizing the dataset from [9], we conducted a comparative analysis between their PCB mining approach and ours. While their method focused solely on the main thread, our approach successfully gathered PCB information for an average of 113 threads per application, demonstrating enhanced coverage. Our approach achieved a zero duplicate rate and a miss rate of less than 1%, compared to their reported 82% duplicate rate and nearly 53% miss rate. These results highlight our method's superior performance and effectiveness in capturing accurate PCB information. Refer to Table 2 for an overview of the distinguishing features of our mining approach compared to existing methods [8,9] (Table 1).

Table 1. Comparative Analysis: Dataset Statistics from Our Proposed Mining Approach and a Previous Method [9].

	Xinning et al. [9]	Our Dataset
Number of Samples	2550	5117
AVG Number of Threads	1	113
Main process Context Switches - AVG	5647	5578
Total Context Switches - AVG	5647	18431
Miss Ratio - AVG	53%	0.57%
Miss Ratio - SD	7.8	0.01
Duplication Ratio - AVG	82%	0%
Duplication Ratio - SD	8.1	0

Table 2. Comparative Analysis: Our Proposed Approach and Prior Methods

	Shahzad et al. [8]	Xinning et al. [9,10]	Our Approach
Android Version	Android 2.3.6	-	Android 8.1
Kernel Version	2.6.35.7	3.4.0	4.4.78
Data Collection Scope	Main process	Main process	Main process & all children
Requires Kernel Modification	Yes	No	Yes
Data Collection is Requested by	Userspace Reading Loop	Userspace Reading Loop	Kernel CPU-Scheduler Context Switching
Data Collection is Triggered by	Target process name found	Target pid found	Launching the matching target process name/ tgid
Data Collection & Transfer Rate	Every 1ms	Every 0.75ms	Every context switching \ variable transfer rate
Data Collection & Transfer Size	One PCB record	One PCB record	PCB patches
User & Kernel-space Communication Channel	System calls	Kernel module	Kernel module

6 Conclusion

In conclusion, our paper introduces two innovative methods: one for dynamically mining PCB information and the other for Android malware detection using deep learning. Operating at the kernel level and synchronized with CPU usage, our approach meticulously monitors PCB parameter changes for all application threads, facilitating thorough analysis of application behavior and detection of malicious activities by non-main threads. Our results showcase the robustness of our approach, capturing over 99% of context switches for the majority of tested applications and detecting malware at various execution stages using a compact sequence size of 12 PCBs, achieving an impressive accuracy of 95.6%. Additionally, we propose a closed framework for concurrent testing of multiple Android devices, enhancing scalability and efficiency. Comparative analysis with related research underscores the significant improvements we achieved in PCB data retrieval and utilization for malware detection.

Looking ahead, future work could explore integrating our approach into the Android operating system to provide real-time malware protection. Online classifier training could ensure continuous updates to combat emerging threats. Furthermore, combining our dynamic approach with static-based detection could yield a more comprehensive and robust malware detection system.

Acknowledgments. We acknowledge the VirusTotal for providing us with access to their valuable malware dataset. Their collaboration was crucial to the success of our research. We appreciate their support of our contribution to the cybersecurity community.

References

1. AV-TEST: Malware statistics (2021). https://www.av-test.org/en/statistics/malware/
2. urwithajit9: Classification of malware PE headers (2019). https://github.com/urwithajit9/ClaMP

3. Euh, S., Lee, H., Kim, D., Hwang, D.: Comparative analysis of low-dimensional features and tree-based ensembles for malware detection systems. IEEE Access **8**, 76796–76808 (2020)
4. Sangal, A., Verma, H.K.: A static feature selection-based android malware detection using machine learning techniques. In: International Conference on Smart Electronics and Communication (ICOSEC), pp. 48–51. IEEE (2020)
5. Ahmed, Y.A., Koçer, B., Huda, S., Al-rimy, B.A.S., Hassan, M.M.: A system call refinement-based enhanced minimum redundancy maximum relevance method for ransomware early detection. J. Netw. Comput. Appl. **167**, 102753 (2020)
6. Enck, W., et al.: TaintDroid: an information-flow tracking system for Realtime privacy monitoring on smartphones. ACM Trans. Comput. Syst. (TOCS) **32**(2), 1–29 (2014)
7. Eltaher, A., Abu-juma'a, D., Dania, H., Alawneh, H.: Design and implementation of a malware detection tool using network traffic analysis in android-based devices. In: 2023 IEEE Jordan International Joint Conference on Electrical Engineering and Information Technology (JEEIT) (2023)
8. Shahzad, F., Shahzad, M., Farooq, M.: In-execution dynamic malware analysis and detection by mining information in process control blocks of lInux OS. Inf. Sci. **231**, 45–63 (2013)
9. Wang, X., Li, C., Song, D.: CrowdNet: identifying large-scale malicious attacks over android kernel structures. IEEE Access **8**, 15823–15837 (2020)
10. Wang, X., Li, C.: Android malware detection through machine learning on kernel task structures. Neurocomputing **435**, 126–150 (2021)
11. Alawneh, H., Umphress, D., Skjellum, A.: Android malware detection using neural networks & process control block information. In: 2019 14th International Conference on Malicious and Unwanted Software (MALWARE), pp. 3–12 (2019)
12. Shi, X., Chen, Z., Wang, H., Yeung, D.Y., Wong, W.K., Woo, W.C.: Convolutional LSTM network: a machine learning approach for precipitation nowcasting. In: Advances in Neural Information Processing Systems, pp. 802–810 (2015)
13. Raccoon - APK downloader. https://raccoon.onyxbits.de/
14. LineageOS: Lineageos android distribution. https://lineageos.org/
15. Monkey - android developers. https://developer.android.com/studio/test/monkey
16. Inetsim project. https://www.inetsim.org/
17. Alawneh, H. Alkofahi, H.: Dataset for android intrusion detection using process control block information (2023). https://doi.org/10.17632/s87267tpp3.1

Integrating Pretrained VGG19 and Bi-LSTM for Violence Detection in Video

Pablo Negre[1(✉)], Ricardo S. Alonso[2,3], Javier Prieto[1], and Paulo Novais[4]

[1] BISITE Research Group, Universidad de Salamanca, Salamanca 37007, Spain
{pablo.negre,corchado,javierp}@usal.es
[2] AIR Institute, Av. Santiago Madrigal, 39, Salamanca 37003, Spain
ralonso@air-institute.com
[3] UNIR (International University of La Rioja), Av. de la Paz, 137, Logrono, Spain
ricardoserafin.alonso@unir.net
[4] Departamento de Informática, Universidade do Minho, Braga, Portugal
pjon@di.uminho.pt

Abstract. Physical aggression presents a pervasive challenge worldwide, disrupting various aspects of individuals' lives and societal functioning. This phenomenon stems from difficulties in emotional regulation, interpersonal conflicts, and socio-economic factors. Women and minors are particularly vulnerable demographics, facing high rates of violence in both intimate and public settings. Despite concerted efforts, violence detection remains a crucial issue with implications for public safety and social well-being. In this study, the challenge of violence detection in videos is addressed utilizing a combination of pre-trained VGG19 and Bi-LSTM layers. While promising results have been demonstrated in previous research utilizing VGG16, the potential effectiveness of VGG19 in this context has not been thoroughly investigated, although a larger number of convolutional layers should mean a better understanding of the scene. Moreover, the use of Bi-LSTM layers has been shown to be superior to the use of LSTM layers by up to 3%. A broad range of hyperparameter combinations is explored to optimize model performance. Positive results are obtained through experiments, with accuracies of 97%, 90%, and 73% achieved on the Hockey Fights dataset, Violent Flow Dataset, and RWF-200, respectively. Although the model does not surpass state-of-the-art approaches utilizing VGG16, it exhibits promise when compared to other proposals within the field. Overall, our study contributes to advancing violence detection methodologies, emphasizing the importance of leveraging deep learning techniques for improving public safety and social well-being.

Keywords: Violence detection · Video Surveillance · Pretrained VGG19 · Bi-LSTM · Fine-tuning

R. Chinthaginjala et al. (Eds.): DCAI 2024, LNNS 1259, pp. 139–149, 2025.
https://doi.org/10.1007/978-3-031-82073-1_14

1 Introduction

Acts of physical aggression pose a significant challenge in our society, exerting a global influence. This phenomenon disrupts various aspects of life, affecting not only the immediate targets and their psychological well-being [15], but also reverberating through families, communities, and the overall functioning of nations (including mobility, tourism, and commerce) [13]. The origins of aggressive conduct stem from difficulties in emotional regulation, interpersonal conflicts, and the socio-economic and demographic characteristics of communities [7].
Among the most affected demographics are women and minors. Concerning women, as per a report by the World Health Organization (WHO), roughly one in three women globally has encountered either physical or sexual violence from an intimate partner or non-partner during their lifetimes [19]. Research conducted by ActionAid revealed that 79% of Indian women, 86% of Thai women, 89% of Brazilian women, and 75% of women in London face harassment or violence in public spaces [25]. Regarding children, findings published by [10] in the American Academy of Pediatrics indicate that in 2015, at least half of children in Asia, Africa, and North America experienced violence, with over 1 billion children globally, aged 2–17, being subjected to such treatment.

In summary, physical aggression presents a pervasive challenge worldwide, impacting nearly every aspect of individuals' lives and entire societies. Protecting the physical well-being of people globally should be a fundamental priority and a universal right.

2 Violence Detection State of the Art

In this Section it will present the most widely used datasets of violence videos in the state of the art, as well as the different types of violence detection algorithms that have been used in the state of the art.

Various approaches to addressing violent behaviors within societies have been explored. Beginning with more indirect and enduring strategies, numerous studies endeavor to comprehend the circumstances or contexts that precipitate acts of violence, with the aim of rectifying them [15]. Conversely, other investigations, exerting a more immediate impact, scrutinize the correlation between crime rates and the urban settings in which they manifest [24]. However, the final line of defense for victims of violence lies in the real-time detection of such occurrences to promptly alert authorities for conflict resolution, as well as the capture of video or imagery that can subsequently aid in identifying the individuals involved [2,11,23].

Violence detection algorithms in videos typically follow the same basic steps [18]; these steps are shown in Fig. 1. First, the input video is processed to extract key frames, which are representative images of different moments in the video. Then, preprocessing is done on the frames before introducing the data to the algorithm. Next, the violence detection algorithm employs feature extraction techniques to capture relevant information from the frames and a training process on violent and non-violent videos. The algorithm classifies each frame as

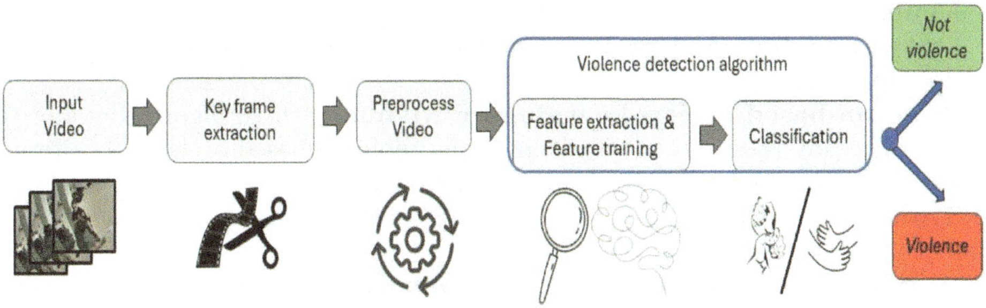

Fig. 1. Basic violence detection steps

"violent" or "non-violent" based on the extracted features and the trained classification model, thereby providing violence detection in the video. Considering that identifying violence constitutes an atypical task and that instances of physical aggression can vary significantly (in terms of participant count, type of aggression, severity level, etc.), ensuring an ample volume of data for algorithm training is paramount. Within the realm of contemporary research, numerous datasets have surfaced in recent years, albeit a select few are consistently utilized to enable comparison with findings from prior studies. Some of the most frequently employed datasets include [18]: *Hockey Fights* [4], which comprises footage from the National Hockey League (NHL); *Action Movies* [4], housing sequences from action-oriented films; *Violent Flow* [9], alternatively known as *Violent Crowd*, encompassing YouTube videos depicting authentic crowded scenarios with accompanying audio; and *Real World Fight-2000 (RWF-2000)* [5], comprising altercations captured by surveillance cameras in real-world.

A wide variety of algorithms is employed for violence detection, traditionally categorized into: *Conventional Methods*, which rely on manual feature extraction and traditional Machine Learning algorithms; and *Deep Learning Methods*, which leverage Deep Learning techniques, [16,23]. However, such categorization proves overly broad; hence, we present recent endeavors in video violence detection using artificial intelligence, classified by the algorithms' nature.

- **CNN**: approaches based on CNNs for violence detection extract spatial features from video frames and utilize them for classification decisions. The CNN adjusts its weights during training to extract features. Several studies demonstrate improved accuracy by employing pre-trained CNNs on extensive image datasets and utilizing fine-tuning techniques, leveraging the pre-trained network weights [2,11].
- **LSTM**: methods based on LSTMs extract temporal features of violence, capturing temporal patterns. LSTMs modify their weights during training to extract features. The combination of CNNs and LSTMs stands as the most prevalent approach in recent literature [18].

- **Manual Feature**: algorithms wherein feature extraction or training, or both, are conducted on mathematical principles unrelated to machine learning or deep learning [12].
- **Skeleton-based (Deep Learning or Manual)**: these algorithms aim to identify body positions in videos and infer violence based on these positions, employing mathematical or deep learning techniques [20].
- **Transformer**: this category employs Transformer-based architectures for violence detection in video content [3].
- **Audio-based (Deep Learning or Manual)**: encompassing studies utilizing video audio for violence detection, although not exclusively dependent on audio as the algorithm's input, emphasizing its significance within these works [14].

The combination of CNN and LSTM for violence detection is the most widely used architecture in the recent state of the art [18]; where some of the works have shown outstanding results [1,16].

3 Objectives and Model Arquitecture

This section presents the objectives and architecture of the model proposed for this work.

One of the most widely used pre-trained CNNs is VGG-16, obtaining excellent results in violence detection in combination with LSTM and Bi-LSTM layers [1,16]. The use of VGG-19 is lower, and the results have not been as good [17], although the use of a greater number of convolutional layers should provide a greater capacity to extract complex patterns. It is also the case that violence detection studies do not use the same datasets, so comparing their results is costly [1]. Finally, in the works that combine the use of CNN and LSTM, it is not highlighted that a certain number of neurons or layers of LSTM or Fully Connected Layers leads to better results in the prediction of violence [8]. Therefore, the aim of this work is the use of pre-trained VGG19 combined with Bi-LSTM layers; the model will be trained and tested with 3 of the most used datasets in the state of the art, and the optimal number of neurons for the Bi-LSTM and Fully Connected Layers will be searched for in a wide range of hyperparameters.

The architecture of the model proposed for this work is shown in Fig. 2 First of all, the frames are pre-processed, since VGG-19 expects to receive an array of shape (224,224,3), which indicates 224 pixels wide and high for each RGB colour channel. VGG-19 consists in going through a series of convolutional blocks, which are made up of consecutive convolutional layers. Between each convolutional block there are interspersed Max Pool Layers. Once the image has passed through all the convolutional blocks, it is passed to three densely connected layers that act as a classifier; in total 16 convolutional layers, 5 pooling layers and 3 fully connected layers. VGG19 pre-trained on the ImageNet dataset is used, taking advantage of its weights for spatial feature extraction. After feature extraction by VGG19 pre-trained features that are obtained from the output of its last convolutional layer, they are pooled and passed through a 2D Global Average Pooling layer. They are then fed into two Bi-LSTM layers that extract spatial features and finally into three Fully connected layers that act as a classifier between whether the video is violent or not.

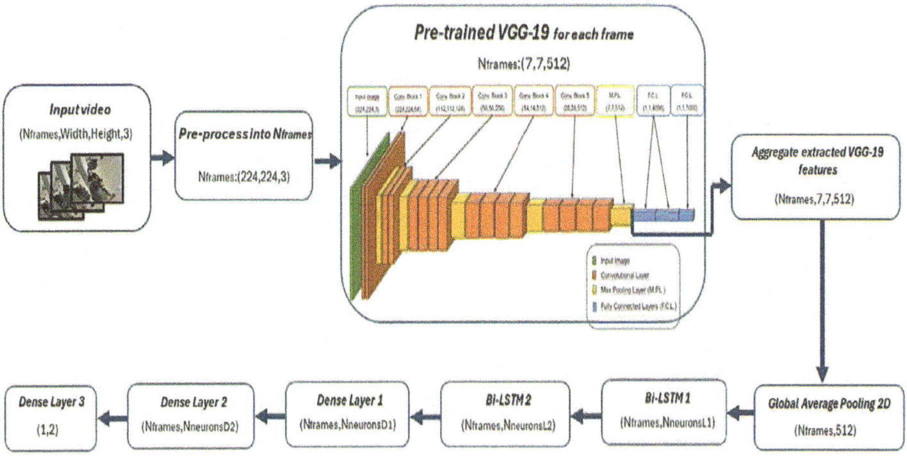

Fig. 2. Proposed Pretrained VGG19 and Bi-LSTM violence detection model

4 Experiments and Results

The results obtained by training and testing the violence detection model shown in Fig. 2 are presented. The model is trained and tested with three of the most widely used physical aggression datasets in the literature, already exposed in Sect. 1: *Hockey Fights* [4] with 1000 videos and 50 median frame number, *Violent Flow* with 247 videos and 107 median frame number [9] and *Real World Fight-2000 (RWF-2000)* [5] with 2000 videos and 150 median frame number. Only Violent Flow has videos with different lengths and since our model requires a fixed length of the analysed video, we take the median as a fixed value of the videos, repeating the last frame in case the video in question has less frames and trimming from the end of the video in case the original video has more than 107 frames.

4.1 Violence Detection Model Training

Since VGG-19 is already pre-trained on ImageNet, all that remains is to train the two Bi-LSTM layers and three fully connected layers. These layers are trained by introducing the extracted spatial features, after grouping them for each video in the form $(Nframes, 7, 7, 512)$ and after applying the Global Average Pool.

During the training, multiple hyperparameter values will be tried. The learning rate tried are: 10^{-2}, 10^{-3}, 10^{-4}, Bi-LSTM and fully connected layers neuron number tried values are: 64, 128, 256, 512 and 1024. It is permitted during training to experiment with varying the number of neurons in the first and second layers of Bi-LSTM and fully connected layers, in order to assess how altering the number of neurons affects the results.

The three datasets will undergo training for a total of 150 epochs. The optimizer utilized is *Adam*, and the loss function is *Binary Cross Entropy.* Due to the extensive search range, the number of potential combinations is significantly high. Keras Tuner is employed to handle this task. A total of 200 combinations are evaluated, representing approximately 10% of the total 1875 possible combinations. Given that our aim during training is to maximize the Validation Accuracy metric, the top three models identified by Keras Tuner in its hyperparameter optimization process are preserved. This is crucial because among models achieving similar Validation Accuracy, the one selected may not necessarily generalize best to unseen test data.

The results of the three best models obtained for the three selected datasets are shown in Table 1, where the Validation Accuracy value obtained and the values of Learning Rate and number of neurons for the Bi-LSTM and Fully Connected Layers are shown for each model.

Table 1. Training results of the best three models with Bi-LSTM layers hyperparameters combinations for the three selected datasets

Dataset	Model	Validation Accuracy	Learning Rate	Bi-LSTM 1	Bi-LSTM 2	F.C.L 1	F.C.L 2
Hockey Fights	1	0.97	0.01	64	64	64	512
	2	0.97	0.01	512	64	1024	512
	3	0.97	0.01	64	256	64	256
RWF-2000	1	0.87	0.01	512	64	256	64
	2	0.87	0.001	512	256	64	256
	3	0.87	0.001	128	128	64	128
Violent Flow	1	0.96	0.001	1024	512	64	256
	2	0.96	0.01	256	256	256	128
	3	0.96	0.01	128	512	128	256

The three models stored for each dataset obtain the same validation accuracy during the training process. In the case of the Hockey Fights and Violent Flow dataset, the results are excellent, obtaining 96% validation accuracy, unlike the RWF-2000 dataset, which obtains 87% validation accuracy. It is understood that this is due to the greater complexity and variety contained in the RWF-2000 dataset videos, as well as greater complexity in the identification of the scene due to poorer lighting conditions, distant violent scenes, etc.

4.2 Violence Detection Model Testing

This Section presents the results obtained during the testing of the top 3 models for the architecture that, based on the spatial features extracted by the pre-trained VGG-19 network, utilizes Bi-LSTM layers for spatial feature extraction and densely connected layers for classification between violence and non-violence. Since Accuracy is the most commonly used metric in the state of the art [18], it will be the basis for choosing the best model among the three best stored by Keras Tuner for each of the three selected datasets.

Table 2 contains the test results of the architecture employing Bi-LSTM layers. In case there are two models with equal testing accuracy, the model with the highest selection by Keras Tuner is chosen. For the Hockey Fights dataset, the best model is the second one, with a testing accuracy of 96%. In the case of the RWF-2000 dataset, the best model is the second one, with a testing accuracy of 72%. Lastly, for the Violent Flow Dataset, the best model is the first model, with a testing accuracy of 86%.

Table 2. Testing results of the best three models for each of the three selected datasets

Dataset	Model	Test Accuracy
Hockey Fights	1	0.93
	2	0.95
	3	0.97
RWF-2000	1	0.73
	2	0.73
	3	0.68
Violent Flow	1	0.90
	2	0.82
	3	0.88

4.3 Comparison with State-of-the-Art

This section presents a comparison of results with other state-of-the-art work on video violence detection. In the table, recent state-of-the-art papers that have used a combination of VGG-16 and VGG-19 in conjunction with some form of RNN are highlighted in blue in the *Cite* column. In orange are two articles, one of them using a CNN in combination with an Bi-LSTM and another article using only a pre-entry CNN (without combining it with an RNN). It can be seen how articles using VGG-16 pre-workout get better results compared to our model. Regarding the study using pre-trained VGG-19 together with more CNN and Bi-LSTM layers, our model obtains better results in the Hockey Fights Dataset, but not in the other two datasets used. With respect to the other state-of-the-art studies, our algorithm obtains better results in the Hockey Fights Dataset and similar results in the RWF-2000 dataset (Table 3).

Table 3. Comparison of results with other state of the art articles on violence detection

Cite	CNN	LSTM	Hockey Fight Acc.	Violent Flow Acc.	RWF-2000 Acc.
[1]	PT VGG-16 (ImageNet)	LSTM	99.1	X	X
[16]	PT VGG-16 (ImageNet)	Bi-Conv-LSTM	99.1	98.4	92.4
[8]	PT VGG-16 (ImageNet)	LSTM/ Bi-LSTM	97.6/ 98.8	92.2/ 95.1	X
[22]	PT VGG16 (INRA)	Bi-GRU	98	95.5	X
[17]	PT VGG-19 (ImageNet) + extra CNN layers	Bi-LSTM	91.29	90.47	90.47
[21]	CNN	LSTM	94.9	92.2	77.31
[6]	CNN PT	N	94.1	X	72
Proposed model	PT VGG-19 (ImageNet)	Bi-LSTM	97	90	73

All in all, the use of pre-trained VGG-19 with Bi-LSTM layers, testing a wide range of hyperparameters of learning rate and number of neurons of the Bi-LSTM and Fully connected layers has given good results, especially in the Hockey Fights Dataset, although not better than the results obtained by other works using pre-trained VGG-16. Still, it is competitive with other state-of-the-art papers using other architectures.

5 Conclusions and Future Work

This work addresses the challenge of violence detection in videos, a crucial problem with significant implications for public safety and social well-being [13,15]. In the state of the art, there are multiple works that contribute different kinds of algorithms for violence detection in videos using artificial intelligence [12,20], of which the most commonly used is the combination of Convolutional Neural Networks along with Long Short-Term Memory [1,16].

The aim of this work has been the development of a violence detection algorithm using pre-trained VGG19 in combination with Bi-LSTM layers, since the use of VGG-16 has yielded excellent results in violence detection, although VGG-19 has been considerably less utilized. Additionally, in theory, a higher number of convolutional layers could imply a better understanding of the scene, but a clear improvement in the use of certain hyperparameter values has not been found in the literature. Therefore, in this work, a wide range of combinations for the hyperparameters of learning rate and number of neurons in the Bi-LSTM and Fully Connected Layers has been used.

The results obtained have been positive, with accuracies of 96%, 86%, and 72% in the Hockey Fights dataset, Violent Flow Dataset, and RWF-2000, respectively. Although these are positive results, especially in the use of the Hockey Fights Dataset, the model did not achieve better results than other state-of-the-art works using pre-trained VGG-16 [1,8,16,22], even after training our model over a wide range of learning rate values and Bi-LSTM and Fully connected layers neuron numbers. Nonetheless, compared to other proposals from the state of the art, the results are positive [6,21].

As future work, the study of how the training of some layers of pre-trained VGG16 CNN combined with Bi-LSTM layers may improve the results of the state of the art works. Additionally, the application of XAI techniques for understanding why the algorithm is incorrectly predicting certain scenes and whether the mispredicted scenes follow any pattern is proposed.

Acknowledgements. This research has been supported by the project "Smart, Sustainable and coheSive Digitalization conceived as a Digital Innovation Hub (DIGIS3)", Grant Agreement 101083898. Co-financed by the European Union (EU) and the Ministry of Industry, Trade and Tourism and the EOI Foundation of the Government of Spain, in the framework of the Recovery and Resilience Mechanism financed by the Next Generation funds of the EU. The views and opinions expressed are solely those of the author(s) and do not necessarily reflect those of the EU, Ministry of Industry, Trade and Tourism or the EOI Foundation. Neither the EU nor the granting authority can be held responsible for them.

References

1. Aarthy, K., Nithya, A.A.: Crowd violence detection in videos using deep learning architecture. In: 2022 IEEE 2nd Mysore Sub Section International Conference (MysuruCon), pp. 1–6. IEEE (2022)
2. Ahmed, M., et al.: Real-time violent action recognition using key frames extraction and deep learning (2021)
3. Aktı, Ş., Ofli, F., Imran, M., Ekenel, H.K.: Fight detection from still images in the wild. In: Proceedings of the IEEE/CVF Winter Conference on Applications of Computer Vision, pp. 550–559 (2022)
4. Bermejo Nievas, E., Deniz Suarez, O., Bueno García, G., Sukthankar, R.: Violence detection in video using computer vision techniques. In: Real, P., Diaz-Pernil, D., Molina-Abril, H., Berciano, A., Kropatsch, W. (eds.) CAIP 2011. LNCS, vol. 6855, pp. 332–339. Springer, Heidelberg (2011). https://doi.org/10.1007/978-3-642-23678-5_39

5. Cheng, M., Cai, K., Li, M.: RWF-2000: an open large scale video database for violence detection. In: 2020 25th International Conference on Pattern Recognition (ICPR), pp. 4183–4190 (2021)
6. Cheng, S.-T., Hsu, C.-W., Horng, G.-J., Jiang, C.-R.: Video reasoning for conflict events through feature extraction. J. Supercomput. **77**(6), 6435–6455 (2021). https://doi.org/10.1007/s11227-020-03514-5
7. Enaifoghe, A., Dlelana, M., Durokifa, A.A., Dlamini, N.P.: The prevalence of gender-based violence against women in South Africa: a call for action. Afr. J. Gend. Soc. Develop. **10**(1), 117 (2021)
8. Gupta, H., Ali, S.T.: Violence detection using deep learning techniques. In: 2022 International Conference on Emerging Techniques in Computational Intelligence (ICETCI), pp. 121–124. IEEE (2022)
9. Hassner, T., Itcher, Y., Kliper-Gross, O.: Violent flows: real-time detection of violent crowd behavior. In: 2012 IEEE Computer Society Conference on Computer Vision and Pattern Recognition Workshops, pp. 1–6. IEEE (2012)
10. Hillis, S., Mercy, J., Amobi, A., Kress, H.: Global prevalence of past-year violence against children: a systematic review and minimum estimates. Pediatrics **137**(3) (2016)
11. Jayasimhan, A., Pabitha, P.: A hybrid model using 2D and 3D convolutional neural networks for violence detection in a video dataset. In: 2022 3rd International Conference on Communication, Computing and Industry 4.0 (C2I4), pp. 1–5. IEEE (2022)
12. Lohithashva, B.H., Aradhya, V.N.M.: Violent video event detection: a local optimal oriented pattern based approach. In: Mahmud, M., Kaiser, M.S., Kasabov, N., Iftekharuddin, K., Zhong, N. (eds.) AII 2021. CCIS, vol. 1435, pp. 268–280. Springer, Cham (2021). https://doi.org/10.1007/978-3-030-82269-9_21
13. Long, D., Liu, L., Xu, M., Feng, J., Chen, J., He, L.: Ambient population and surveillance cameras: the guardianship role in street robbers' crime location choice. Cities **115**, 103223 (2021)
14. Mahalle, M.D., Rojatkar, D.V.: Audio based violent scene detection using extreme learning machine algorithm. In: 2021 6th International Conference for Convergence in Technology (I2CT), pp. 1–8. IEEE (2021)
15. Muarifah, A., Mashar, R., Hashim, I.H.M., Rofiah, N.H., Oktaviani, F.: Aggression in adolescents: the role of mother-child attachment and self-esteem. Behav. Sci. **12**(5) (2022)
16. Mugunga, I., Dong, J., Rigall, E., Guo, S., Madessa, A.H., Nawaz, H.S.: A frame-based feature model for violence detection from surveillance cameras using convLSTM network. In: 2021 6th International Conference on Image, Vision and Computing (ICIVC), pp. 55–60. IEEE (2021)
17. Mumtaz, N., Ejaz, N., Aladhadh, S., Habib, S., Lee, M.Y.: Deep multi-scale features fusion for effective violence detection and control charts visualization. Sensors **22**(23), 9383 (2022)
18. Omarov, B., Narynov, S., Zhumanov, Z., Gumar, A., Khassanova, M.: State-of-the-art violence detection techniques in video surveillance security systems: a systematic review. PeerJ Comput. Sci. **8**, e920 (2022)
19. de la Salud, O.M.: Violence against women (2021)
20. Su, Y., Lin, G., Zhu, J., Wu, Q.: Human interaction learning on 3D skeleton point clouds for video violence recognition. In: Vedaldi, A., Bischof, H., Brox, T., Frahm, J.-M. (eds.) ECCV 2020. LNCS, vol. 12349, pp. 74–90. Springer, Cham (2020). https://doi.org/10.1007/978-3-030-58548-8_5

21. Talha, K.R., Bandapadya, K., Khan, M.M.: Violence detection using computer vision approaches. In: 2022 IEEE World AI IoT Congress (AIIoT), pp. 544–550. IEEE (2022)
22. Traoré, A., Akhloufi, M.A.: 2D bidirectional gated recurrent unit convolutional neural networks for end-to-end violence detection in videos. In: International Conference on Image Analysis and Recognition, pp. 152–160. Springer (2020)
23. Ullah, F.U.M., et al.: AI-assisted edge vision for violence detection in IoT-based industrial surveillance networks. IEEE Trans. Ind. Inf. **18**(8), 5359–5370 (2021)
24. Vomfell, L., Härdle, W.K., Lessmann, S.: Improving crime count forecasts using twitter and taxi data. Decis. Support Syst. **113**, 73–85 (2018)
25. Wilkinson, S.: Meet the heroic campaigners making cities safe for women. In: Action Aid, Global Street Harassment-Making Street Safer (2016)

Automated Machine Learning and Explainable Artificial Intelligence in Predictive Maintenance: An MLJAR Framework Review

Bruno Mota, Pedro Faria(✉), and Carlos Ramos

GECAD - Research Group on Intelligent Engineering and Computing for Advanced Innovation and Development, LASI - Intelligent Systems Associate Laboratory, Polytechnic of Porto, Rua Dr. António Bernardino de Almeida 431, 4200-072 Porto, Portugal
{bamoa,pnf,csr}@isep.ipp.pt

Abstract. Predictive Maintenance (PdM) through Machine Learning (ML) has become an essential strategy for industries to reduce redundant maintenance activities. Even so, many companies lack the technical knowledge to implement these PdM systems, and more often than not, do not trust ML models for their lack of transparency and interpretability. To mitigate these issues, the present paper explores and implements an Automated Machine Learning (AutoML) and Explainable Artificial Intelligence (XAI) framework designated as MLJAR. The framework is evaluated for its AutoML and XAI capabilities in a widely used synthetic PdM dataset in the literature. Promising results were found as the framework was able to outperform most works in the literature by up to 26.6% in recall score, with the only work surpassing MLJAR by up to 3.2%, yet having the drawbacks of 58.3% worse precision score, and no AutoML or XAI capabilities. Overall, the MLJAR framework was able to, on average, provide 13.2% and 2.2% better scores in recall and accuracy, respectively.

Keywords: Automated Machine Learning · Explainable Artificial Intelligence · Predictive Maintenance

1 Introduction

Predictive Maintenance (PdM) strategies that employ Machine Learning (ML) have become more efficient in evaluating machines' conditions when compared to conventional approaches, which focus on time-based preventative maintenance. This trend may be a result of the increasing complexity and data availability in industries that ML models can handle when compared to conventional strategies [1]. PdM, using ML models, allows for higher cost savings in companies by avoiding redundant maintenance activities, even after accounting for investment expenses PdM still outperforms reactive and conventional preventive maintenance strategies in terms of cost savings [2]. Historical maintenance data and/or machine health indicators are evaluated in PdM to develop high-performance ML models capable of predicting machines' failure state, allowing for optimal maintenance activities scheduling [3].

R. Chinthaginjala et al. (Eds.): DCAI 2024, LNNS 1259, pp. 150–159, 2025.
https://doi.org/10.1007/978-3-031-82073-1_15

Nonetheless, ML model implementation is challenging for companies as it demands high levels of technical understanding not just to deploy the models but also to fine-tune them. Automated Machine Learning (AutoML) aims to overcome this issue by automatically pre-processing, exploring, and optimizing hyperparameters on ML models without the need for an expert [4]. This strategy saves money and time, by enabling more reliable and faster deployments of ML systems in industrial environments.

As useful as AutoML can be, complex ML models, which provide better predictions, still suffer from low transparency and interpretability due to their black-box characteristics [5]. Accordingly, trustworthiness in these systems is very small, with users often becoming skeptical of the results provided by ML models, which distances users from the decision-making process since understanding a model's behaviors becomes an almost impossible task for a human [6]. However, breakthroughs in explainable Artificial Intelligence (XAI) strive to solve this issue by providing explanations for an ML model's predictions that are simple and brief enough to be understood by humans [7]. Nevertheless, there is still a lack of research that focuses on PdM with XAI, with the majority of PdM works employing non-interpretable ML models [8].

The purpose of the present paper is to investigate the framework MLJAR [9], which provides AutoML and XAI capabilities, in a PdM problem. To achieve this, the present paper implements, explores, and compares the MLJAR framework to other state-of-the-art ML methodologies for PdM. Performance evaluation of the framework AutoML capabilities is carried out using an unbalanced PdM dataset [10], which is widely used in the literature, and comparing it to other works that also employ the aforementioned dataset such as in [11–19]. For XAI, a detailed analysis is performed of the explanations generated by the framework.

2 Proposed Methodology

To explore the open-source MLJAR framework, it is taken into account a synthetic PdM dataset from the literature, available in [10]. The cited dataset is used in a variety of works, being an ideal dataset to evaluate MLJAR framework capabilities in handling PdM problems when compared to other methodologies in the literature. Another reason is that it is an extremely unbalanced dataset, composed of 10000 data samples with only 339 of them represing machine failures which equals to approximately a ratio of 1:28 of failures to non-failures, hence replicating very well PdM problems. From the dataset, the following five features are considered: air temperature in Kelvin (K), process temperature in Kelvin (K), rotational speed in revolutions per minute (rpm), torque in Newton-meters (Nm), and tool wear in minutes used (min). Also from the dataset, a target class is taken into account: machine failure with 0 for non-failure and 1 for failure.

Implementation is done utilizing the ML-supervised framework [9], in MLJAR, for binary classification. For AutoML, the framework is able to preprocess data (i.e., imputation and converting categoricals), feature engineering (i.e., feature construction and selection), and ML algorithm choosing with hyperparameter tuning. When building an ML model, the framework has a pool of eight algorithms for binary classification: Baseline, Linear, Random Forest (RF), Decision Tree (DT), Light Gradient-Boosting Machine (LightGBM), Extreme Gradient Boosting (Xgboost), CatBoost, and Artificial

Neural Network (ANN). Regarding XAI, while the framework provides three different levels of interpretability, only the highest one is employed in the present paper, as it provides permutation feature importance, Shapley Additive Explanations (SHAP), tree plots (only for DTs), and coefficients (only for linear models).

3 Implementation of MLJAR Framework

The present paper implements version 1.1.3 of the MLJAR framework for supervised ML in [9], for the Python programming language. The framework can be easily implemented with a few lines of code, useful for non-expert users in ML.

First the automl.AutoML method is used to define the different parameters for training and building the ML models. There are up to thirty parameters that can be defined in the AutoML method, with all of them having a default value. Despite that, only the following parameters were considered relevant to define:

- mode – defines the overall objective of the AutoML. It has four possible values: 'Explain' for better explainability, 'Perform' for a balance between ML interpretability and performance, and 'Compete' for better performance;
- ml_task – defines the type of ML problem. Although the framework allows for binary or multiclass classification, as well as, regression, in the present paper, only binary classification is considered as it portrays the problem at hand;
- eval_metric – defines the evaluation metric when comparing ML models for early stopping. For binary classification, the framework has the following metrics: log loss, area under curve, f1-score, average precision, and accuracy. Nevertheless, the framework allows for custom evaluation metrics, essential as from the abovementioned metrics only the average precision is appropriate for PdM problems that suffer from unbalanced datasets. Accordingly, through custom evaluation metrics it was possible to implement the f2-score, f0.5-score, and recall. The f2- and f0.5-scores are implemented using the metrics.fbeta_score method (parameter beta defined as 2 for f2-score, and 0.5 for f0.5-score), while the recall is obtained using the metrics.recall_score method, both methods from the Scikit-learn library [20];
- explain_level – defines the degree of interpretability, using XAI techniques, for ML models. There are three different levels, that is, levels 0, 1, and 2. Level 0 provides no explanations, while level 1 provides permutation feature importance, tree plots, and linear coefficients. The last level, level 2, provides all XAI techniques from level 1 plus SHAP explanations;
- validation_strategy – defines the type of validation strategy, from train/test split to k-fold cross-validation. In all scenarios of the present paper train/test split strategy is taken into account, as it is more consistent for comparison with other methodologies, and allows SHAP to give global explanations instead of explanations per fold;
- train_ensemble – defines if an ensemble method, composed of the best ML models created by the framework, is generated. This functionality is turned off because the ensemble model is always only slightly better or equal to the best ML model found by the framework while not providing any explanations. In addition, it makes comparisons with other ML models from other works in the literature more complex, as an ensemble model from MLJAR can be composed, for example, of fifty different ML models of repeated algorithms but with different hyperparameters.

After defining and executing the automl.AutoML method, an MLJAR model is obtained, which can then be trained using the fit method. For this method, three parameters are needed: a DataFrame containing the features (i.e., air temperature, process temperature, rotational speed, torque, and tool wear), another DataFrame with the target class (i.e., machine failure), and the sample weights. It is through the sample weights parameter that the bias for non-machine failures, due to the dataset being highly unbalanced towards this class (i.e., a ratio of 1:28 of failures to non-failures), can be mitigated. This is achieved, by assigning each sample a weight corresponding to its target class value, with higher weights penalizing more an ML model for incorrectly predicting the corresponding class. For instance, an ML model during training is significantly less penalized in a sample corresponding to a non-failure with an assigned weight value of around 0.52, when compared to a failure sample with a weight value of around 14.75. To obtain the class weights, it is used the utils.class_weight.compute_class_weight method, from the Scikit-learn library [20], with a class_weight parameter defined as 'balanced'. Also worth mentioning is that documentation, such as performance metrics and XAI data, is automatically saved in the user's system.

4 Results and Discussion

The MLJAR framework is validated for its AutoML and XAI capabilities, applied to PdM, by considering the dataset in [10], described in Sect. 2. To evaluate the framework's capacity in dealing with PdM datasets, which tend to be very unbalanced, four evaluation metrics (i.e., parameter eval_metric) are considered for early stopping, taking into account different industry-specific scenarios: average precision, for when the positive class, that is, machine failure, is more relevant than the negative class in an industry scenario; f2-score, for when false negatives, that is, unsuccessful failure predictions, are more detrimental to the industry; f0.5-score, for when false positives, that is, false alarms, are more detrimental; and recall, for when false negatives need to be minimized as much as possible.

Also worth noting, for the parametrization of the MLJAR framework in the following scenarios, the ml_task is always defined as binary classification, explain_level as level 2 for maximum ML model interpretability, validation_strategy specified as a split 80% training and 20% test strategy for more consistent comparisons and SHAP global explanations, and the train_ensemble parameter is False since the ensemble model does not provide any XAI explanations and are often compromised of too many ML models. In addition, the MLJAR framework sets by default its random state, which corresponds to the parameter that controls randomness, to a fixed value of 1234. Consequently, all MLJAR executions with the same dataset and parametrization give the same results, allowing for reproducibility of the results as only one execution per scenario is needed.

4.1 MLJAR Automated Machine Learning

In order to mitigate the bias of the dataset for non-failure data samples, which in turn makes trained models have low recall scores, sample weights need to be defined. Employing the compute_class_weight method, described in Sect. 3, non-failure samples obtained a weight of around 0.52, while failure samples obtained a weight of around 14.75. As a result, the obtained performance, using the aforementioned weight samples, MLJAR parametrization, and PdM dataset, is described in Table 1.

Table 1. MLJAR framework performance results with sample weight consideration. (*the color green represents the best overall performing model found by the MLJAR framework*)

MLJAR Parameter		Performance Metric of the Best Algorithm				AutoML Metrics	
Mode	Evaluation Metric	Recall	Precision	F1-score	Accuracy	Training Time (seconds)	Models Explored
Explain	Average Precision	0.91	0.31	0.47	92.90%	44	5
	F2-score	0.85	0.42	0.56	95.45%	44	5
	F0.5-score	0.85	0.42	0.56	95.45%	43	5
	Recall	0.94	0.28	0.43	91.65%	44	5
Perform	Average Precision	0.91	0.52	0.66	96.85%	372	48
	F2-score	0.94	0.36	0.52	94.05%	311	41
	F0.5-score	0.87	0.37	0.52	94.55%	380	53
	Recall	0.94	0.28	0.43	91.65%	372	54
Compete	Average Precision	0.90	0.62	0.73	97.75%	877	120
	F2-score	0.90	0.44	0.59	95.75%	891	128
	F0.5-score	0.85	0.42	0.56	95.45%	900	126
	Recall	0.94	0.23	0.37	89.15%	783	124

It demonstrates that the 'Explain' mode is useful to quickly obtain working ML models, in around 44 s, to explore its predictions and interpret the model through XAI. Nevertheless, if more time and models are explored, such as in the 'Perform' method, results get much better, with all scenarios having an increase of up to 10.6%, 67.7%, 40.4%, and 4.2% to the recall, precision, f1-score, and accuracy, respectively. Also from Table 1, it is possible to conclude that there are diminishing returns, with the difference between modes 'Perform' and 'Compete' netting fewer gains in performance, and in some cases, loss in performance, for example, when comparing the evaluation metric recall between the two mentioned modes, all performance metrics decreased or stayed the same in the 'Compete' mode. The best overall ML model found by the MLJAR framework, considering that recall is the most important metric in PdM followed by precision, based on Table 1, was obtained using the 'Perform' mode, with sample weight balanced, and f2-score evaluation metric. It boasts a recall score of 0.94, precision of 0.36, f1-score of 0.52, and an accuracy of 94.05%. The model employs an Xgboost algorithm, trained in 4 s, with the following hyperparameter values: 'objective' with binary:logistic, 'eta' of 0.1, 'max_depth' of 8, 'min_child_weight' of 1, 'subsample' of 0.9, and a 'colsample_bytree' of 1. From the Xgboost model, only 4 machine failures were undetected (i.e., false negatives), and 115 false alarms were given (i.e., false positives).

It is worth noting that, no sample weight definition was also considered, however, recall scores were much worse. For instance, considering only the mode 'Explain', recall scores achieved a maximum of 0.54, a decrease of 42.5% compared to Table 1.

4.2 MLJAR Predictive Maintenance Literature Performance Comparison

To evaluate the performance of the MLJAR framework compared to other state-of-the-art methodologies, nine works from the literature were selected due to their exceptional performance (e.g., high recall and precision scores) and validation strategies (e.g., by not using the same data for training and testing as it overfits performance results). Accordingly, Table 2 presents a quantitative comparison of the performance metrics (recall, precision, f1-score, and accuracy) between the MLJAR framework's best overall performing model found, as described in 4.1, and nine works from the literature. It demonstrates that the MLJAR framework, with few lines of code and high usability for non-expert users in ML, can outperform works such as in [11, 14–16, 18], and [19] by 21.3%, 3.2%, 26.6%, 23.4%, 3.2%, and 24.5%, respectively, in recall score. Nonetheless, these works tend to trade recall for precision, having higher values of precision aiming to reduce false alarms, which is not considered in the present study as the most important metric, because undetected failures, calculated through recall, are seen as more detrimental problems for the industry.

Furthermore, the works [13] and [17] are the few that have really good recall scores, with 0.94 and 0.92, respectively, however, they still fail to outperform the MLJAR framework both in recall and precision. From all the works in Table 2, the PdM methodology presented in [12] is without a doubt the most successful in increasing recall, with an increase between 1.1% and 3.2% compared to the recall from the MLJAR framework's model. Its biggest drawback is the significant precision score loss when compared to the MLJAR model, as it loses between 16.7% and 58.3% of the precision score. Overall, when comparing the MLJAR model to the average of the cited works, it was possible to create a model with 13.2% and 2.2% better results in the recall and accuracy scores, with the trade-off of being 29.4% and 7.1% worse in precision and f1-score, respectively. It is noteworthy that, from all the works in Table 2, none implemented AutoML, requiring users to be experienced in ML in order to preprocess data, fine-tune models, and evaluate ML models. In addition, only the works in [18] and [19] implemented XAI techniques, with the MLJAR framework also implementing XAI while also providing better performance results.

4.3 MLJAR Explainable Artificial Intelligence

The MLJAR framework provides the essential tools to interpret all of its generated models globally. To accomplish this, permutation feature importance and SHAP plots (i.e., bar, decision and dependence plots) are employed for almost all models, while tree plots and coefficients are only provided to DT and linear models, respectively. Hence, MLJAR

Table 2. Comparison of the MLJAR framework with state-of-the-art methodologies applied to the same dataset. (*citations sorted by decreasing publication year*)

Methodology Information					Performance Metric			
Citation	Validation	Auto ML	XAI	Method[1]	Recall	Precision	F1-score	Accuracy
[11]	0.2-split test	No	No	Xgboost + FA	0.74	0.88	0.80	–
				Xgboost + GIFA	0.73	0.91	0.81	–
[12]	0.2–split test	No	No	RF	0.95	0.30	0.46	93.15%
				ANN	0.97	0.15	0.27	83.65%
[13]	0.2-split test	No	No	Xgboost	0.94	0.24	0.38	–
				CatBoost	0.93	0.25	0.39	–
[14]	3-fold	No	No	ctGAN + CatBoost	0.83	–	–	87.08%
				SmoteNC + ctGAN + CatBoost	0.91	–	–	88.83%
[15]	0.2-split test	No	No	ANN	0.69	–	–	98.50%
[16]	0.3-split test	No	No	DT	0.72	0.84	0.78	–
				RF	0.61	0.83	0.70	–
[17]	0.2-split test	No	No	GB	0.87	0.34	0.49	94.55%
				SVM	0.92	0.24	0.38	91.00%
[18]	5-fold	No	Yes	RUSBoost Trees	0.91	0.31	0.46	92.74%
[19]	5-fold	No	Yes	Bagged DT	0.71	0.87	0.78	98.34%
Average					0.83	0.51	0.56	91.98%
MLJAR	0.2-split test	Yes	Yes	Xgboost	0.94	0.36	0.52	94.05%

[1] FA: Firefly Algorithm, GIFA: Genetically Inspired FA, ctGAN: Conditional Tabular Generative Adversarial, SmoteNC: Synthetic Minority Over-Sampling Technique for Nominal and Continuous Network, GB: Gradient Boosting, SVM: Support Vector Machine, RUSBoost: Random Undersampling Boosting

only offers global interpretations of the models, if users require local explanations other Python libraries need to be implemented and integrated.

For the permutation feature importance, the MLJAR framework provided the following importance values for each feature: air temperature with 0.3028, torque with 0.2531, rotational speed with 0.1808, tool wear with 0.1165, and process temperature with 0.0102. Consequently, it is considered that air temperature and torque were the features found to be the most important when building the model, while the rotational speed and tool wear had less but still significant importance. Also, process temperature was found to be almost negligible, having the lowest importance with a wide margin. To a non-expert user, these values easily show the features the model found to be the most important to predict a machine failure, as higher values indicate high importance.

SHAP also gives feature importance values, with the rotational speed having a value of 0.1919, torque at 0.1347, tool wear at 0.1343, air temperature at 0.0666, and process temperature at 0.0080. In this case, SHAP measured that the rotational speed, torque, and tool wear were the most important features that impacted the model, whereas air and process temperatures had the lowest SHAP importance values. When compared to the previously mentioned permutation feature importance values, SHAP measured rotational speed to be significantly more important while the opposite was done to air temperature, which decreased from first to fourth most important. Moreover, torque stayed as the second most important feature both in permutation and SHAP, and process temperature also stayed as the least important feature. The SHAP decision plot comprised of the best machine failure decisions is portrayed in Fig. 1. From the plot, it can be concluded that for each feature the higher the value it assumes, the higher the tendency the model has to classify it as a machine failure. For a non-expert user, Fig. 1 provides a clear understanding of the model's feature behavior, allowing for transparency and higher levels of interpretability of a PdM model's performance.

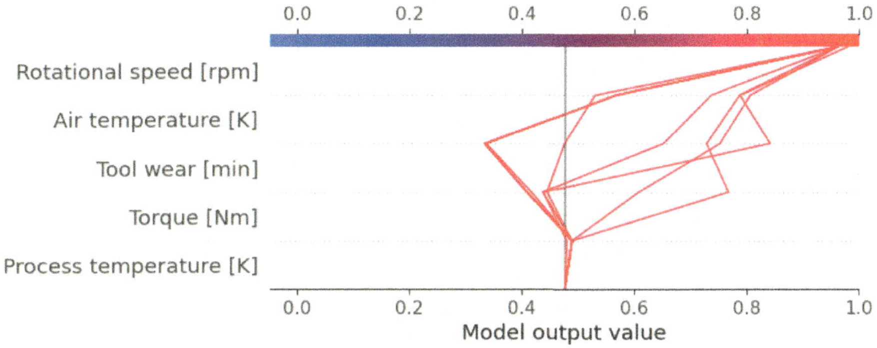

Fig. 1. SHAP decision plot for the best machine failure predictions.

When compared with the works in [18] and [19], that also implemented XAI, MLJAR has the major limitation of only considering the global interpretation of ML models, while the literature also includes local explanations. Despite that, there are plans in the MLJAR roadmap to implement local explanations, allowing non-expert users to easily understand pertinent predictions. In view of this, the trustworthiness and affordability of PdM systems would improve as, through the MLJAR framework, it would be possible for a non-expert user to set up a simple PdM model with great performance and XAI capabilities, making companies more comfortable investing in PdM.

5 Conclusions

When compared to PdM state-of-the-art methodologies that used the same dataset, the MLJAR framework was capable of outperforming almost all of them by up to 26.6% in recall score. Only an RF and ANN models in the literature had higher recall scores by about 1.1% and 3.2%, respectively, more than the MLJAR model. Nevertheless, these models traded recall for precision since precision scores decreased by 16.7% and 58.3% compared to the MLJAR model. On average in the literature, the MLJAR framework provided a model with 13.2% and 2.2% higher scores in recall and accuracy, but 29.4% and 7.1% worse scores in precision and f1-score.

One major limitation of the MLJAR framework is its lack of oversampling and undersampling techniques, with sample weights being the only alternative provided by the framework to handle unbalanced datasets. In addition, XAI techniques are only available at a global interpretation level, no local explanations are provided by MLJAR.

In future work, the MLJAR framework will be explored for multi-class classification, of multiple machine failures, using the same dataset in the present paper.

Acknowledgments. This work has received funding from FEDER Funds through COMPETE program and from National Funds through (FCT) under the scholarship 2023.03776.BD. The authors acknowledge the work facilities and equipment provided by GECAD research center (UIDB/00760/2020), DOI:10.54499/UIDB/00760/2020 to the project team.

Disclosure of Interests. The authors have no competing interests to declare that are relevant to the content of this article.

References

1. Zonta, T., da Costa, C.A., da Rosa Righi, R., de Lima, M.J., da Trindade, E.S., Li, G.P.: Predictive maintenance in the industry 4.0: a systematic literature review. Comput Ind Eng. **150**, 106889 (2020). https://doi.org/10.1016/j.cie.2020.106889
2. Florian, E., Sgarbossa, F., Zennaro, I.: Machine learning-based predictive maintenance: a cost-oriented model for implementation. Int. J. Prod. Econ. **236**, 108114 (2021). https://doi.org/10.1016/j.ijpe.2021.108114
3. Theissler, A., Pérez-Velázquez, J., Kettelgerdes, M., Elger, G.: Predictive maintenance enabled by machine learning: use cases and challenges in the automotive industry. Reliab. Eng. Syst. Saf. **215**, 107864 (2021). https://doi.org/10.1016/j.ress.2021.107864
4. Truong, A., Walters, A., Goodsitt, J., Hines, K., Bruss, C.B., Farivar, R.: Towards automated machine learning: evaluation and comparison of AutoML approaches and tools. In: 2019 IEEE 31st International Conference on Tools with Artificial Intelligence (ICTAI), pp. 1471–1479. IEEE (2019). https://doi.org/10.1109/ICTAI.2019.00209
5. Adadi, A., Berrada, M.: Peeking inside the black-box: a survey on explainable artificial intelligence (XAI). IEEE Access. **6**, 52138–52160 (2018). https://doi.org/10.1109/ACCESS.2018.2870052
6. Hassija, V., et al.: Interpreting black-box models: a review on explainable artificial intelligence. Cognit Comput. **16**, 45–74 (2024). https://doi.org/10.1007/s12559-023-10179-8

7. Minh, D., Wang, H.X., Li, Y.F., Nguyen, T.N.: Explainable artificial intelligence: a comprehensive review. Artif. Intell. Rev. **55**, 3503–3568 (2022). https://doi.org/10.1007/s10462-021-10088-y
8. Vollert, S., Atzmueller, M., Theissler, A.: Interpretable machine learning: a brief survey from the predictive maintenance perspective. In: 2021 26th IEEE International Conference on Emerging Technologies and Factory Automation (ETFA), pp. 01–08. IEEE (2021). https://doi.org/10.1109/ETFA45728.2021.9613467
9. Płońska, A., Płoński, P.: MLJAR: State-of-the-art Automated Machine Learning Framework for Tabular Data (2021). https://github.com/mljar/mljar-supervised
10. Matzka, S.: AI4I 2020 Predictive Maintenance Dataset: UCI Machine Learning Repository. https://doi.org/10.24432/C5HS5C. Accessed 02 Mar 2024
11. Bozovic, A., et al.: Metaheuristic optimized extreme gradient boosting milling maintenance prediction. Presented at the (2024). https://doi.org/10.1007/978-981-99-9037-5_28
12. Mota, B., Faria, P., Ramos, C.: Machine learning applied to industrial machines for an efficient maintenance strategy: a predictive maintenance approach. In: Jørgensen, B.N., da Silva, L.C.P., Ma, Z. (eds.) Energy Informatics. EI.A 2023, vol 14467, pp. 289–299. Lecture Notes in Computer Science. Springer, Cham (2024). https://doi.org/10.1007/978-3-031-48649-4_17
13. Dilbaz, Ü., Cingiz, M.Ö.: Predictive maintenance on industrial data using soft voted ensemble classifiers. In: García Márquez, F.P., Jamil, A., Eken, S., Hameed, A.A. (eds.) Computational Intelligence, Data Analytics and Applications. ICCIDA 2022. Lecture Notes in Networks and Systems, vol. 643, pp. 370–384. Springer, Cham (2023). https://doi.org/10.1007/978-3-031-27099-4_29
14. Chen, C.-H., Tsung, C.-K., Yu, S.-S.: Designing a hybrid equipment-failure diagnosis mechanism under mixed-type data with limited failure samples. Appl. Sci. **12**, 9286 (2022). https://doi.org/10.3390/app12189286
15. Kamel, H.: Artificial intelligence for predictive maintenance. J. Phys. Conf. Ser. **2299**, 012001 (2022). https://doi.org/10.1088/1742-6596/2299/1/012001
16. Vuttipittayamongkol, P., Arreeras, T.: Data-driven industrial machine failure detection in imbalanced environments. In: 2022 IEEE International Conference on Industrial Engineering and Engineering Management (IEEM), pp. 1224–1227. IEEE (2022). https://doi.org/10.1109/IEEM55944.2022.9989673
17. Mota, B., Faria, P., Ramos, C.: Predictive maintenance for maintenance-effective manufacturing using machine learning approaches. Presented at the (2023). https://doi.org/10.1007/978-3-031-18050-7_2
18. Torcianti, A., Matzka, S.: Explainable artificial intelligence for predictive maintenance applications using a local surrogate model. In: 2021 4th International Conference on Artificial Intelligence for Industries (AI4I), pp. 86–88. IEEE (2021). https://doi.org/10.1109/AI4I51902.2021.00029
19. Matzka, S.: Explainable artificial intelligence for predictive maintenance applications. In: Proceedings - 2020 3rd International Conference on Artificial Intelligence for Industries, AI4I 2020, pp. 69–74. Institute of Electrical and Electronics Engineers Inc. (2020). https://doi.org/10.1109/AI4I49448.2020.00023
20. Pedregosa, F., et al.: Scikit-learn: machine learning in python. J. Mach. Learn. Res. **12**, 2825–2830 (2011)

FuzzTheREST: An Intelligent Automated Black-Box RESTful API Fuzzer

Tiago Dias(✉), Eva Maia, and Isabel Praça

Research Group on Intelligent Engineering and Computing for Advanced Innovation and Development (GECAD), Porto School of Engineering (ISEP), 4200-072 Porto, Portugal
{tiada,egm,icp}@isep.ipp.pt

Abstract. Software's pervasive impact and increasing reliance in the era of digital transformation raise concerns about vulnerabilities, emphasizing the need for software security. Fuzzy testing is a dynamic analysis software testing technique that consists of feeding faulty input data to a System Under Test (SUT) and observing its behavior. Specifically regarding black-box RESTful API testing, recent literature has attempted to automate this technique using heuristics to perform the input search and using the HTTP response status codes for classification. However, most approaches do not keep track of code coverage, which is important to validate the solution. This work introduces a black-box RESTful API fuzzy testing tool that employs Reinforcement Learning (RL) for vulnerability detection. The fuzzer operates via the OpenAPI Specification (OAS) file and a scenarios file, which includes information to communicate with the SUT and the sequences of functionalities to test, respectively. To evaluate its effectiveness, the tool was tested on the Petstore API. The tool found a total of six unique vulnerabilities and achieved 55% code coverage.

Keywords: Fuzzy Testing · RESTful APIs · Vulnerability Detection · Code Security · Artificial Intelligence

1 Introduction

Software has had a positive impact over the years and its adherence continues to grow. The ubiquity of information technology, coupled with the seamless availability of web services, such as REST, has made software much more accessible in today's interconnected landscape. However, considering its prevalence, the existence of software vulnerabilities is worrying. Nowadays, people depend on technology, and as such, poor-quality software can have detrimental effects. Considering that software is the backbone of many modern operations, ensuring software reliability and security is crucial [10,18,23].

Software testing is one of the main contributors to software quality capable of answering the validation and verification standard [16]. Software testing encompasses a very extensive taxonomy of methods, each focusing on improving

R. Chinthaginjala et al. (Eds.): DCAI 2024, LNNS 1259, pp. 160–169, 2025.
https://doi.org/10.1007/978-3-031-82073-1_16

different software quality attributes [9]. One such method is vulnerability detection, which consists of spotting and addressing weaknesses in software code. Vulnerabilities are typically spotted following a certain type of analysis: (i) static and/or (ii) dynamic. The first one consists of analyzing the source code itself, whilst the second consists of analyzing the behavior of the SUT by feeding it varied input data [10,18,21,23,28]. Fuzzy testing is a dynamic analysis vulnerability detection technique that consists of testing a SUT with faulty input and monitoring its behavior. In recent research, multiple works have attempted to reduce the search space of the inputs for efficient fuzzing, with most approaches relying on heuristics [26]. However, most works do not measure code coverage in this setting.

This paper introduces a black-box RESTful API fuzzy testing tool that leverages RL to perform efficient fuzzy testing. The fuzzer interacts with the SUT in an interoperable manner by using the OAS file and allows the configuration of scenarios, which are functionality execution sequences. The role of the RL in this context is to guide the input generation, by learning the context of the SUT relying on its HTTP responses. A vulnerability report is provided along with the RL agent's performance on the SUT making the intelligent system highly explainable as well.

This paper is organized as follows. Section 2 provides an analysis of related works. Section 3 details the fuzzer's implementation. Section 4 presents the experimental setup, demonstrates the suitability of the tool in the case study API, and discusses the results. Finally, Sect. 5 summarises and outlines the key takeaways of the work, describing the current ongoing improvements to the fuzzer.

2 State-of-the-Art

Fuzzy testing is a vulnerability detection technique that has shown remarkable results in identifying software defects, particularly when automated [3,8,12,22]. This technique involves generating faulty input to feed the functionalities of a SUT and monitoring its behavior according to the actual requirements and performance of the system [23,28]. Since it consists of trying different inputs to find vulnerabilities, this method is prone to low code coverage due to randomness, and inefficient execution due to redundant tests [22] leading to a higher rate of missed vulnerabilities. However, it is generally more accurate and has fewer false alarms than static analysis [28].

Nowadays, web APIs are the de-facto standard for Software integration [19], being most software made available via RESTful APIs. As RESTful APIs gain momentum, so does their testing. Recent advances in the literature show that black-box testing of RESTful APIs has outputted effective results and is capable of contributing to the validity, reliability, and correctness of these systems.

Atlidakis et al. [2] presented RESTler, a REST API fuzzy testing tool. The tool analyses the API specifications and generates a sequence of requests by inferring order dependencies and analyzing dynamic feedback of prior tests. Their tool was capable of detecting 28 bugs in various cloud services, which were confirmed and tested by service owners. Ed-douibi et al. [7] also presented an automated REST API testing tool that relied on OpenAPI Specifications (OAS) to generate specification-based test cases to ensure APIs meet the requirements. Their tool was validated on 91 APIs, and the results showed that the generated test cases cover, on average, 76.5% of the elements included in the definitions, with 40% of the tested APIs failing. Karlsson et al. [14] attempted automatic property-based tests generated from OAS documentation, where the input was randomly generated. Haraldsson et al. [11] collected public data and made it faulty to produce valuable test cases.

More recently, research has attempted to leverage RL to solve the search-based problem of finding the right input and combination of use cases. Sameer et al. [20] utilized RL to enhance valid test input generation, addressing the challenge of generating valid inputs for parameterized test drivers like web APIs. They introduce a tabular guide based on the Monte Carlo Control (MCC) learner, which constrains the input space within a random generator. While their system may face scalability issues due to the complexity of the MCC algorithm, their tool, demonstrated improved average time and reliability in bug detection in comparison to other tools, though it was unable to detect all bugs. Similarly, Zheng et al. [27] introduced WebExplor, an automated web testing framework employing curiosity-driven reinforcement learning to generate action sequences for end-to-end web testing. The framework adopts a Use Case testing approach, recognizing that a sequence of system functionalities can reveal unexpected defects. To address local optima, they incorporate a Deterministic Finite Automaton to track RL exploration. WebExplor successfully identified 3466 exceptions and errors in a real-world commercial web application.

Although these approaches can diminish the input space, and produce realistic faulty test data leading to the discovery of vulnerabilities, most do not provide code coverage metrics to validate the solutions and only describe the errors or vulnerabilities found in numbers, without explaining them. This work tackles these issues by presenting a novel fuzzer that uses a custom RL-based approach to guide input generation.

3 FuzzTheREST: A TestLab Module

This work presents FuzzTheREST, an automated black-box RESTful API fuzzer that leverages RL to strategically generate input values. The tool integrates the TestLab ecosystem, which is an automated software testing framework that integrates a diverse range of testing approaches across various levels and viewpoints [6]. FuzzTheREST's approach utilizes as main data source the OAS file for interoperability purposes. Several mutation methods were implemented to generate faulty input for the fuzzy tests. When successful, these result in the SUT's anomalous response, which the RL algorithm learns to improve input generation. In addition, the fuzzer documents the findings and performance of the RL, granting visibility into the process. This fuzzer sets itself apart through its novel RL algorithmic approach, explainability, adaptability, and interoperability.

By eliminating the need for in-depth source code knowledge, it streamlines the testing process, making it more accessible.

3.1 Reinforcement Learning Design

This section describes meticulously the design and execution of the novel RL algorithmic approach employed in the fuzzer for improved input generation. Fig 1 depicts the custom environment and the agent characteristics, interactions, and architecture. The following subsections describe the data acquisition process, RL environment, and Agent's design, showcased in Fig 1.

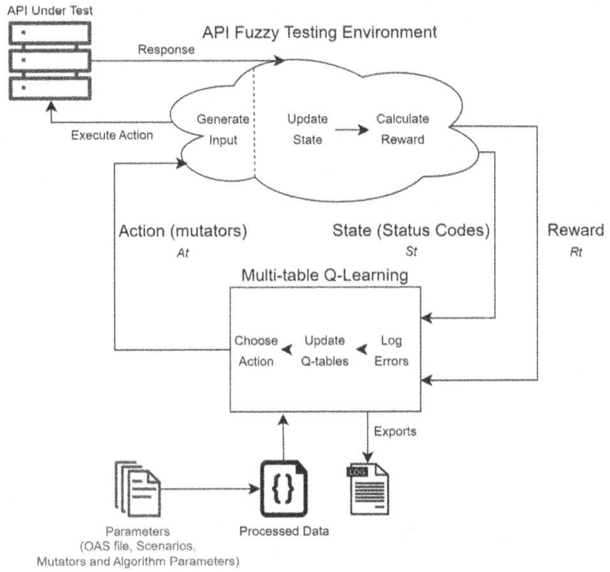

Fig. 1. Reinforcement Learning Environment and Agent Overview

Data Sources and Processing. The fuzzer requires four main sources of information to operate: (i) the OAS file, containing information to communicate with the SUT, (ii) the scenarios file, containing functionality execution sequences, (iii) the mutators, responsible for generating the faulty input, and (iv) the algorithm parameters, which will impact the learning process of the agent, therefore affecting its performance and the results obtained during testing.

The OAS file is divided into three main sections: (i) API information, (ii) API paths and operations, and (iii) reusable components for data schemas and security schemes. Since the OAS file contains a huge corpus of unnecessary data, before training the RL algorithm, this file was processed for dimensional reduction.

```
Function = {
        path,
        content-type,
        HTTP_method,
        input_parameters,
        input_body
}

Input_parameters = {
        name,
        location,
        schema,
        sample
}

Input_body = {
        schema,
        sample,
        schema_name
}
```

Fig. 2. Proposed Fuzzer's Data Schema

The extracted data is the one shown in Fig. 2, following a JSON notation. The `Function` contains the request type, header, and payload information of each API function. The `Input_parameters` store each function's query parameters and the `Input_body` the request body schema. The fuzzer is implemented to deal currently with five different datatypes: (i) integer, (ii) float, (iii) boolean, (iv) string, (v) and byte. The `sample` field in both `Input_parameters` and `Input_body` represents an instantiated schema. This value can be randomly generated or mutated via the mutation methods. Eight mutation methods were implemented inspired in successful fuzzers like AFL and libFuzzer: (i) bit flips, (ii) byte shuffling, (iii) byte injection/deletion, (iv) bytes substitution, (v) truncation, (vi) dictionary, (vii) arithmetic operations, and (viii) random generation.

Reinforcement Learning Environment and Agent. The RL environment is composed of the observation and action space. The states of the observation space are the HTTP status codes. The action space is defined by the actions that the agent can take. In this context, it corresponds to the mutation methods previously mentioned. Considering the nature of the mutation methods, there is a relationship of one-to-many between the datatypes and actions.

The RL agent interacts with the environment, by acting according to the current state. In this environment, the action is making an HTTP request with the faulty input to the SUT, which generates a response containing the HTTP status code. The status is used to assign a reward, which can be positive or negative depending on the outcome, to the agent. In this work, rewards were modeled around HTTP status codes: 1XX rewards 0 points, 2XX and 3XX rewards 5 points, 4XX takes 20 points, and 5XX, grants 10 points. This action and reward system allows the agent to learn over time. Nonetheless, it is important to choose a suitable agent to interact with such a specific environment.

An agent is an algorithm that learns from the environment and makes decisions to maximize cumulative rewards. This is done by figuring out the best possible actions to take given its current state. The agent is a Multi-table Q-Learning algorithm [13] that is well-suited to the environment developed. The Multi-table approach was motivated by the relationship between datatypes and actions, assigning each datatype its own Q-Table. To optimize learning, an epsilon-greedy policy with exploration decay was implemented to balance exploitation and exploration.

Figure 1 depicts the interactions between the agent and the environment during the training process. For a parameterized amount of episodes, the agent engages with the environment, recording outcomes and updating Q-values in the Q-tables. These values measure the success of each action in a given state and are calculated using the Bellman Equation. The agent then applies the policy to decide the next action. In each training episode, the agent can perform a parameterized amount of steps in the environment. However, it may end sooner if a 5XX HTTP status code is encountered. In this work, each agent only tests one single function. As such, if the function being tested consists of a creational one (e.g.: create user), then the agent uses substring identification and word inflection to store the unique identifiers that are generated so that others may use that information, ensuring the dependency between different functions.

4 Petstore API Case Study

To demonstrate the proposed fuzzer's effectiveness, a case study was conducted on the Petstore API [1]. The Petstore API is an illustrative Java-based RESTful API, designed around the concept of a pet store business, which includes three main entities: (i) Pets, (ii) Orders, (iii) and Users. For each entity, there are several variants of Create, Read, Update, and Delete (CRUD) methods. For the case study, three distinct scenarios targeting the functions of each entity were crafted in the CRUD order. This API is often used by many literature works in the testing field [4,5,15,17], and as such it was considered most relevant for the benchmark.

The Petstore API was deployed to a Docker container, to simulate a production environment, and a JaCoCo agent was set up for real-time code coverage. The API was tested using the fuzzer with the parameters described in Table 1. In bold are the parameters that had the best performance.

Table 1. Agent's Hyperparameters

Parameter name	Value
Number of max. steps	5 or **10**
Number of episodes	50 or 100 or 200 or **500**
Exploration rate	**1** or 0.8
Exploration decay rate	0.01

Harnessing the parameters detailed in Table 1, a suite of 18 agents, one for each tested function, were deployed and trained. The results show that agents testing functions where no vulnerabilities were found quickly reach a plateau. However, this was not true for all functions. For instance, the results achieved in the GetPetById function, depicted in Fig. 3, which represent the convergence of Q-Values and the number of received HTTP status codes that range from 1XX to 5XX, show that the agent is exploring the environment and is just barely making any successful requests, with most responses being in the 4XX range. Nonetheless, after a few training iterations, the agent learns which input data works best to interact with the SUT, leading to an increase in success and therefore receiving multiple HTTP responses with status 2XX and 5XX. This pattern replicates in multiple functions, exhibiting the learning process of the agents. A deeper analysis showed that initially, the agent was trying to fetch non-existing pets, and only when it learned that it should choose valid unique identifiers did it start to successfully communicate with the SUT. Since unique identifiers are integers, the action distribution plot corroborates with this observation, reinforcing the assertion that the automated requests are indeed yielding productive outcomes.

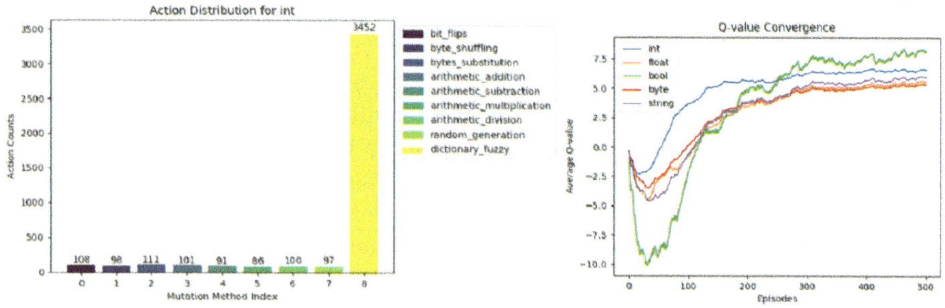

Fig. 3. GetPetById Agent performance

The results also show that depending on the complexity of functions the agent takes longer to learn better, as its interaction requires learning more details. An example of this was the findPetsByStatus which requires the pet status attribute to be of a specific range of values.

From an analysis of the vulnerability reports extracted from each function, out of the 2549 requests that resulted in a 5XX status code, six unique vulnerabilities were identified. Table 2 summarizes the findings.

The reports highlight critical issues within the SUT. Firstly, there's an erroneous string-to-number conversion in the source code. Additionally, multiple functionalities generate 5XX error codes due to potentially malicious content in URLs. Three vulnerabilities found show issues with illegal character combinations resulting in 5XX and surprisingly one 2XX. This is because malformed inputs were recorded but are unable to be fetched. The report shows this problem stems from a deficiency in the *jackson.core* framework. Similarly, the pets recorded with a status outside of the three possible values were unable to be retrieved.

The Petstore API vulnerabilities documented in other works [17] were also identified by the proposed fuzzer. However, unlike the proposed tool, most scientific works that present testing tools, do not describe the errors found and mostly focus on the number of received 5XX HTTP error code [5,15,24,25].

Table 2. Petstore Vulnerabilities Found

Vulnerability	Status Code	Description	Faulty Framework
String to Number	500	Conversion error from string input to number.	Not applicable
Malicious Content	500	Issue with input validation or security checks.	Spring boot security framework
Illegal Surrogate Pair	500	Error due to illegal surrogate pair.	com.fasterxml.jackson.core
Unmatched Surrogate Pair	500	Second part of surrogate pair missing.	com.fasterxml.jackson.core
Invalid White Space	200	Invalid white space makes fetching certain objects impossible.	com.fasterxml.jackson.core
Fetch Operation Error	500	Internal server error due to unexpected status for pet records.	Not applicable

To validate the effectiveness of the exploration-exploitation inherent to the RL approach implemented, the code coverage was also collected during the testing.

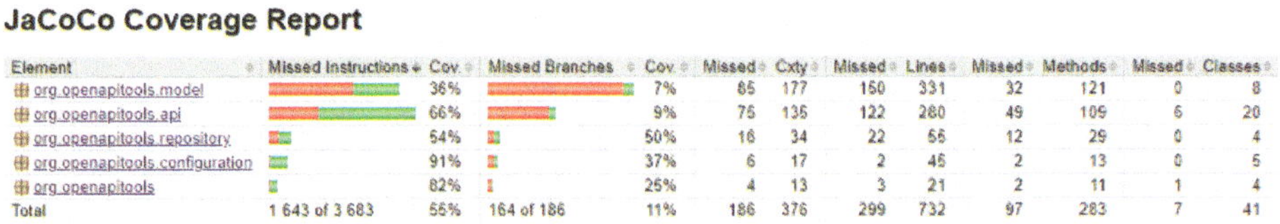
JaCoCo Coverage Report

Element	Missed Instructions	Cov.	Missed Branches	Cov.	Missed	Cxty	Missed	Lines	Missed	Methods	Missed	Classes
org.openapitools.model		36%		7%	85	177	150	331	32	121	0	8
org.openapitools.api		66%		9%	75	135	122	280	49	109	6	20
org.openapitools.repository		54%		50%	16	34	22	56	12	29	0	4
org.openapitools.configuration		91%		37%	6	17	2	45	2	13	0	5
org.openapitools		82%		25%	4	13	3	21	2	11	1	4
Total	1 643 of 3 683	55%	164 of 186	11%	186	376	299	732	97	283	7	41

Fig. 4. JaCoCo Coverage Report

The code coverage report depicted in Fig. 4, shows that 55% of code was traversed. Naturally, due to the nature of fuzzy testing, some paths may have been missed, which of course reflects on the code coverage achieved. Nonetheless, after further analysis it was noticed that several classes that had defined methods, which are included in the coverage calculation, are never utilized, meaning that such code will never be executed nor tested but will still affect the coverage. As such, if the unused code was removed, the coverage would increase as many methods contain multiple instructions that are never traversed.

5 Conclusions

This work presents an RL-based black-box RESTful API fuzzer that tackles input generation, exhaustive testing, and function dependency problems. The system was evaluated in a case study API and was able to uncover six unique vulnerabilities whilst achieving considerable code coverage, highlighting its learning capability. The experiment also shows that the HTTP status codes cannot be trusted alone for identifying vulnerabilities, which has been used in many literature works. Notably, the tool's usage did not impose significant time constraints on the testing process, rendering it a valuable asset for software testing. Further improvements include increasing data types, mutators, and algorithms, and improvements to the unique identifier storing method to make it suitable for all APIs. Incorporating coverage as a decision metric during testing could provide valuable guidance for balancing exploration and exploitation.

Acknowledgements. The present work has received funding from UIDP/00760/2020 and UIDB/00760/2020.

References

1. Swagger UI — petstore.swagger.io. https://petstore.swagger.io/
2. Atlidakis, V., et al.: RESTler: Stateful REST API fuzzing. In: 2019 IEEE/ACM 41st International Conference on Software Engineering (ICSE), pp. 748–758 (2019). https://doi.org/10.1109/ICSE.2019.00083
3. Böhme, M., Szekeres, L., Metzman, J.: On the reliability of coverage-based fuzzer benchmarking. In: Proceedings of the 44th International Conference on Software Engineering. ACM (2022). https://doi.org/10.1145/3510003.3510230
4. Corradini, D., et al.: Restats: a test coverage tool for restful APIs. In: Proceedings - 2021 IEEE International Conference on Software Maintenance and Evolution, ICSME 2021, pp. 594–598 (2021).https://doi.org/10.1109/ICSME52107.2021.00063
5. Corradini, D., et al.: Automated black-box testing of nominal and error scenarios in restful APIs. Softw. Test. Verification Reliab. **32**, e1808 (2022). https://doi.org/10.1002/STVR.1808
6. Dias, T., et al.: Testlab: an intelligent automated software testing framework. In: Mehmood, R., et al. (eds.) Distributed Computing and Artificial Intelligence, pp. 355–364. Springer, Cham (2023)
7. Ed-Douibi, H., et al.: Automatic generation of test cases for rest APIs: a specification-based approach. In: 2018 IEEE 22nd International Enterprise Distributed Object Computing Conference (EDOC), pp. 181–190 (2018)
8. Fioraldi, A., et al.: Dissecting American fuzzy lop – a fuzzbench evaluation (2022). https://doi.org/10.13140/RG.2.2.13803.82722
9. Ganney, P.S., et al.: Software Engineering, pp. 131–168. Elsevier (2020). https://doi.org/10.1016/b978-0-08-102694-6.00009-7
10. Ghaffarian, S.M., et al.: Software vulnerability analysis and discovery using machine-learning and data-mining techniques: a survey. ACM Comput. Surv. **50**(4), 1–36 (2017). https://doi.org/10.1145/3092566
11. Haraldsson, S.O., et al.: The use of automatic test data generation for genetic improvement in a live system. In: 2017 IEEE/ACM 10th International Workshop on Search-Based Software Testing (SBST), pp. 28–31 (2017). https://doi.org/10.1109/SBST.2017.10
12. Hazimeh, A., et al.: Magma. Proc. ACM Measur. Anal. Comput. Syst. **4**(3), 1–29 (2020). https://doi.org/10.1145/3428334

13. Kantasewi, N., et al.: Multi q-table q-learning. In: 10th International Conference on Information and Communication Technology for Embedded Systems, IC-ICTES 2019 - Proceedings (2019). https://doi.org/10.1109/ICTEMSYS.2019.8695963
14. Karlsson, S., et al.: QuickREST: property-based test generation of OpenAPI-described restful APIs. arXiv preprint arXiv:1912.09686 (2019)
15. Liu, Y., et al.: Morest: model-based RESTful API testing with execution feedback, pp. 1406–1417. Association for Computing Machinery (2022). https://doi.org/10.1145/3510003.3510133
16. Altaie, M.A., et al.: Verification and validation of a software: a review of the literature. Iraqi J. Comput. Inf. **46**(1), 40–47 (2020). https://doi.org/10.25195/ijci.v46i1.249 https://doi.org/10.25195/ijci.v46i1.249 https://doi.org/10.25195/ijci.v46i1.249 https://doi.org/10.25195/ijci.v46i1.249
17. Mahmood, R., et al.: A framework for automated API fuzzing at enterprise scale. In: Proceedings - 2022 IEEE 15th International Conference on Software Testing, Verification and Validation, ICST 2022, pp. 377–388 (2022). https://doi.org/10.1109/ICST53961.2022.00018
18. Malhotra, R., et al.: A systematic review on application of deep learning techniques for software quality predictive modeling. In: 2020 International Conference on Computational Performance Evaluation (ComPE). IEEE (2020). https://doi.org/10.1109/compe49325.2020.9200103
19. Martin-Lopez, A., et al.: Test coverage criteria for restful web APIs, pp. 15-21. Association for Computing Machinery (2019). https://doi.org/10.1145/3340433.3342822
20. Reddy, S., et al.: Quickly generating diverse valid test inputs with reinforcement learning, pp. 1410–1421. Association for Computing Machinery (2020). https://doi.org/10.1145/3377811.3380399
21. Shafiq, S., et al.: A literature review of using machine learning in software development life cycle stages. IEEE Access **9**, 140896–140920 (2021). https://doi.org/10.1109/access.2021.3119746
22. Shen, Q., et al.: A systematic review of fuzzy testing for information systems and applications, pp. 156–162. IEEE (2021). https://doi.org/10.1109/CECIT53797.2021.00035
23. Shen, Z., et al.: A survey of automatic software vulnerability detection, program repair, and defect prediction techniques. Secur. Commun. Netw. **2020**, 1–16 (2020). https://doi.org/10.1155/2020/8858010
24. Tokos, A.: Evaluating fuzzing tools for automated testing of rest APIs using OpenAPI specification (2023). https://www.doria.fi/handle/10024/187395
25. Viglianisi, E., et al.: RestTestGen: automated black-box testing of restful APIs, pp. 142–152. IEEE (2020). https://doi.org/10.1109/ICST46399.2020.00024
26. Vitorino, J., et al.: Constrained adversarial learning and its applicability to automated software testing: a systematic review. arXiv preprint arXiv:2303.07546 (2023)
27. Zheng, Y., et al.: Automatic web testing using curiosity-driven reinforcement learning, pp. 423–435. IEEE (2021). https://doi.org/10.1109/ICSE43902.2021.00048
28. Zhu, Y., et al.: The application of neural network for software vulnerability detection: a review. Neural Comput. Appl. **35**(2), 1279–1301 (2022). https://doi.org/10.1007/s00521-022-08046-y

TRUNC: A Transfer Learning Unsupervised Network for Data Clustering

Rita Xavier[1(✉)], John Peller[2(✉)], and Leandro N. de Castro[1,2(✉)]

[1] Graduate Program in Technology, Universidade Estadual de Campinas, São Paulo, Brazil
r223447@dac.unicamp.br

[2] U.A. Whitaker College of Engineering, Florida Gulf Coast University, Fort Myers, USA
{jpeller,ldecastrosilva}@fgcu.edu

Abstract. Clustering aims to group similar data objects together while keeping dissimilar objects apart. Various bioinspired algorithms have been developed to address different challenges in clustering tasks. One promising approach is the use of self-organizing neural networks, which can adapt and learn the underlying patterns in the data. Transfer Learning (TL) has also gained attention for its ability to leverage knowledge from one domain to improve learning in another. In this context, a Transfer Learning Unsupervised Network (TRUNC) is proposed, integrating a self-organizing network with TL to enhance clustering performance. This paper introduces TRUNC, presents a sensitivity analysis of the algorithm to the transfer learning term, and an evaluation of its effectiveness when applied to synthetic data.

Keywords: Unsupervised Network · Self-Organizing Neural Network · Transfer Learning · Data Clustering · Bioinspired Computing

1 Introduction

Clustering primary objective is to organize a set of data objects into meaningful groups, or clusters, such that objects within the same cluster are more similar to each other than to those in other clusters [1]. It plays a crucial role in various fields, including pattern recognition, image analysis, and bioinformatics, where understanding the underlying structure of data is essential.

Over the years, numerous clustering algorithms have been developed, each with its strengths and weaknesses, to address different challenges in clustering tasks. One promising approach in clustering is the use of self-organizing networks, which mimic the organization and learning processes of the cortex. These networks can adapt to the data and learn the underlying patterns, making them suitable for complex clustering tasks [21]. Another emerging learning technique is Transfer Learning (TL), which aims to transfer knowledge from one domain to another to improve learning efficiency and performance [9]. By leveraging knowledge from a source domain, TL can enhance learning in a target domain with limited labeled data.

R. Chinthaginjala et al. (Eds.): DCAI 2024, LNNS 1259, pp. 170–180, 2025.
https://doi.org/10.1007/978-3-031-82073-1_17

In this context, a Transfer Learning Unsupervised Clustering Network (TRUNC) is proposed. TRUNC integrates a self-organizing neural network with TL to enhance clustering performance, particularly in scenarios where labeled data is scarce or unavailable. By leveraging knowledge from a source domain, TRUNC aims to improve the clustering performance in a target domain, making it a promising approach for various clustering tasks.

This paper introduces TRUNC and analyzes its effectiveness in synthetic data. The methodology used for evaluation, including the experimental setup and evaluation metrics, is detailed to provide a clear understanding of the experimental process. The results are discussed, highlighting the strengths and weaknesses of TRUNC and its potential for enhancing clustering performance in various domains.

This paper consists of six sections, including the introductory first section. Section 2 presents the concepts of TL for clustering and places this proposal in the context of the literature. Section 3 presents the unsupervised network and introduces TRUNC. Section 4 describes the experiments and results obtained. Section 5 concludes the paper and provides discussions for future works.

2 Transfer Learning for Clustering

TL is a machine learning approach where knowledge learned from one task can be reused to improve performance in another task, what may significantly increase learning efficiency [1–4]. TL can be categorized based on learning and transfer strategies [1], including inductive, transductive, and unsupervised approaches. These methods improve the efficiency and effectiveness of machine learning models in various applications. Clustering is often unsupervised, akin to TL, where past knowledge improves a different task, even with varied nature or domain distributions [2].

TL can also be categorized based on transfer approaches, called transfer strategies [3]. Like traditional TL, clustering TL also falls into four categories: instance transfer, that reuses and re-weights source domain instances for better clustering; the feature transfer, that learns effective feature representations; the parameter transfers, that reuses shared parameters or distributions between domains; and relational knowledge transfer, that leverages similarities in relationships between instances [1].

In TL, as shown in Fig. 1, data sampling and acquired knowledge are typically from the source domain. Due to the abundance of data and limited (or absent) labeling, and the differing distribution, direct clustering of data is not feasible [5].

Fig. 1. How to transfer useful knowledge using Transfer Learning.

Several techniques that can be implemented in TL, such as Manifold TL [6], Self-Adaptive TL [7], Two-Level TL [8], and Prototype-Based TL [5, 9–11]. Prototype-Based TL (PBTL) is a widely used technique, especially in clustering algorithms. This approach utilizes clusters identified in a source domain as representative models to assist in clustering data in a target domain. It proved particularly beneficial when data in the target domain is scarce or of lower quality, as it enables more effective data analysis across various application domains, like in DE-TRC algorithm [9]. The approach to be introduced here follows this same paradigm of PBTL, in which the prototypes obtained for the clusters in the source domain help guiding the search for the prototypes in the target domain.

3 TRUNC: A Transfer Learning Unsupervised Network

This section introduces TRUNC, a Transfer Learning Unsupervised Network for Data Clustering, its cost function, weight update rule, and further details.

Self-Organizing Network for Clustering

A self-organizing network, also known as an unsupervised network, is a type of artificial neural network (ANN) designed for unsupervised learning tasks [12–14]. In these networks, training does not rely on labeled data, instead, the neurons compete among themselves to represent the input patterns, making the network competitive in nature.

The architecture employed here is a single-layer feedforward network, where postsynaptic neurons are arranged along the nodes of a one-dimensional grid [15]. The network is trained using a winner-takes-all competitive learning scheme. The output neurons adapt to various input patterns or groups of patterns through unsupervised learning. The network begins by initializing the weights and then, two essential processes occur: competition and adaptation.

Let $\mathbf{x} = [x_1, x_2,\ldots, x_m]$ be an input vector (training pattern) presented in parallel to all neurons $i = 1,\ldots, o$ in the network. The complete dataset $\mathbf{X}$ consists of N patterns, $\mathbf{X} \in \mathbb{R}^{N\times m}$. The weight vector of unit i is $\mathbf{w}_i = [w_{i1}, w_{i2},\ldots, w_{im}]$, and the weight matrix is $\mathbf{W} \in \mathbb{R}^{o\times m}$, where o is the number of output (post-synaptic) units in the network.

In the competition phase, each input pattern triggers a competition among the neurons, and the winning neuron i is the one whose weight vector w_i is closest to the input pattern x, based on a certain metric $\| \cdot \|$, typically the Euclidean distance:

$$i(\mathbf{x}) = \underset{j}{argmin}|\mathbf{x} - \mathbf{w}_j|, \quad j = 1, \ldots, o \tag{1}$$

The neuron $i(x)$ that satisfies this equation is known as the best matching unit, or *winner*. The update process for the winner weight vector is:

$$\mathbf{w}_i(t + 1) = \mathbf{w}_i(t) + \alpha(t)[\mathbf{x}(t) - \mathbf{w}_i(t)] \tag{2}$$

where the learning rate $\alpha(t)$ decreases over time, $0 < \alpha(t) < 1$.

Prototype-based Transfer Process

Prototype-based clustering is a well-known clustering process that typically involves minimizing an objective function J that takes into account a partition matrix **U** and a set of prototypes W [5]:

$$\min J = \min f(\mathbf{U}, W). \tag{3}$$

Assuming that the final weight vectors of the trained network are the cluster prototypes, the problem solved by the unsupervised network for clustering (UNC) can be expressed as:

$$\text{UNC} : \min_{\mathbf{W}} J_{\text{UNC}} = \sum_i \sum_j \|\mathbf{x}_j - \mathbf{w}_i\|^2 \tag{4}$$

where J_{UNC} is the cost function to be optimized (minimized), $\mathbf{x}_j$ is the j-th input pattern, $j = 1,\ldots,N$, $\mathbf{w}_i$ is the i-th prototype (weight vector of neuron i), $i = 1, \ldots, o$, and $\|.\|$ is a distance function, usually Euclidean, that quantifies the distance between the weight vector and the input data.

TL can be used in many ways to assist this learning process, but there is a problem of understanding which information is relevant and what knowledge should be transferred from one domain problem to another, or vice versa. In the PBTL technique, the most valuable information is the location of the prototypes (or centroids) of the group. Formally, the PBTL mechanism is represented as [5, 9]:

$$\Delta\left(\tilde{W}_S, W_T\right) = \sum_{k=1}^{o} \|\widetilde{\mathbf{w}_k} - \mathbf{w}_k\|^2 \tag{5}$$

where $\widetilde{W}_S$ is the set of prototypes for the source domain, W_T is the set of prototypes for the target domain, $\widetilde{\mathbf{w}}_c$ is the c-th cluster centroid (network weight vector) for the source domain, $\mathbf{w}_c$ is the c-th cluster centroid for the target domain, and o is the number of clusters in the source and target domains. Note that in this unsupervised network case, each neuron weight vector represents one cluster centroid.

Equation (5) states that the prototype-based TL process must consider the centroids for the source domain and move the centroids of the target domain in their direction as a way of transferring the knowledge from one domain to another.

TRUNC: Transfer Learning Unsupervised Network for Clustering

The objective of the learning process of a traditional self-organizing network was presented in Eq. (4) and corresponds to the process of finding the most compact clusters, that it, those whose distance between the input patterns and their corresponding centroids are minimal.

The objective function of the Transfer Learning Unsupervised Network for Clustering, TRUNC, uses the target domain data X_T and the cluster centers $\widetilde{W}_S = \{\widetilde{\mathbf{w}}_k\}$, $k = 1 \ldots, o$, of the source domain as auxiliary knowledge. The cluster centers of the source domain are the weight vectors of the network trained for the source data. The new objective function is:

$$\text{TRUNC} : \min_{\mathbf{W}} J_{\text{UNC}} = \sum_i \sum_j \|\mathbf{x}_j - \mathbf{w}_i\|^2 + \lambda \sum_k \sum_i \|\widetilde{\mathbf{w}}_k - \mathbf{w}_i\|^2 \tag{6}$$

where $i = 1, \ldots, c_T$ is the index of neurons (prototypes) in the target domain, $j = 1, \ldots, N$ is the index of the input patterns (objects), $k = 1, \ldots, c_s$ is the index of neurons in the source domain, and λ is a trade-off parameter responsible for weighing the influence of the source domain in the learning of the target domain. In the present case, $c_s = c_T = o$, that is, both networks have the same number of output neurons.

Note that the first term in Eq. (6) measures the compactness of the groups in the target domain, whilst the second term measures the similarity between the weight vectors (prototypes) in the source and target domains. This integration of TL introduces new avenues for enhancing clustering performance by leveraging prototype-based TL strategies, as summarized in Fig. 2.

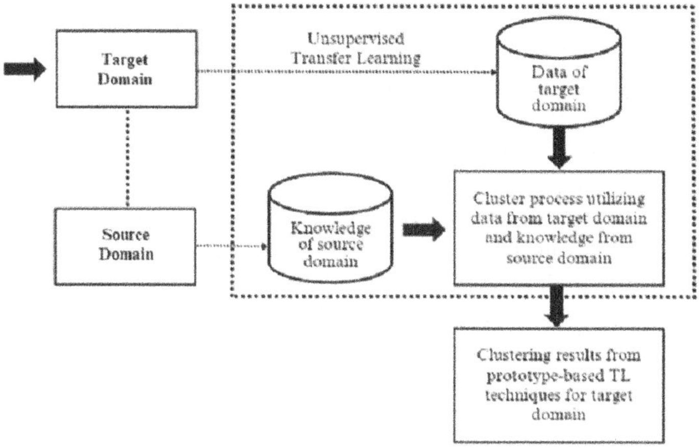

Fig. 2. Flow of prototype-based transfer learning in a single-layer self-organizing network

Based on this proposal, to adapt the learning process and incorporate TL into the self-organizing network, it is necessary to adjust the weight update presented in Eq. (2) so as to accommodate the prototype-based transfer mechanism:

$$\mathbf{w}_i(t+1) = \mathbf{w}_i(t) + \alpha(t)[\mathbf{x}(t) - \mathbf{w}_i(t)] + (t)[\tilde{\mathbf{w}}_k(t) - \mathbf{w}_i(t)] \tag{7}$$

where $\lambda(t)$ is the TL weight parameter, which also decreases over time, and $\tilde{\mathbf{w}}_k$ is the weight vector trained for the source data that is closest to the winner neuron i. The TRUNC algorithm is summarized in Algorithm 1.

Algorithm 1. TRUNC: Transfer Learning Unsupervised Network for Clustering

```
function TRUNC (data, n_prototypes, mode, source_prototypes, source_weight, max_iter,
                learning_rate, max_iter, convergence_threshold)
  Initialize prototypes randomly
  for each interation to i up to max_iter do
    for each point in data do
      Compute distances between the point and the prototypes
      Identify the closest prototype
      Compute the movement direction to the closest prototype
        if mode is target_tl and there is source_prototypes then
          Compute movement direction from source_prototype
          Add movement direction from source_prototypes weighted by source_weight
          end if
      Update the closest prototype in the calculated direction with learning_rate
    end for
    Compute convergence by comparing prototype movements with the previous iteration
    if all movements are below convergence threshold, then
      Break
      end if
   end for
   return final prototypes and a log with the movements for each iteration
end function
```

4 Performance Evaluation

To assess the performance of TRUNC, some experiments will be run with four synthetic datasets created by [9]. These experiments were conducted as an initial validation of the proposal based on tests made by [9]. Three external clustering measures were used: *Accuracy* (ACC); *Normalized Mutual Information* (NMI); and *Rand Index* (RI).

TRUNC was run with prototypes randomly initialized with input data, 100 maximum iterations, $\alpha \in [0.1, 1.0]$, and $\lambda \in [0.1, 1.0]$. 10 experiments were performed for each configuration. TRUNC was developed in Python, using libraries Numpy, Pandas, Matplotlib, Seaborn, Time, IterTools, JSon, Scikit-Learn, and Scipy. The tests were conducted in the Google Colaboratory environment.

The datasets are two-dimensional and are designed to assess the effectiveness of TL for clustering. The three datasets were generated as summarized in Table 1, where S2 to S5 are the source domains, and T2 to T5 are the target domains. In the experiments performed in [9], λ was used to understand the role of TL during the execution of TRUNC Table 1 illustrates the sensitivity of TRUNC to $\lambda = [0.10, 0.15, 0.20, 0.25, 0.30, 0.35, \ldots, 1.0]$.

With the execution of these experiments, two peculiarities were observed. The closer to the extremes of λ values, the greater the sensitivity of TRUNC in the clustering process. The fact that TRUNC performs well with both low and high λ values suggests that it can adapt well to distinct levels of TL. This may indicate that the algorithm is flexible enough to manage various TL scenarios. It also means that small changes in these values can result in significant changes in the clusters generated, which indicates that choosing the right λ is crucial for obtaining optimal clustering results (Fig. 3).

Based on the sensitivity analysis performed $\lambda = 0.1$ was chosen to run the final experiments and present the results. Figure 4 illustrates the clustering results in UNC

Table 1. Synthetic dataset characteristics.

Datasets	Cluster	Source			Target		
		Mean	Covariance	Size	Mean	Covariance	Size
Dataset 1: S2 and T2	C1	[− 5 5]	[5 0;0 5]	200	[− 5 5]	[8 0;0 8]	10
	C2	[0 15]	[5 0;0 5]	200	[0 15]	[8 0;0 8]	10
	C3	[5 5]	[5 0;0 5]	200	[5 5]	[8 0;0 8]	10
Dataset 2: S3 and T3	C1	[2 4]	[10 0;0 10]	250	[3 4]	[10 0;0 11]	125
	C2	[9 15]	[25 0;0 7]	250	[10 12]	[25 0;0 7]	125
	C3	[8 30]	[30 0;0 20]	250	[9 29]	[30 0;0 20]	125
Dataset 3: S4 and T4	C1	[2 4]	[10 0;0 10]	250	[3 4]	[10 0;0 11]	125
	C2	[9 15]	[25 0;0 7]	250	[10 12]	[25 0;0 7]	125
	C3	[8 30]	[30 0;0 20]	250	[9 29]	[30 0;0 20]	125
	Noise				**[20 20]**	**[4 0;0 4]**	**35**
Dataset 4: S5 and T5	C1	[10 5]	[5 0;0 5]	450	[10 5]	[5 0;0 5]	40
	C2	[10 15]	[5 0;0 5]	300	[10 15]	[5 0;0 5]	40
	C3	[15 –5]	[2 0;0 1]	150	[15–5]	[2 0;0 1]	10

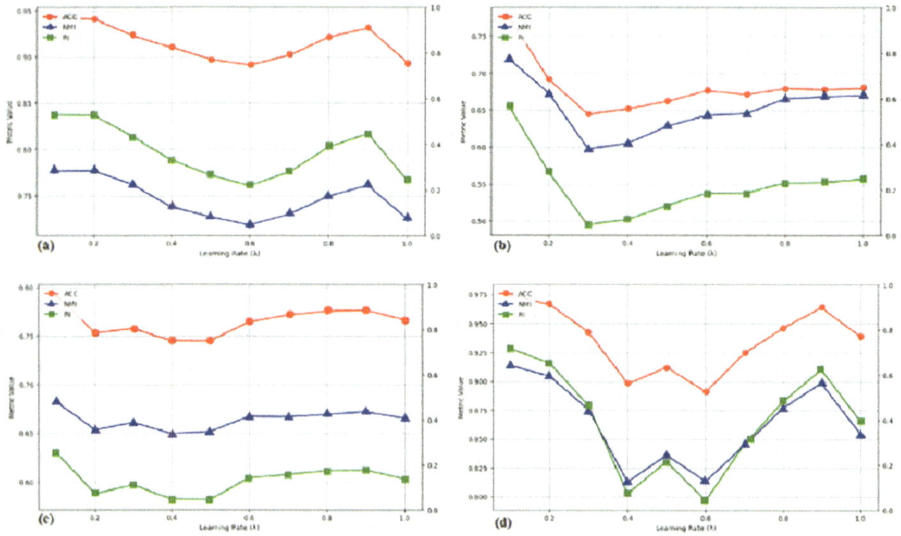

Fig. 3. Sensitivity analysis of TRUNC in the target data (T2 to T5).

for $\lambda = 0.1$, and Fig. 5 shows the same datasets but displaying the clustering results with TRUNC for $\lambda = 0.1$.

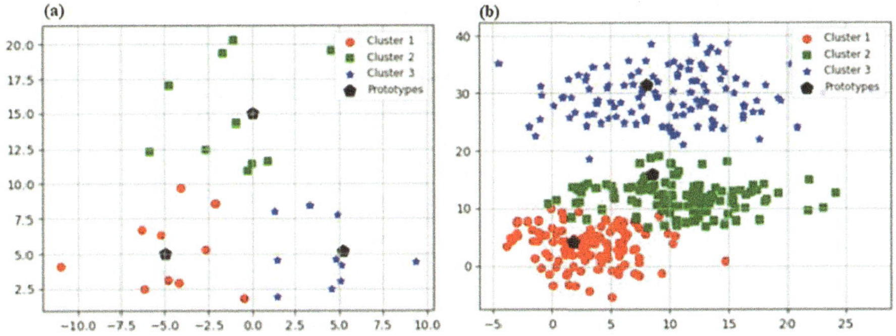

Fig. 4. Clustering for UNC with $\lambda = 0.1$. (a) T2. (b) T3.

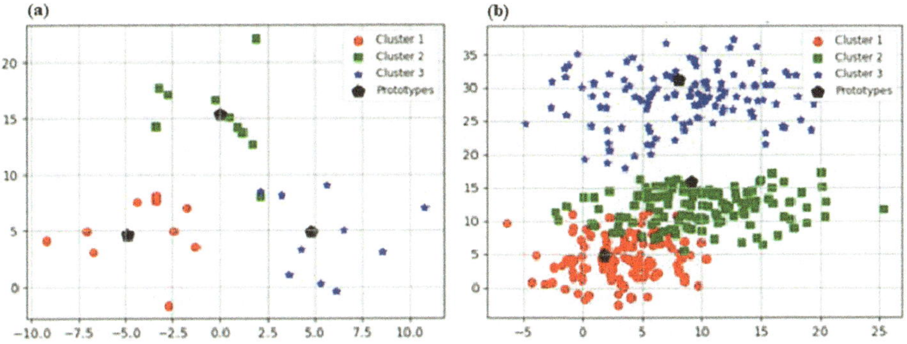

Fig. 5. Clustering for TRUNC with $\lambda = 0.1$. (a) T2. (b) T3.

Comparing UNC and TRUNC across datasets reveals some trends. For T2, while UNC performed slightly better in terms of ACC, NMI, and RI, TRUNC achieved comparable results, suggesting TRUNC's viability for T2. In T3, although UNC outperformed TRUNC in ACC, NMI, and RI, the difference was insignificant, indicating TL's potential value, particularly with limited labeled data. For T4, results were mixed. UNC performed better in ACC and RI, while TRUNC showed slightly better NMI, suggesting TL's effectiveness may depend on dataset characteristics and clustering algorithms. Lastly, for T5, TRUNC achieved similar ACC and RI results to UNC, with a slight NMI improvement, indicating potential TL benefits for T5.

In Table 2 it is possible to observe the mean and standard deviation of ACC, NMI, and RI in the clustering process with different learning rates. These experiments were performed using the source weight parameter between 0.55 and 1.0 and exploring the λ values between 0.1 and 1.0 .

The results show that the TRUNC algorithm achieved high ACC on datasets T2 and T5, indicating its effectiveness in correctly classifying instances. However, the algorithm's performance was lower on datasets T3 and T4, with lower ACC scores. The NMI results suggest that TRUNC achieved moderate to good performance in capturing the clustering structure similarity between predicted and true clusters. It performed

Table 2. Performance of TRUNC for synthetic datasets.

Datasets	ACC	NMI	RI
	Av ± Std	Av ± Std	Av ± Std
T2	0.85 ± 0.14	0.67 ± 0.15	0.70 ± 0.20
T3	0.55 ± 0.13	0.55 ± 0.20	0.41 ± 0.18
T4	0.69 ± 0.18	0.58 ± 0.23	0.51 ± 0.23
T5	0.82 ± 0.22	0.73 ± 0.23	0.70 ± 0.31

Table 3. Performance comparisons of DE-TRC, UNC and TRUNC algorithms on synthetic datasets.

Datasets	ACC			NMI			RI		
	DE-TRC	UNC	TRUNC	DE-TRC	UNC	TRUNC	DE-TRC	UNC	TRUNC
	Av ± std	Av ± std	Av ± std	Av ± std	Av ± std	Av ± std	Av ± std	Av ± std	Av ± std
T2	1.00 ± 0.00	1.00 ± 0.00	1.00 ± 0.00	1.00 ± 0.00	1.00 ± 0.00	1.00 ± 0.00	1.00 ± 0.00	1.00 ± 0.00	1.00 ± 0.00
T3	0.87 ± 0.00	0.96 ± 0.83	0.96 ± 0.13	0.87 ± 0.00	0.86 ± 0.10	0.86 ± 0.19	0.87 ± 0.00	0.89 ± 0.10	0.89 ± 0.18
T4	0.93 ± 0.00	0.80 ± 0.16	0.79 ± 0.18	0.92 ± 0.00	0.68 ± 0.20	0.66 ± 0.23	0.92 ± 0.00	0.64 ± 0.20	0.66 ± 0.24
T5	1.00 ± 0.00	0.55 ± 0.19	0.97 ± 0.22	1.00 ± 0.00	0.39 ± 0.21	0.90 ± 0.24	1.00 ± 0.00	0.24 ± 0.27	0.91 ± 0.32

best on dataset T5, indicating a strong alignment between predicted and true clusters. TRUNC achieved the highest RI mean on dataset T5, indicating a high level of similarity between predicted and true clusters.

Table 3 shows a comparison between the standard approach (UNC), TRUNC, and the algorithm proposed in [9] called DE-TRC. For comparison, the results here use the same parameters as in [9], with transfer and source rates equal to 0.9.

The comparison of algorithms sheds light on their relative performances. For dataset T2, all algorithms achieved perfect scores, indicating effective handling. However, TRUNC and UNC showed comparable results, suggesting limited advantage for TL. In T3, TRUNC slightly outperformed UNC in ACC and NMI, while DE-TRC, while performing well, showed variability in RI. For T4 and, TRUNC and DE-TRC performed similarly, with TRUNC surpassing UNC, indicating TL's potential to enhance clustering in complex datasets.

5 Discussion and Future Trends

The TRUNC algorithm, incorporating TL, shows promise for enhancing clustering performance across various synthetic datasets. While traditional algorithms like UNC and DE-TRC performed well on certain datasets, TRUNC demonstrated competitive or even superior results in some cases.

For datasets like T2, where all algorithms achieved perfect scores, TRUNC proved to be a viable alternative to traditional methods. In datasets like T4 and T5, TRUNC showcased its effectiveness, outperforming UNC and demonstrating similar performance to DE-TRC. This suggests that TL can be particularly beneficial in datasets with complex structures or with high unlabeled data.

While further optimizations and refinements to the TRUNC algorithm are needed, especially in handling dataset characteristics, TL presents itself as a valuable tool for improving clustering performance in various applications. Future research can explore additional enhancements to TRUNC and further investigate its applicability in real-world scenarios.

In the future, efforts are planned to further enhance the TRUNC algorithm by integrating a particle swarm strategy and conducting comparative studies with other state-of-the-art algorithms identified through a systematic review. These future efforts aim to refine the algorithm and explore its full potential in enhancing clustering performance across different domains.

Acknowledgments. This study was financed in part by the Coordenação de Aperfeiçoamento de Pessoal de Nível Superior of Brasil (CAPES). Finance Code 001 and Grant 2021/11905-0 of the São Paulo Research Foundation (FAPESP).

Disclosure of Interests. The authors have no competing interests to declare that are relevant to the content of this article.

References

1. Pan, S.J., Yang, Q.: A survey on transfer learning. IEEE Trans. Knowl. Data Eng. IEEE **22**(10), 1345–1359 (2009)
2. Dai, W., Yang, Q., Xue, G.-R., Yu, Y.: Self-taught clustering. In: Proceedings of the 25th International Conference on Machine Learning. [S.l.], pp. 200–207 (2008)
3. Zhuang, F., et al.: A comprehensive survey on transfer learning. Proc. IEEE IEEE **109**(1), 43–76 (2020)
4. Weiss, K., Khoshgoftaar, T.M., Wang, D.: A survey of transfer learning. J. Big Data Springer Open **3**(1), 1–40 (2016)
5. Deng, Z., et al.: Transfer prototype-based fuzzy clustering. IEEE Trans. Fuzzy Syst. **24**(5), 1210–1232 (2015)
6. Jiang, M., et al.: A fast dynamic evolutionary multiobjective algorithm via manifold transfer learning. IEEE Trans. Cybern. **51**(7), 3417–3428 (2020)
7. Zhang, X., Jin, Y., Qian, F.: A self-adaptive dynamic multi-objective optimization algorithm based on transfer learning and elitism-based mutation. Neurocomputing **559**, 126761 (2023)
8. Ma, X., Chen, Q., Zhu, Z.: A two-level transfer learning algorithm for evolutionary multitasking. Front. Neurosci. **13**, 493604 (2020)
9. Zhao, F., Wang, C., Liu, H.: Differential evolution-based transfer rough clustering algorithm. Complex Intell. Syst. Springer **9**(5), 5033–5047 (2023)
10. Jiao, L., et al.: TECM: transfer learning-based evidential c-means clustering. Knowl.-Based Syst. **257**, 109937 (2022)
11. Medina, C., Devos, A., Grossglauser, M.: Self-supervised prototypical transfer learning for few-shot classification. arXiv preprint:2006.11325 (2020)
12. Aliu, O.G., Imran, A., Imran, M.A., Evans, B.: A survey of self-organization in future cellular networks. IEEE Commun. Surv. Tutorials IEEE **15**(1), 336–361 (2012)

13. Guthikonda, S.M.: Kohonen self-organizing maps. Wittenberg University, 98 (2005)
14. Kohonen, T.: Essentials of the self-organizing map. Neural Netw. **37**, 52–65 (2013)
15. Klaine, P.V., Imran, M.A., Onireti, O., Souza, R.D.: A survey of machine learning techniques applied to self-organizing cellular networks. IEEE Commun. Surv. Tutorials IEEE **19**(4), 2392–2431 (2017)

An Architectural Proposal for Injury Assessment in Occupational Medicine

Mariana Fernandes and Hugo Peixoto and José Machado(✉)

LASI/Algoritmi, University of Minho, Braga, Portugal and and
jmac@di.uminho.ptand and

Abstract. Occupational medicine focuses on protecting workers' health, through the prevention of work-related illnesses and diseases, ensuring safety and well-being to perform their functions. This reflects in more productivity and resilience that comes with healthy workers, who experience fewer sick days and have more job satisfaction. Ultimately, this benefits both employees and the companies they work for, contributing to the overall performance and good functioning of the workplace. However, there is a gap in terms of existing software for corporal damage assessment of workplace injuries. Therefore, it is proposed the development of a web application designed to help professionals in the occupational medicine field with the evaluation of diseases and injuries sustained in the workplace. This solution aims to be more precise, effective, and timesaving, thereby enhancing the efficiency of assessments in this field. Moreover, the data collection of injury types and disability coefficients aims to establish a dataset for future machine learning applications. This initiative holds promise for advancing occupational health practices and promoting safer work environments for all.

Keywords: Web Application · Occupational Medicine · Injury Assessment · Machine Learning

1 Introduction

In today's dynamic work environment, occupational medicine plays a vital role in protecting the health and well-being of workers across various industries. This specialized area is dedicated not only to the prevention of work-related diseases and illnesses but also to creating a culture of well-being and safety in the workplace. By proactively identifying and mitigating occupational hazards, occupational medicine not only reacts to health issues but also promotes a safe and healthy work environment, resulting in increased productivity and organizational success.

This importance becomes even more evident when considering the significant number of work accidents in Portugal. In 2021, 166,028 work-related accidents

This work has been supported by FCT (Fundacão para a Ciência e Tecnologia) within the R&D Units Project Scope: UIDB/00319/2020.

R. Chinthaginjala et al. (Eds.): DCAI 2024, LNNS 1259, pp. 181–190, 2025.
https://doi.org/10.1007/978-3-031-82073-1_18

were recorded, marking an increase of 6.4% when compared to the previous year. These accidents led to over 4 million days of work lost [8]. These numbers highlight the crucial link between worker's health and productivity, making it clear that healthy employees are more resilient and productive. Furthermore, experience fewer sick days and have more job satisfaction, significantly contributing to the performance and good functioning of the workplace.

This proposal seeks to address occupational medicine needs by introducing an architecture designed to optimize practices and improve outcomes for both workers and employers.

1.1 Occupational Medicine in Portugal

During the Industrial Revolution, in the first half of the 19th century, rapid industrialization and increasing demand for production led workers to face extremely inhumane working conditions. These adverse conditions seriously affected the health and quality of life of workers, making the need for intervention imperative. Occupational medicine emerged as a response to these needs, aiming to assess, prevent, and treat occupational health problems [11]. In Portugal, occupational medicine has emerged as a significant tool since the 1990s in preventing occupational diseases among workers, ultimately contributing to the improvement of employees' well-being.

In occupational medicine, accurately assessing workplace injuries and disabilities is crucial. A key tool for this, in Portugal, is the *Tabela Nacional de Incapacidades* (National Disability Table), providing a standardized framework for evaluating bodily harm and determining disability levels. This table, addressed in Decree-Law No. 352/2007, of October 23rd, is a crucial instrument for assessing bodily harm in Portugal. It plays a vital role in determining changes in individuals' physical and mental health, a complex process due to factors such as the difficulty in interpreting sequelae, the subjectivity of some damages, and the need for accurate medical assessments. It plays a crucial role by enabling doctors to assess injuries and assign a disability coefficient to each of them. This coefficient, within the range established by the table, reflects the severity of the injury and the extent to which it impacts the individual's ability to perform work activities. Ultimately, all coefficients converge to an overall degree of disability for work, reflected in the percentage that will appear on the certificate signed by the doctors.

2 Related Work

The integration of technology into healthcare has revolutionized health monitoring and occupational health management. This section explores the recent advancements in digital solutions and artificial intelligence (AI) applications within these domains.

2.1 Applications for Health Monitoring and Occupational Health

In recent years, digital solutions have emerged in the healthcare sector, offering innovative tools for health monitoring and occupational health management. This section covers several implemented applications, addressed at both patients and health professionals, focusing on occupational health.

SNS 24 [17] is a health app developed by the Portuguese National Health Service, available for all citizens. It offers patients access to various digital health services, including access to personal health information, vaccination records, sick leave records, medication prescriptions, test results, and consultation appointments. It has facilitated a significant shift in healthcare delivery by empowering individuals to take a more active role in managing their health. Furthermore, it has been instrumental in optimizing the efficiency of healthcare services, reducing administrative burdens, and promoting a more patient-centric healthcare ecosystem.

Safemed [15] is a software designed for managing occupational health and safety requirements, available for service providers and companies that organize these services internally. There are several modules available, including the Occupational Medicine module. This software offers several essential features for clinical professionals and clients, such as customizable clinical records, digital fitness forms, appointment scheduling, and appointment history, among others.

A multidisciplinary group of professors from different departments of the University of Aveiro (UA) [2] created an initiative, e.cuidHaMUs. The program was developed to support the Occupational Medicine Office of the UA, contributing to the monitoring and epidemiological surveillance of employee health at higher education institutions. In the medium term, it aims to provide users with quick and easy access to their clinical history. Some of their features include an individual questionnaire for registration, with personalized feedback to adjust their health behaviors and promote healthy lifestyles.

Connolly et al. [4] studied the development of Sterling Connect, a web-based Management Information System. It is a personalized web platform, developed by Sterling HealthCare, a scottish company that offers occupational health services, in partnership with a local university and financing by the UK Technology Strategy Board. This system aims to facilitate referral and management of employee health services. It is made up of several subsystems, including Customers, Patients, and Appointments; Occupational Health Reports; Human Resources; Security; and Electronic Health Records. The core functionality of the system revolves around the ability of authorized personnel, such as Human Resources managers and technicians, to forward employees for various types of assessments and medical examinations. Additionally, the system assists in legal compliance, providing guidance to managers and Human Resources for reporting injuries and illnesses, as required by regulations in the UK (RIDDOR). The first version of the system works in a closed environment, without data transferring from or to another external system. Despite that, it is developed so that, in the future, it can integrate other external systems, to evolve according to change and technological advancements. For EHR management, the system uses openEHR.

While the existing applications offer essential functionalities in occupational medicine, such as consultation appointments and injury reporting, practitioners have identified specific gaps in meeting their needs. This paper aims to address one of these unmet needs.

2.2 Artificial Intelligence in Occupational Medicine

Several studies have been conducted in the medical field with the use of artificial intelligence. Some examples include prediction of COVID-19 diagnosis [14], mortality and occurrence of complications for gastric cancer patients [12], hospital readmission for diabetic patients [13], as some examples. Other applications involve natural language processing for extraction of drug-drug interactions [9], for example.

The use of AI in occupational health is present in numerous applications such as safety enhancement, accident prevention, risk assessment, stress prediction, noise level evaluation, and worker fatigue determination. Several studies have been conducted in this area, demonstrating the growing importance and potential of AI in occupational health [10].

Respiratory illnesses, among the most common occupational diseases, benefit from AI algorithms' ability to analyze medical images. These algorithms exhibit high precision and accuracy in detecting abnormalities indicative of lung conditions in chest X-rays, computed tomography scans, and magnetic resonance imaging scans. This aids in improving decision-making, especially in cases more challenging for radiologists [16].

In a study by Choi et al. [3], predictive machine learning models were applied to analyze fatal accidents at construction sites, utilizing national public data. The purpose of the study is to enhance the efficiency of on-site safety management practices. The dataset included various workers' parameters, such as age, gender, length of service, and type of construction. The study achieved high accuracy values in its predictions.

Ajayi et al. [1] proposed a big data architecture for health hazards analytics. The architecture comprises multiple layers: an application layer, an analytics and functional model layer, a semantic layer, and a data storage layer. The data storage layer contains the data sources (finance and health and safety risks) needed for the development of the machine learning models. Due to the diverse nature of the stored data, a NoSQL dataset is implemented. The semantic layer provides the data exchange formatting and data provisioning to the application layer. The analytics and functional model layer includes one functional model (health and safety visualization) and three analytics models (exploratory, predictive, and prescriptive analytics). The end-user interface, represented by the application layer, consists of APIs for integration with the system. The results obtained indicated that the proposed architecture can identify relevant variables and improve preliminary prediction accuracy and explanatory capacities.

AI-based work management systems have extended their application to mental health monitoring, such as the assessment of psychological distress levels. In a study conducted by Doki et al. [6], the goal was to use AI for predicting psychological distress among workers by considering various objective factors, including sociodemographic, lifestyle patterns, and sleep data. The study compared the predictive capabilities of AI with those of psychiatrists. The created machine learning model was able to accurately assess 65.2% of workers with moderate psychological distress and 89.9% for severe psychological distress, higher than the accuracy of psychiatrists.

Another application in the area of mental health was studied by Estevez-Mujica and Quintane [7] to identify employees at risk of experiencing burnout. To achieve this, the authors used e-mail communication and burnout data collected through an online survey. The results demonstrated that the created model can distinguish between employees at a higher and lower risk of burnout. With this, concluded that the main goal of the study, which is to propose a cost-effective early warning system for burnout, was achieved.

3 Architecture Proposal

This section presents an overview of the proposed architecture aimed at addressing challenges in occupational health management. It offers a detailed description of the architecture's components, accompanied by a use case diagram to aid comprehension. Additionally, it outlines the architecture's implementation across three distinct layers, emphasizing the integration and processing of data extracted from the official document serving as the project's foundation.

3.1 Description of the Components

Doctors frequently evaluate injuries sustained by workers during their tasks. Currently, professionals acknowledge the necessity of a platform to assist them in this calculation, which is based on the official Portuguese document, the *Tabela Nacional de Incapacidades* (National Disability Table). With this in mind, this research proposes a solution to address this issue. To achieve this goal, a web application is proposed with three distinct modules: Registration, Corporal Damage Assessment, and Monitoring.

- **Registration Module**: The platform should allow detailed registration of the patient's information, including data related to occupational activity.
- **Corporal Damage Assessment Module**: The platform should integrate the National Disability Table, allowing the doctor to select all injuries sustained and determine the final range of impairment. Additionally, it should enable the doctor to input the final value, submit the obtained result, and store it in the database.
- **Monitoring Module**: The platform should facilitate the easy creation, editing, and deletion of monitoring dashboards. It should also gather relevant collected data and present it in a clear and organized manner.

This platform addresses the critical issue of the absence of a digital method for the calculation of total disability, providing a crucial solution for healthcare professionals. Offering a digital tool for assessing bodily damage and determining impairment levels using the National Disability Table, marks a significant advancement in occupational health management.

For a clearer grasp of how this innovative tool operates in a real-world scenario, a use case diagram is provided in Fig. 1. Furthermore, to provide a practical perspective, an example will be presented to illustrate how this process is applied in real-world scenarios, such as after an industrial accident when a worker is brought to the hospital for evaluation. The attending doctor, using the Corporal Damage Assessment tool, selects the specific injuries sustained by the patient, such as fractures or contusions, and assigns the appropriate impairment values based on the severity of each injury. Meanwhile, the system calculates the overall degree of incapacitation, considering all selected injuries. Once the assessment is complete, the doctor submits the final evaluation, which is then stored in the hospital's database for reference in the worker's medical record. This practical example illustrates how the Corporal Damage Assessment tool streamlines the evaluation process for healthcare professionals, ensuring accurate and efficient injury assessment in real-world scenarios.

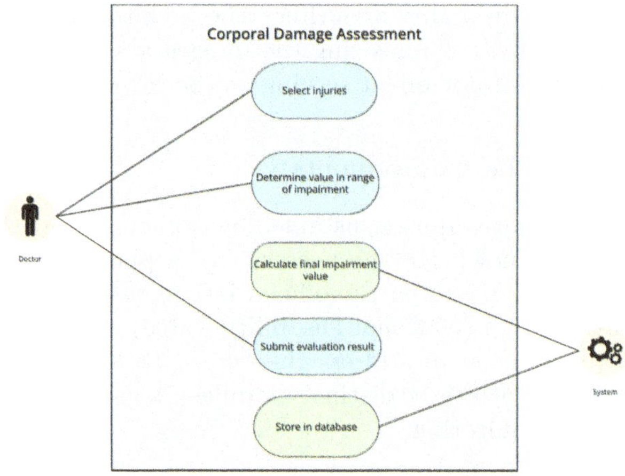

Fig. 1. Use case diagram of the Corporal Damage Assessment process.

3.2 Description of the Architecture

The proposed architecture, as depicted in Fig. 2, adopts a 3-tier model, with distinct layers for presentation, application, and data handling. In the presentation layer, the frontend is developed in ReactJS, a JavaScript library. Libraries such as Material UI, and Bootstrap offer pre-designed components and styles for UI development. OpenEHR is integrated for Electronic Health Records (EHR) management, providing a standardized approach to storing and exchanging health data [5].

The application layer, situated between the presentation and data layers, is implemented using Node.js and Express.js. Node.js provides a robust runtime environment for server-side JavaScript execution, while Express.js simplifies the development of RESTful APIs for communication between the frontend and backend. Within this layer, Axios is utilized for handling HTTP requests/responses, facilitating seamless communication between the frontend and backend components.

At the data layer, MongoDB serves as the database management system for storing and managing data. MongoDB's document-oriented model aligns well with the flexible data structures required for healthcare records. Additionally, Mongoose, a Node.js library, is employed for data modeling and interaction with MongoDB, enhancing the efficiency of database operations within the backend.

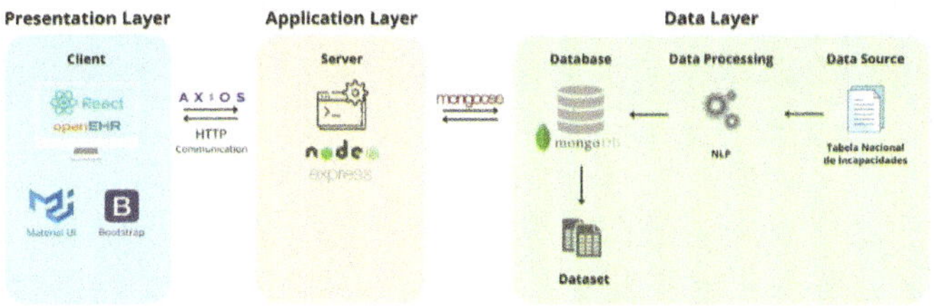

Fig. 2. Schematic representation of the proposed architecture.

In terms of database, the official document, *Tabela Nacional de Incapacidades*, depicted in Fig. 3, serves as the data source.

1. Spine

1.1 Sprains, fractures and dislocations:

1.1.1. Spinal injuries without fracture, or with healed fractures without deformation or with insignificant deformation:

a) Asymptomatic. 0,00

Fig. 3. Excerpt from the official *Tabela Nacional de Incapacidades* (translated).

Data processing techniques were applied, specifically natural language processing to extract relevant information from the document using regular expressions and incorporating it into the database system in a structured manner.

The analysis process begins with data extraction from a structured document that follows a layered hierarchy, transitioning from general to specific terms. Incorporating regular expressions into the extraction process streamlined the identification of specific patterns or formats within the document. Each element of the document is individually captured, including its ID, description, and, when present, associated value.

Following extraction, the data is organized based on IDs and the presence or absence of values. Elements lacking an associated value indicate they are not final injuries and may contain supplementary information. Such information is grouped with 'parent' elements through shared IDs. For instance, IDs 1.1. and 1.2. share the parent ID 1.

Following this approach, the extracted data was transformed into a JSON file, presenting the information from the original PDF document in a more accessible format. This facilitated ease of use and further analysis. A snippet of the resultant JSON file, as visualized in Fig. 4, corresponds to the structured data previously depicted in Fig. 3.

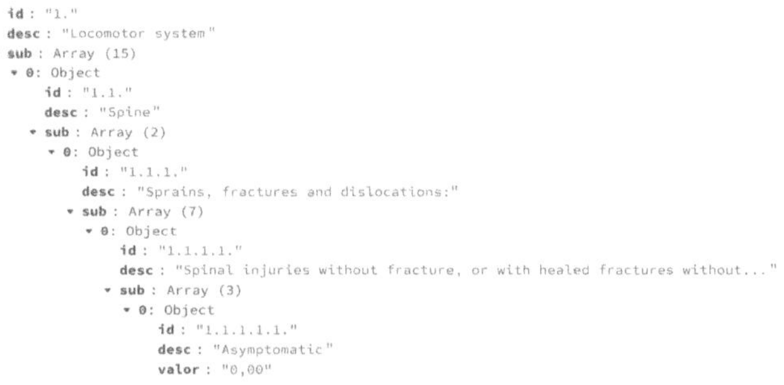

Fig. 4. Excerpt from the final JSON incorporating the extracted informations from *Tabela Nacional de Incapacidades*, extracted using MongoDB Compass tool (translated).

Following each medical evaluation, data regarding the description of the damages suffered, corresponding coefficients, and final evaluation, are stored in the database. The final goal is to establish a high-quality dataset, suitable for application in machine learning and artificial intelligence algorithms.

The integration of the bodily harm assessment tool is crucial as it provides healthcare professionals with a digital, automated, and user-friendly solution for their assessments and calculations. This eliminates the necessity for manual searches and computations, optimizing the process and enhancing efficiency.

Additionally, the platform's capability to store detailed records of bodily damage creates a valuable database that can be utilized for various purposes, including the training of machine learning algorithms. Machine learning can significantly enhance the proposed platform's functionality and effectiveness in occupational health management. By analyzing historical data and patterns, machine learning algorithms can predict potential risks, assist in diagnosing injuries, develop personalized treatment plans, monitor patients' health in real time, and optimize resource allocation. These capabilities empower healthcare professionals to make more informed decisions, improve patient outcomes, and enhance overall workplace safety and productivity. This integration of technology not only improves efficiency in disability assessment but also lays the foundation for future advancements in data-driven healthcare practices.

4 Conclusions and Future Work

The proposed project addresses a common concern in occupational health: assessing injuries sustained in the workplace. This evaluation is important to understand the gravity of occupational risks, identify areas of safety improvement, and guarantee the safety of workers. The proposed solution benefits professionals in this field by introducing a transition from the traditional method to a digital and innovative approach. This tool offers a more precise and objective evaluation, reducing the margin of error associated with traditional methods.

Machine learning can be useful in this field because, due to its capacity to analyze extensive datasets, it can identify patterns and trends that traditional methods may fail to capture. Additionally, it can predict potential safety risks based on historical data of the injuries and optimize preventative measures to reduce them. Moreover, another potential application is automatic injury classification, facilitating the analysis and decision-making.

To evaluate the platform's performance and effectiveness, conducting a study with field experts, such as doctors, to gather feedback on the system"s utility would be an ideal approach. This feedback ensures continuous improvement and refinement of the platform to better serve the needs of occupational health professionals.

References

1. Ajayi, A., et al.: Big data platform for health and safety accident prediction. World J. Sci. Technol. Sustain. Develop. **16**, 2–21 (2019). https://doi.org/10.1108/WJSTSD-05-2018-0042
2. Brandão, M.P., Sa-Couto, P., Gomes, G., Beça, P.: Description of an integrated e-health monitoring system in a Portuguese higher education institution: the e.cuidHaMUstm program. Global Health Promot. **29**(1), 65–73 (2022). https://doi.org/10.1177/1757975920984222
3. Choi, J., Gu, B., Chin, S., Lee, J.S.: Machine learning predictive model based on national data for fatal accidents of construction workers. Autom. Constr. **110**, 102974 (2020). https://doi.org/10.1016/J.AUTCON.2019.102974
4. Connolly, T.M., MacPhail, A., Scott, D.G.W.: Developing a web-based mis system for occupational health. In: Proceedings of the 6th International Conference on Web Information Systems and Technology (2010). https://doi.org/10.5220/0002792100810088
5. Dias, M., Sousa, R., Duarte, J., Peixoto, H., Abelha, A., Machado, J.: Enhancing data science interoperability: an innovative system for managing openEHR structures. Commun. Comput. Inf. Sci. **1942 CCIS**, 288–299 (2023). https://doi.org/10.1007/978-981-99-7969-1_21

6. Doki, S., et al.: Comparison of predicted psychological distress among workers between artificial intelligence and psychiatrists: a cross-sectional study in Tsukuba science city, Japan. BMJ Open **11**, e046265 (2021). https://doi.org/10.1136/BMJOPEN-2020-046265
7. Estévez-Mujica, C.P., Quintane, E.: E-mail communication patterns and job burnout. PLOS ONE **13**, e0193966 (2018). https://doi.org/10.1371/JOURNAL.PONE.0193966
8. Gabinete de Estratégia e Planeamento: Acidentes de trabalho 2021 - síntese (2023). http://www.gep.mtsss.gov.pt/documents/10182/26338/at2021sint.pdf/8f4fb65c-1a90-4746-b701-e63a860364a1
9. Machado, J., Rodrigues, C., Sousa, R., Gomes, L.M.: Drug-drug interaction extraction-based system: an natural language processing approach. Expert Syst., e13303 (2023). https://doi.org/10.1111/EXSY.13303
10. Maheronnaghsh, S., Zolfagharnasab, H., Gorgich, M., Duarte, J.: Machine learning in occupational safety and health - a systematic review: review. Int. J. Occup. Environ. Saf. **7**, 14–32 (2023). https://doi.org/10.24840/2184-0954_007-001_001586 https://doi.org/10.24840/2184-0954_007-001_001586 https://doi.org/10.24840/2184-0954_007-001_001586 https://doi.org/10.24840/2184-0954_007-001_001586
11. Mendes, R., Dias, E.C.: Da medicina do trabalho à saúde do trabalhador. Rev. de Saúde Pública **25**(5), 341–349 (1991). https://doi.org/10.1590/s0034-89101991000500003
12. Neto, C., Brito, M., Lopes, V., Peixoto, H., Abelha, A., Machado, J.: Application of data mining for the prediction of mortality and occurrence of complications for gastric cancer patients. Entropy **21** (2019). https://doi.org/10.3390/E21121163
13. Neto, C., et al.: Different scenarios for the prediction of hospital readmission of diabetic patients. J. Med. Syst. **45** (2021). https://doi.org/10.1007/S10916-020-01686-4
14. Oliveira, D., Ferreira, D., Abreu, N., Leuschner, P., Abelha, A., Machado, J.: Prediction of COVID-19 diagnosis based on openEHR artefacts. Sci. Rep. **12**(1), 1–13 (2022).https://doi.org/10.1038/s41598-022-15968-z
15. Safemed solutions: Safemed solutions: software segurança no trabalho (2022). https://www.safemed.solutions/safemed-2/
16. Shah, I.A., Mishra, S.D.: Artificial intelligence in advancing occupational health and safety: an encapsulation of developments. J. Occup. Health **66**, 1–12 (2024). https://doi.org/10.1093/JOCCUH/UIAD017
17. SNS24 (2023). https://www.sns24.gov.pt/sobre-nos/

Dynamic Programming for Training DNN in Distributed Computing

Julia García Cabello[1(✉)] and S. Carbó-García[2]

[1] Department of Applied Mathematics, Andalusian Research Institute in Data Science and Computational Intelligence, University of Granada, Granada, Spain
cabello@ugr.es

[2] Department of Computer Science and Artificial Intelligence, Andalusian Research Institute in Data Science and Computational Intelligence, University of Granada, Granada, Spain
carbo@ugr.es

Abstract. In recent years, the architecture and structure of Deep Neural Networks (DNNs) have become progressively more complex in order to respond to the increasing complexity of real problems. A strategy to deal with this complexity when it affects training would be to partition DNN training in some way: for example, by distributing it among different components of a computer network. For this, training (which is in essence the minimization of the loss function) should be performed through separated "smaller pieces". This paper offers an alternative to the gradient-based DNN training from a Dynamic Programming (DP) point of view (DP is an optimisation methodology supported by the division of a complex problem into many problems of lower complexity). To do so, conditions which enable the DNN minimization algorithm to be solved under a DP perspective are studied here.

In this line, in this work is proved that *any* artificial neural network ANN (and thus also DNNs) with monotonic activation is separable. Furthermore, whenever ANNs are considered as a dynamical system in the form of a network (known as coupled cell networks CCNs), we show that the transmission function is a separable function assuming that the activation is non-decreasing.

Keywords: Dynamic Programming · Separable function · DNN training

1 Introduction

Learning skills of Artificial Neural Networks (ANNs) are indisputable. Amongst the ANNs, Deep Neural networks (DNNs), which are learning models with multiple hidden layers, stand out because they require fewer neurons in function approximation compared to shallow networks. The DNN complex structure enables a deeper processing more suitable for certain contexts, with results into

R. Chinthaginjala et al. (Eds.): DCAI 2024, LNNS 1259, pp. 191–200, 2025.
https://doi.org/10.1007/978-3-031-82073-1_19

a better performance in the determination of the parameters that define the learning process.

DNNs is a term that evokes the network *architecture* but includes different varieties of processing (recurrent neural networks or convolutional neural networks). In recent years, such architecture have become progressively more complex in order to respond to the increasing complexity of real problems. A strategy to deal with complex architectures when it affects DNN training would be to partition the training in some way: for example, by distributing it amongst different computational units of a network in order to lower the computational cost. For this, training (which is in essence a minimization process) should be divided into 'smaller pieces" and processed sequentially.

Dynamic Programming (DP) [1] is a crucial branch of Optimisation Theory. The major contribution of DP is the idea of decomposing complex optimisation problems into many problems of lower complexity, to which the Bellman's equation can be applied. The theoretical assumptions are set in the Principle of Optimality (PO), [4], which is enabled provided that the function to be optimized (objective) is separable in the additive sense (classical DP) or, in more general contexts, separable under a definition for which there is no unified version in the literature. Surprisingly, very few papers have addressed the potential application of the DP mode to neural network training. The most representative is [2] where, amongst other things, a complete review of separability is performed. Previously, works such as [9] and [11] provides formulas for minimizing the loss function by determining the solution at each stage.

The main target of this work is to analize those requirements under which the ANN learning process can be approached from a DP point of view (i.e., through simpler optimization programs that run sequentially) as an alternative solution to mitigate the DNN processing drawbacks. To find such requirements (those that allow to use PO) we shall analyse them in general scenarios. After an exhaustive review of the different versions of the notion in [2], this paper articulates a general definition of separability applied to the objective function. With this, we show that those ANNs which have non-decreasing activation function are separable. Furthermore, we also show that the transmission function is separable when ANNs are considered as a networked dynamical system (also called coupled cell network, see [8]).

This article is structured as follows. The main problem is stated in Sect. 2. Section 3 contains different views of ANNs and the UAT property. In Sects. 4 and 5, the intermediate and main contributions are shown, respectively. Section 6 gathers the conclusions.

2 Problem Definition

The main objective is to find sufficient conditions under which the minimization algorithm which underlies any ANN can be addressed under a DP view. The rigurous definition of a Dynamic Programming problem with a single-objective but multiple stages is as follows:

$$\begin{aligned} &\underset{(u,x)}{Max} \qquad g(x,u) \\ &\text{subject to } x(t+1) = f(x(t), u(t), t), \quad t = 0, \dots, T-1 \end{aligned} \tag{1}$$

where $x(0) = x_0 \in \mathbb{R}^n$ is the initial condition, $\mathbf{x} = (x(0), \dots, x(T))$, $\mathbf{u} = (u(0), \dots, u(T-1))$ are the state and control variables respectively and $(\mathbf{x}^*, \mathbf{u}^*)$ denotes a solution (i.e., $(\mathbf{x}^*, \mathbf{u}^*) \in \underset{\mathbf{x},\mathbf{u}}{\mathbf{arg\,max}} g(\mathbf{x}, \mathbf{u})$). In a compact form, a Dynamic Programming problem with multiple stages, written as $(g, f, \underset{0 \le t \le T}{\{X_t\}}, U, T)$, is a tuple consisting of

- $g : \mathbb{R}^{n \times (T+1)} \times \mathbb{R}^{m \times T} \longrightarrow \mathbb{R}, \quad (\mathbf{x}, \mathbf{u}) \longmapsto g(\mathbf{x}, \mathbf{u})$ represents the objective function. By means of the isomorphisms $\mathbb{R}^{n \times (T+1)} \cong \mathfrak{M}_{n \times (T+1)}(\mathbb{R})$ and $\mathbb{R}^{m \times T} \cong \mathfrak{M}_{m \times T}(\mathbb{R})$, we will use both vector and matrix notation for $\mathbf{x}, \mathbf{u}$:
$$\mathbf{x} = \begin{pmatrix} x_1 \\ \vdots \\ x_n \end{pmatrix} = \begin{pmatrix} x_1(0) & \dots & x_1(T) \\ \vdots & \vdots & \vdots \\ x_n(0) & \dots & x_n(T) \end{pmatrix} = (x(0), \dots, x(T)),$$
$$\mathbf{u} = \begin{pmatrix} u_1 \\ \vdots \\ u_m \end{pmatrix} = \begin{pmatrix} u_1(0) & \dots & u_1(T-1) \\ \vdots & \vdots & \vdots \\ u_m(0) & \dots & u_m(T-1) \end{pmatrix} = (u(0), \dots, u(T-1)),$$
- $f : \mathbb{R}^n \times \mathbb{R}^m \times \mathbb{N} \to \mathbb{R}^n$ is the dynamics.
- $X_{t+1} = \{x(t+1) = f(x(t), u(t), t) | \mathbf{x} \in X_t, \mathbf{u} \in U, t \in \mathbb{N}\} \subseteq \mathbb{R}^n$ and $U \subseteq \mathbb{R}^m$ contain those $x(t) \in X_t$ and $u(t) \in U$, known as feasible and control variables respectively.

The DP cornerstone is the Principle of Optimality (PO) ([1]) that states that the solutions of an optimal solution of the problem are themeselves optimal solutions for the corresponding sub-problems in which the original problem is decomposed. In classical Dynamic Programming, $g(x, u)$ (the function to be optimized) is considered as separable in the additive sense, i.e.:

$$g(\mathbf{x}, \mathbf{u}) = \sum_{t=0}^{T-1} c_t(x(t), u(t)) + c_T(x(T)), \text{with } c_t(x(t), u(t)) \text{ single time-stage,}$$

such that, under this assumption, the solution $F(x, t)$ is found by using the Bellman's equation in a recursive manner. In some contexts, separability applied to the objective function is a requirement to enable the Principle of Optimality. Generally speaking, in classical Dinamic Programming both separability and monotonicity are sufficient conditions to enable the Principle of Optimality. The following is our definition of separability:

Definition 1. ([7]) *Let us consider a function* $g : X_0 \times \ldots \times X_T \times \overbrace{U \times \ldots \times U}^{T} \to \mathbb{R}$. *We say that* g *is a separable function as long as there are real-valued functions*

$\{g_t : X_t \times U \times Img_{t+1} \to \mathbb{R}\}_{0 \le t \le T-1}$ *and* $g_T : X_T \to \mathbb{R}$ *in such a way that* g *can be written as composition of* $\{g_t\}_{t=0}^{T}$*, as follows:*

$$g(\mathbf{x}, \mathbf{u}) = g_0\left(x(0), u(0), g_1\left(x(1), u(1), \ldots g_T(x(T)) \ldots\right)\right),$$

where every g_t *is non-decreasing in the last argument: i.e., for* $z, w \in Img_{t+1}$*, thus* $z \ge w \Rightarrow g_t(x, u, z) \ge g_t(x, u, w)$.

We analyse this definition in low dimensions:

- $T = 1$: $g : X_0 \times X_T \times U \to \mathbb{R}$ is separable as long as there are functions

$$\begin{aligned} &g_0 : X_0 \times U \times Im\, g_T \to \mathbb{R} \quad \text{non-decrasing in the last argument} \\ &g_T : X_T \to \mathbb{R} \\ &\text{where} g(\mathbf{x}, \mathbf{u}) = g(x(0), x(T), u(0)) = \\ &\qquad = g_0(x(0), u(0), g_T(x(T))). \end{aligned}$$

- $T = 2$: $g : X_0 \times X_1 \times X_T \times U \times U \to \mathbb{R}$ is separable as long as there is a sequence of functions

$$\left.\begin{aligned} g_0 &: X_0 \times U \times Im\, g_1 \to \mathbb{R} \\ g_1 &: X_1 \times U \times Im\, g_T \to \mathbb{R} \end{aligned}\right\} \text{which must be non-decrasing in the last argument}$$

$$\begin{aligned} &g_T : X_T \to \mathbb{R} \\ \text{so that } g(\mathbf{x}, \mathbf{u}) &= g(x(0), x(1), x(T), u(0), u(1)) = \\ &= g_0(x(0), u(0), g_1(x(1), u(1), g_T(x(T)))). \end{aligned}$$

3 Definitions of Neural Networks. Universal Approximators

Focusing on the architecture, ANNs can be mathematically defined in different ways.

CCN. CNN is the acronym which stands coupled cell network, a mathematical object consisting on cells connected by links. Rigorously (see [3]), a coupled cell network is a network representation of a dynamical system so that the dynamics are cohesive with the network structure: i.e., the network nodes gather the dynamical system dynamics while edges represent linkages between nodes. This definition is particularly useful when the DNN functioning is described by a dynamic system with transmission of signals.

Composition of Parametric Functions. The original definition of a multi-layered $n - L - m$ ANN as a CCN formed by n inputs, L hidden layers and m outputs is shown to be equivalent to the following one:

Proposition 1. *A* $n - L - m$ *ANN is a function* $F : \mathbb{R}^n \to \mathbb{R}^m$ *depending on parameters* $W^{[\cdot]}, b^{[\cdot]}$*, which can be decomposed into functions as follows:* $F = \varphi \circ f_{L+1} \circ \varphi \circ f_L \circ \ldots \circ \varphi \circ f_1$ *where* $f_i : \mathbb{R}^{k_i} \to \mathbb{R}^{r_i}, i = 1, \ldots, L+1$ *are affine*[1] *maps* $f_i(x^i) = W^{[i+1]}x + b^{[i+1]}$, $W^{i+1} \in \mathfrak{M}_{r_i \times k_i}, x^i \in \mathbb{R}^{k_i}, b^{i+1} \in \mathfrak{M}_{r_i \times 1}$*, and* φ *is a non-linear function.*

[1] An affine function is a function of the form: $f(x) = Wx + b$ for matrices $W \in \mathfrak{M}_{m \times n}(\mathbb{R})$ and $b \in \mathfrak{M}_{m \times 1}(\mathbb{R})$.

Proof. By means of mathematical induction on L, the results follows. Actually, $n-1-1$ ANNs (called perceptron) verifies the property. By assuming that the result is true for $n-L-m$ ANN, note that the result is also true for $n-(L+1)-m$ by composing the $n-L-m$ ANN with the perceptron.

In this paper, we will focus on multilayered $n-L-1$ ANNs.

Corollary 1. *Any $n-L-1$ ANN is a function $F:\mathbb{R}^n\rightarrow\mathbb{R}$ with parameters $W^{[\cdot]}, b^{[\cdot]}$ resulting from the composition of a even number of functions, as follows: $F=\varphi\circ f_{L+1}\circ\varphi\circ f_L\circ\ldots\circ\varphi\circ f_1$, $f_i:\mathbb{R}^{k_i}\rightarrow\mathbb{R}^{r_i}, i=1,\ldots,L, f_i(x)=W^{[i+1]}x+b^{[i+1]}, i=1,\ldots,L$, $f_{L+1}:\mathbb{R}^{r_{L+1}}\rightarrow\mathbb{R}$ affine mappings and φ a non-linear function.*

The advantage of this definition is that, under this perspective, ANNs can be analyzed through their components, affine and activation functions.

CCN Revisited. According to the previous view of ANNs as parametric function, the first definition in terms of CCNs can now be revisited:

Definition 2. *A $n-L-m$ ANN is a CCN where the transmission of signals (i.e., the dynamics) is given by $x^{i+1}=\varphi\circ f_i(x^i)=\varphi(W^{i+1}x^i+b^{i+1})$ from layer i (input x^i) to layer $i+1$ (output x^{i+1}), where $f_i, i=1,\ldots,L+1$ and φ are affine mappings and a non-linear function, respectively.*

Remark 1. Importantly, the layer counter is equal to the counter of transition of states of neurons from one layer to the next one. In other words, ANNs can be viewed as multistage sequential decision processes that can be considered under DP mode.

Universal Approximators. The key property of ANNs -that explains their high learning capacity- is contained into the UAT (Universal Approximation Theorem) according to which ANNs are approximators of continuous functions (see the Hilbert's thirteenth problem[2]). The similarities between the decomposition of the objective function into continuous components and that of separable functions provides additional support for addressing the minimization ANN algorithm under a DP mode.

Theorem 1. (UAT, [5]). *α denote a continuous function with compact domain $K\subset\mathbb{R}^n$. Then, there exists an ANN F which approximates the given α (called the target function): $||F-\alpha||_\infty = sup_{x\in K}|F(x)-\alpha(x)|<\epsilon$, for some $\epsilon>0$, (under smooth requirements on the its activation).*

Therefore, we can assume that the loss (which is minimized in the ANN processing) has a very specific form:

Corollary 2. *The loss is a linear function of the ANN (considerd as a parametric function, see proposition 1).*

[2] The Hilbert's thirteenth problem posed whether a continuous function of two variables can be written in terms of single-variable continuous functions.

4 Intermediate Results

In order to show the central result of this work ((theorem 4 in Sect. 5), a series of intermediate results must be proven. This section is devoted to this together with a synthesis of their proofs. For further details, see [2].

4.1 General Results

Theorem 2. *Let ϕ be a non-decreasing function and g a separable function. Thus, the (left) composition of both $\phi \circ g$ is separable.*

Proof. Consider a separable function $g : X_0 \times \ldots \times X_T \times U^T \to \mathbb{R}$. According to definition 1, there is a sequence of functions $\{g_t : X_t \times U \times Img_{t+1} \to \mathbb{R}\}_{0 \leq t \leq T-1}, g_T : X_T \to \mathbb{R}$ which verify that g may be written in terms of the functions $\{g_t\}_{t=0}^{T}$. Then, the functions defined as $\phi_0 = \phi \circ g_0, \phi_t = g_t, \quad 1 \leq t \leq T-1, \phi_T = g_T$ is the sequence that shows that the composition $\phi \circ g$ is a separable function. Actually, :

$$\phi \circ g(\mathbf{x}, \mathbf{u}) = \phi \circ g_0\left(x(0), u(0), g_1\left(x(1), u(1), \ldots g_T(x(T)) \ldots\right)\right) 0 \leq t \leq T.$$

Importantly, note that functions ϕ_t are non-decreasing in the last argument since g_t are. Thus, the property is satisfied for $1 \leq t \leq T$. In the case $t = 0$, by applying the chain rule:

$$\frac{\partial(\phi \circ g_t)}{\partial g_{t+1}(\cdot)} = \underbrace{\phi'(g_t(x(t), u(t), g_{t+1}(\cdot)))}_{\geq 0} \cdot \underbrace{\frac{\partial g_t}{\partial g_{t+1}(\cdot)}(x(t), u(t), g_{t+1}(\cdot))}_{\geq 0} \geq 0,$$

we have shown that the property also holds true.

Theorem 3. *Let $g : X_0 \times \ldots \times X_T \times U^T \to \mathbb{R}$ be a function that is separable in the sense of definition 1. Thus, its partial derivative on the last argument is non-negative (g is non-decreasing in the last argument).*

Proof. On one hand, there is a sequence $\{g_t : X_t \times U \times Img_{t+1} \to \mathbb{R}\}_{0 \leq t \leq T-1}$ and $g_T : X_T \to \mathbb{R}$ in such a way that g is written like

$$g(\mathbf{x}, \mathbf{u}) = g_0\left(x(0), u(0), g_1\left(x(1), u(1), \ldots g_T(x(T)) \ldots\right)\right),$$

where each function g_t is non-decreasing in the third argument. On the other hand, g can be expressed in terms of projections p_t and $\{g_t : X_t \times U \times Img_{t+1} \to \mathbb{R}\}_{0 \leq t \leq T-1}, g_T : X_T \to \mathbb{R}$:

$$\begin{aligned} g(\mathbf{x}, \mathbf{u}) &= g_0\left(x(0), u(0), g_1\left(x(1), u(1), \ldots g_T(x(T)) \ldots\right)\right) \\ &= \underbrace{g_0 \circ (p_0 \times p_{T+1} \times g_1 \circ (p_1 \times p_T) \ldots)}_{g}(x(0, \ldots, x(T), u(0), \ldots, u(T-1)) \end{aligned}$$

Note that the jacobian matrix of g is

$$J(g) = (\frac{\partial g_0}{\partial x(0)}, \frac{\partial g_0}{\partial u(0)}, \frac{\partial g_0}{\partial g_1(\cdot)}) \begin{pmatrix} \frac{\partial p_0}{\partial x(0)} \cdots \frac{\partial p_0}{\partial u(0)} \cdots \frac{\partial p_0}{\partial g_2(\cdot)} \\ \frac{\partial p_T}{\partial x(0)} \cdots \frac{\partial p_T}{\partial u(0)} \cdots \frac{\partial p_T}{\partial g_2(\cdot)} \\ \frac{\partial g_1}{\partial x(0)} \cdots \frac{\partial g_1}{\partial u(0)} \cdots \frac{\partial g_1}{\partial g_2(\cdot)} \end{pmatrix}.$$

From this, it can be seen the positive sign of the corresponding partial:

$$\frac{\partial g}{\partial g_2(\cdot)} = (\frac{\partial g_0}{\partial x(0)}, \frac{\partial g_0}{\partial u(0)}, \frac{\partial g_0}{\partial g_1(\cdot)}) \begin{pmatrix} \frac{\partial p_0}{\partial g_2(\cdot)} \\ \frac{\partial p_T}{\partial g_2(\cdot)} \\ \frac{\partial g_1}{\partial g_2(\cdot)} \end{pmatrix} \geq 0.$$

Corollary 3. *Separability is closed under composition.*

Proof. By mathematical induction together with the associative law for composition, it is sufficient to show the property for two functions. Let h_1, h_2 two separable functions such that $h_1 \circ h_2$ exists. According to theorem 3, both functions h_1, h_2 are non-decreasing in their last argument, particularly h_1 is. The result follows by applying theorem 2.

4.2 Results Shown on Affine Functions

Proposition 2. *Let $g : \mathbb{R}^{n\times(T+1)} \times \mathbb{R}^{m\times T} \to \mathbb{R}$ be an additively separable function. Then g is separable.*

Proof. Suppose that $g : \mathbb{R}^{n\times(T+1)} \times \mathbb{R}^{m\times T} \to \mathbb{R}$ is additively separable, i.e., g can be decomposed as

$$g(\mathbf{x}, \mathbf{u}) = \sum_{t=0}^{T-1} c_t(x(t), u(t)) + c_T(x(T)).$$

The sequence of functions defined as $g_t : X_t \times U \times Im\, g_{t+1} \to \mathbb{R}$, $g_T : X_T \to \mathbb{R}$, $g_t(x(t), u(t), z) := c_t(x(t), u(t)) + z$, $g_T = c_T$ shows that g is separable (see [2] for further details).

This result allows us to prove that the affine functions are separable as well:

Proposition 3. *Any affine function is separable.*

Proof. Any affine map $f_i : \mathbb{R}^{k_i} \to \mathbb{R}^{r_i}$ responds to the pattern $f_i(x) = W^{[i+1]}x + b^{[i+1]}$, $x = (x_1, \ldots, x_{k_i})$ for $i = 1, \ldots, l$. Thus, the following decomposition of f shows that f is additively separable and therefore separable:

$$f_i \begin{pmatrix} x_1 \\ \vdots \\ x_{k_i} \end{pmatrix} = \begin{pmatrix} w_{11}^{i+1} & \cdots & w_{k_i 1}^{i+1} \\ \vdots & \ddots & \vdots \\ w_{r_i 1}^{i+1} & \cdots & w_{r_i k_i}^{i+1} \end{pmatrix} \begin{pmatrix} x_1 \\ \vdots \\ x_{k_i} \end{pmatrix} + \begin{pmatrix} b_1^{i+1} \\ \vdots \\ b_{r_i}^{i+1} \end{pmatrix} \quad (2)$$

$$= \begin{pmatrix} w_{11}^{i+1} \\ \vdots \\ w_{r_i 1}^{i+1} \end{pmatrix} x_1 + \ldots + \begin{pmatrix} w_{k_i 1}^{i+1} \\ \vdots \\ w_{r_i k_i}^{i+1} \end{pmatrix} x_{k_i} + \begin{pmatrix} b_1^{i+1} \\ \vdots \\ b_{r_i}^{i+1} \end{pmatrix}$$

$$= g_1(x_1) + \ldots + g_{k_i}(x_{r_i}).$$

5 Main Result

In this section is stated and proved the principal result of this work: that any $n - L - 1$ ANN (considered as function which depends on some parameters) is a separable function under a smooth condition. This result finally demonstrates that the DP philosophy can be applied to the minimisation algorithm of neural networks.

Theorem 4. *Any $n-L-1$ ANN (when seen as a function depending on parameters) is a separable function provided that the corresponding activation function is a non-decreasing function.*

Proof. According to corollary 1, F is $F = \varphi \circ f_{L+1} \circ \ldots \circ \varphi \circ f_1$ with $f_i : \mathbb{R}^{k_i} \to \mathbb{R}^{r_i}, i = 1, \ldots, L, f_{L+1} : \mathbb{R}^{r_{L+1}} \to \mathbb{R}$ affine maps and φ the activation. By the associative law, F can thus be written by means of composite units of the type $(\varphi \circ f_i)$: $F = (\varphi \circ f_{L+1}) \circ (\varphi \circ f_L) \circ \ldots \circ (\varphi \circ f_1)$. Note that these composite units are separable according to theorem 2 since they are the left composition of a non-decreasing function φ with a separable one f_i (as they are affine maps, according to proposition 3). Thus, F is separable by corollary 3.

From corollary 4, next corollay can be obtained as a consequence, where the main result is stated in terms of CCN revisited (see definition 2) with dynamics given by $x^{i+1} = \varphi \circ f_i(x^i) = \varphi(W^{i+1}x^i + b^{i+1})$.

Corollary 4. *Let us consider a $n - L - m$ multilayered ANN as a CCN. Then, the corresponding transition function $\varphi \circ f_i$, $i = 1, \ldots, L$, is a separable function provided that the activation φ is non-decreasing function.*

Table 1 thus sets up the comparison between DP and ANN minimization problems where equation $x^{i+1} = h(x^i, W^{i+1}, i)$ is an abstract form to express $x^{i+1} = \varphi \circ f_i(x^i) = \varphi(W^{i+1}x^i + b^{i+1}) (= h(x^i, W^{i+1}, i))$ for affine maps f_i : $\mathbb{R}^{k_i} \to \mathbb{R}^{r_i}$, $f_i(x^i) = W^{[i+1]}x + b^{[i+1]}$ and φ a non-linear activation.

Table 1. Translation from ANN minimization algorithm into DP terms

ANN minimization algorithm	DP multi-stage minimization algorithm
$\underset{W^t,b^t}{Min}\ E(x) = F(x) - \alpha(x)$ subject to $x^{i+1} = h(x^i, W^{i+1}, i)$ $= \varphi(W^{i+1}x^i + b^{i+1})$	$\underset{(u,x)}{Min}\quad g(x,u)$ subject to $x(t+1) = f(x(t), u(t), t)$
φ is the activation function $x^i \in K \subseteq \mathbb{R}^{k_i}$ $W^{i+1} \in \mathfrak{M}_{r_i \times k_i}, i = 1, \ldots, L$	$x(0) = x_0$ is the initial condition $\mathbf{x} = (x(0), \ldots, x(T))$ $\mathbf{u} = (u(0), \ldots, u(T-1))$ $t = 0, \ldots, T-1$

6 Conclusions

The complexity of DNNs in real scenarios (extremely large or complex models) urges finding alternative which mitigate the serious problem as that classical training present in practice. This paper thus explores the minimal requirements that enable the underlying ANN minimization algorithm to be run under a DP mode (which shall allow partitioning the training procedure). Breaking down the training procedure into manageable parts makes it more feasible to train such models, which might otherwise be impractical due to memory constraints or excessive training times. Other advantages of the DP training mode for DNNs in real contexts include:

- Resource Efficiency. Large DNNs often demands significant computational resources, including memory and processing power. Partitioning the training process allows for better utilization of available resources by distributing the workload across multiple computing units or devices (e.g. distributing computations across multiple GPUs or even across different machines).
- Parallelism. Deep learning frameworks often leverage parallelism to accelerate training. By the DP mode, different parts of the model can be trained simultaneously, exploiting parallel processing capabilities.
- Transfer Learning and Fine-Tuning. The DP mode facilitates transfer learning and fine-tuning since, instead of training an entire model, one can leverage pre-trained partitions and adapt them to the specific task at hand.
- Fault Tolerance and Robustness. In distributed training setups, partitioning the training can enhance fault tolerance and robustness. If one partition fails or experiences issues, the overall training process can continue without being completely disrupted. This is crucial for large-scale deployments where hardware failures or network issues are not uncommon.
- Privacy and Security. In certain scenarios where data privacy is a concern, partitioning can be employed to ensure that sensitive information remains localized and is not exposed during the training process. This is particularly

relevant in federated learning setups, where models are trained across distributed devices without centralizing sensitive data.

Acknowledgement. This work was supported in part by the Spanish Ministry of Universities through the research project "Disruptive group decision making systems in fuzzy context: Applications in smart energy and people analytics" (PID2019-103880RB-I00). Main Investigator: Enrique Herrera Viedma. Also and Junta de Andalucía by "Excellence Groups" (P12.SEJ.2463) and Junta de Andalucía (TIC186). In part was supported by the "MdM" Excellence Unit IMAG CEX2020-001105-M.

Disclosure of Interests. None

References

1. Bellman, R.: Dynamic Programming. Press Princeton, New Jersey (1957). https://doi.org/10.1126/science.153.3731.34
2. García Cabello, J.: Improved deep neural network performance under dynamic programming mode. Neurocomputing **559**, 126785 (2023). https://doi.org/10.1016/j.neucom.2023.126785
3. Golubitsky, M., et al.: Nonlinear dynamics of networks: the groupoid formalism. Bull. Am. Math. Soc. **43**(3), 305–364 (2006). https://doi.org/10.1090/S0273-0979-06-01108-6
4. Hussein, M.L., et al.: Decomposition of multiobjective programming problems by hybrid fuzzy-dynamic programming. Fuzzy Sets Syst. **60**(1), 25–32 (1993). https://doi.org/10.1016/0165-0114(93)90286-Q
5. Leshno, M., et al.: Multilayer feedforward networks with a nonpolynomial activation function can approximate any function. Neural Netw. **6**(6), 861–867 (1993). https://doi.org/10.1016/S0893-6080(05)80131-5
6. Li, D., et al.: The envelope approach for multiobjeetive optimization problems. IEEE Trans. Syst. Man Cybern. **17**(6), 1026–1038 (1987). https://doi.org/10.1109/TSMC.1987.6499313
7. Li, D.: Multiple objectives and non-separability in stochastic dynamic programming. Int. J. Syst. Sci. **21**(5), 933–950 (1990). https://doi.org/10.1080/00207729008910422
8. Nijholt, E., et al.: Center manifolds of coupled cell networks. SIAM Rev. **61**(1), 121–155 (2019). https://doi.org/10.1137/18M1219977
9. Saratchandran, P.: Dynamic programming approach to optimal weight selection in multilayer neural networks. IEEE Trans. Neural Netw. **2**(4), 465–467 (1991). https://doi.org/10.1109/72.88167
10. Stewart, I., et al.: Symmetry groupoids and patterns of synchrony in coupled cell networks. SIAM J. Appl. Dyn. Syst. **2**(4), 609–646 (2003). https://doi.org/10.1137/S1111111103419896
11. Sun, M.: Training multilayer feedforward neural networks using dynamic programming. In: Proceedings of 28th Southeastern Symposium on System Theory, pp. 163–167 (1996). https://doi.org/10.1109/SSST.1996.493491

Designing User Experience in the Context of Human-Centered AI and Generative Artificial Intelligence: A Systematic Review

Carlos Alberto Peláez[1], Andrés Solano[1], Juan M. Núñez V[2](✉), David Castro[3], Juan J. Cardona[3], Juan S. Duque[3], Juan C. Espinosa[1], Ana S. Montaño[1], and Fernando De la Prieta[2]

[1] Universidad Autónoma de Occidente, Cali, Valle del Cauca 760030, Colombia
[2] BISITE Research Group, University of Salamanca, Salamanca 37007, Spain
jmnunez@usal.es
[3] Corporación Talentum, Cali, Valle del Cauca, Colombia

Abstract. This study presents an assessment of the state-of-the-art based on a systematic literature review methodology, focusing on the problematic relationship between User Experience (UX) design, generative artificial intelligence (GenAI), and human-centered design. It examines how UX principles and generative AI technologies can be harmoniously integrated to enhance user experiences, prioritizing human needs and values in the design process. The findings reveal significant insights into ethical considerations, user personalization, the importance of human-centric approaches, trust and perception challenges, and advances in human-AI interaction.

Keywords: HCAI · User experience (UX) · GenAI

1 Introduction

Recent advancements in Generative Artificial Intelligence (GenAI), especially Large Language Models (LLMs) with billions of parameters [1], have spawned a wide range of applications. This surge is evident across different organizations, including Antler, which supports 191 companies in the use of GenAI for diverse IT services [2]. Notably products such as Midjourney and You have gained a global consumer acceptance.

The rise of GenAI-driven AI solutions heralds a new era in Human-Computer Interaction (HCI), evolving to Human-Centered AI (HCAI) [3], focusing on trust, reliability, security, and ethical concerns. These factors crucially affect the User Experience (UX) with GenAI-based solutions [4], as seen in Nielsen Norman Group's exploration of GenAI for UX enhancement [5].

R. Chinthaginjala et al. (Eds.): DCAI 2024, LNNS 1259, pp. 201–209, 2025.
https://doi.org/10.1007/978-3-031-82073-1_20

The discussion on adapting UX design principles, such as the "flexibility-usability tradeoff", to GenAI-based solutions continues. The diversity in applications, from OpenAI's broad offerings to Perplexity's specialized focus on information search [6], highlights concerns about GenAI's capacity to reinterpret and adhere to these core UX principles effectively. The discussion underscores the transformative impact of GenAI democratization, highlighting the essential preparedness of academic, industrial, governmental, and societal sectors. This shift is guided by the HCAI philosophy, which focuses on developing AI applications that enhance rather than replace human capabilities [7].

This review analyzes the merging of HCAI with UX and GenAI, and is structured into the following sections: introduction, background, methodology, results analysis, and discussion.

2 Background

Deep learning, a subset of AI, concentrates on developing multilayered neural networks capable of learning hierarchical data representations [8]. The three primary approaches utilized in deep learning include supervised, unsupervised, and semi-supervised methods [9]. GenAI involves deep learning models, such as Transformers with an attention mechanism, which generate high-quality text, images, and content by learning from data [10]. Unlike recurrent or convolutional architectures, Transformers process sequences in parallel, efficiently managing long-range dependencies [11].

On the other hand, HCAI is a discipline that prioritizes human experiences, satisfaction, and needs. Its objective is to enhance, augment, and refine human performance, ensuring that AI systems are reliable, secure, and trustworthy. This fosters human self-efficacy, stimulates creativity, clarifies responsibilities, and promotes social participation [12]. These elements are deemed empirically significant due to their potential impact on the UX offered by GenAI-based systems. Nevertheless, some authors highlight the necessity for additional research concerning UX associated to AI-assisted chatbots [13]. Various authors express worries about GenAI-based solutions such as ChatGPT due to their ability to produce incorrect or made-up information and bypass systems designed to detect duplicate content, which is crucial in fields where originality is key [14].

LLMs vulnerabilities can degrade UX and HCI by undermining trust through biased [15], inaccurate, or manipulative responses, leading to poor decisions and perceptions of unreliability. They also pose risks of social engineering attacks and privacy breaches, eroding trust in digital platforms and driving users towards safer options [16]. Addressing these issues is key to preserving a positive UX and HCI [15], requiring measures to counteract biases, prevent malicious content, and boost data privacy and security [17]. Furthermore, incorporating decentralized and open-source LLMs could enhance collaboration, innovation, and system resilience [18], despite potential challenges in model consistency and a learning curve for users and developers [19–21].

3 Research Methodology

This systematic review was carried out using the PRISMA protocol, involving four phases: identifying studies from databases, reviewing titles and abstracts, examining full texts of selected studies, and extracting data from the analyzed articles [22].

3.1 Research Questions

The research questions that define the scope of this systematic review are shown in Table 1.

Table 1. Research questions

Code	Question	Motivation
RQ1	What are currently the main factors of concern in UX design for the development of GenAI-based solutions?	A crucial aspect is understanding the current factors of concern regarding UX design for the development of GenAI-based solutions.
RQ2	How does a human-centered design (HCD) approach impact the development of systems based on GenAI?	We aim to understand the potential impact of HCD on the development of GenAI-based solutions.

3.2 Information Resources and Search Strategies

The search was conducted across five databases: Scopus, Springer, ACM Digital Library, IEEE Xplore, and ScienceDirect. The search queries were applied to the titles, abstracts, and keywords of the publications as follows in Table 2.

Table 2. Search queries

Databases	Sentences
Scopus	"UX" OR "User Experience" AND "Generative Artificial Intelligence" OR "Generative AI" - "Human Centered" AND "Generative Artificial Intelligence" OR "Generative AI"
Springer	"User Experience" where the title contains "Generative Artificial Intelligence" - "Human Centered" where the title contains "Generative Artificial Intelligence"
IEEE Xplore	"User Experience" OR "Human Centered" AND "Generative Artificial Intelligence"
ACM DL	"User Experience" OR "Human Centered" AND "Generative Artificial Intelligence"
ScienceDirect	"User Experience" OR "Human Centered" AND "Generative Artificial Intelligence"
arXiv	"User Experience" OR "Human Centered" AND "Generative Artificial Intelligence Open Source"

3.3 Inclusion and Exclusion Criteria

The search spanned five databases: Scopus, Springer, ACM Digital Library, IEEE Xplore, ScienceDirect, and arXiv, targeting titles, abstracts, and keywords of publications as detailed in Table 3.

Table 3. Inclusion and exclusion criteria

Inclusion Criteria	Exclusion Criteria
IC1: Studies focused on UX and GenAI.	EC1: The paper was not available for download.
IC2: Studies published between 2020 and 2024.	EC2: The paper is not available in English or Spanish language.
IC3: Studies focused on human-centered and GenAI	EC3: Studies focused on generic AI.

3.4 Data Recovery

A Microsoft Excel template streamlined data organization and teamwork for our review, classifying initial results by database after search and selection protocols. Data on each article included search protocol, title, authors, year, source, abstract, identifiers, and a brief source analysis. Having established the timeframe between September 2023 and March 2024, our search yielded 223 results, Following the application of the inclusion and exclusion criteria, the results were narrowed down to 44 relevant documents. The methodology outcomes are summarized in Table 4.

Table 4. Results summary

Database	Search Results	Relevant Documents
Scopus	45	11
Springer	13	10
IEEE Xplore	5	2
ACM DL	45	3
ScienceDirect	25	3
arXiv	90	15
Total	**223**	**44**

4 Results Analysis

Liu and Siau's [23] research highlight the necessity for human-centered design in GenAI applications with metaverse. Hassan [24], Brandtzaeg et al. [25], Shin and Ahmad [26] Discuss ethical integration in GenAI development, emphasizing fairness, transparency, and accountability. Sison et al. [27], Weisz et al. [28] advocate for comprehensive HCAI approaches and outline ethical design principles for GenAI. Shah et al. [29], Tlili et al. [30], York [31], Hernandez-Leo [32] and Mørch and Andersen [33] focus on GenAI in education, emphasizing the need for ethical practices and human-centered design to enhance creativity and provide personalized feedback.

Banh and Strobel [34], Peruchini et al. [35], Wang et al. [36], Choi et al. [37], Lo [38], Oniani et al. [39], and Guo et al. [40], discuss the impact of human feedback and the limitations of current GenAI systems on UX, advocating for better communication techniques and customizable solutions.

Mao et al. [41], Fisher [42] and Qadri et al. [43], Highlight the creative potentials of GenAI and the importance of inclusive design. Sun et al. [44], Emphasize the importance of Explainable Artificial Intelligence (XAI) in crucial sectors like medicine and finance, advocating for transparency to improve trust and usability. Kim [45] extends the concepts of data transparency and the role of generative AI in enhancing learning through interactions, highlighting the necessity of discerning between human-derived knowledge and GenAI outputs.

Huang [46] Discusses GenAI's influence on human-computer co-creation over the next decade, stressing the need for more attention to UX and interaction design in GenAI development. Asadi [47] examines the role of human-centered design in enhancing the presentation of information through GenAI. Yoon and Jun [48], and Zhihan [49], explore the cautious approach of UX designers toward bias and justification issues in GenAI models, emphasizing the need for personalized and immersive experiences in virtual environments.

Baek and Kim [50], and Sarraf et al. [51], analyze user motivations for engaging with GenAI systems and the impact of anthropomorphic design on user experience. Li et al. [52], and Nguyen et al. [53], discuss the roles of GenAI in enhancing digital security and user safety, advocating for open-source models to detect and respond to online threats.

The review collectively stresses the vital role of ethical considerations in GenAI development and its impact on UX, emphasizing the need for transparency and explainability. These principles, by clarifying how responses and content are generated, enhance user trust and acceptance, thereby improving UX. It also highlights the importance of fair and unbiased feedback, privacy, and security in UX design, noting that GenAI systems should adapt to and personalize user experiences without compromising data security.

Accessibility and inclusivity are identified as crucial for making GenAI solutions available to a wider audience, ensuring everyone can benefit from such technologies. The review underscores the significance of human-centered design, advocating for continuous user feedback to refine GenAI solutions in alignment with user needs and expectations. This involves engaging users throughout the development lifecycle to validate algorithms, ensure utility, remove biases, and build confidence in the outcomes. User education and training is also recommended for better interaction with GenAI, helping them understand and manage their needs and expectations with these technologies.

5 Discussion

The review emphasizes the necessity of embedding ethical principles, transparency, and social considerations in UX design for GenAI development, ensuring these solutions are equitable, responsible, and human-value oriented. It highlights the importance of tailoring GenAI systems to user-specific needs, improving relevance, utility, and satisfaction. This approach also stresses the critical role of ensuring the privacy and security of user information in enhancing UX perception. Moreover, significant challenges are associated with building trust and managing user expectations regarding the capabilities of GenAI, underscoring the need for clear communication about what these systems can and cannot do.

This research highlights the need for human-centered design and UX as key to the successful implementation of GenAI systems, focusing on users' emotions, behaviors, and socio-cultural contexts. It outlines challenges for the emerging economies that adopt these technologies to boost productivity, education, research, innovation, and governance. These challenges are heightened because many algorithms, particularly large language models, are trained on data mainly from European and North American contexts, potentially limiting their applicability in diverse cultures.

This study underscores the strategic value of UX design in fostering the usage and adoption of GenAI solutions, highlighting the essential role of human-centered design in involving society in the development of these technologies. It notes the complex and interdisciplinary nature of UX design in GenAI, calling for a comprehensive approach that encompasses technical, ethical, and human factors from the onset. This includes addressing user trust in GenAI systems and the potential for uncritical acceptance of AI-generated information and insights, underscoring the importance of critical engagement with such technologies.

Open-Source Large Language Models (LLMs) enhance HCI and UX through customization and global innovation. They boost transparency and trust, reduce costs, and promote open standards. However, their open nature demands strict security and privacy measures. Community-driven problem-solving introduces unpredictability compared to centralized systems.

Based on the literature review, it is evident that there is a gap in the development of specific metrics that can quantify the user experience in contexts where GenAI plays a crucial role. These metrics would help to more objectively evaluate how GenAI-based solutions meet user expectations and needs. While the review highlights the importance of accessibility and inclusivity, there is little research on how to design and develop GenAI solutions that are truly accessible to users with diverse abilities and in different socioeconomic contexts.

One of the main barriers to the adoption of GenAI is user distrust in how machines generate and act upon information. By employing XAI principles, developers can design interfaces that provide visibility into the AI decision-making processes. This not only increases user trust but also allows them to evaluate the relevance and reliability of the AI's responses, which are critical elements in UX.

Acknowledgements. This research is part of the International Chair Project on Reliable Artificial Intelligence and Demographic Challenge within the National Strategy for Artificial Intelligence (ENIA), in the framework of the European Recovery, Transformation and Resilience Plan. Referencia: TSI-100933-2023-0001. This project is funded by the Secretary of State for Digitalization and Artificial Intelligence and by the European Union(Next Generation).

References

1. Carlini, N., et al.: Extracting training data from large language models. In: Proceedings of the 30th USENIX Security Symposium (2021)
2. Antler: Antler Gen-AI Landscape (2023). https://airtable.com/shrBeWpMlxf3e14E8/tblS4TkbJbm0cqT0o. Accessed 21 Apr 2024
3. Xu, W., Dainoff, M., Ge, L., Gao, Z.: Transitioning to human interaction with AI Systems: new challenges and opportunities for HCI professionals to enable human-centered AI. Int. J. Hum.-Comput. Interact. **39**(3), 459–518 (2022)
4. Shneiderman, B.: Bridging the gap between ethics and practice: guidelines for reliable, safe, and trustworthy human-centered AI systems. ACM Trans. Interact. Intell. Syst. (TiiS) **10**(4), 1–31 (2020). https://doi.org/10.1145/3427592
5. Liu, F., Zhang, M.: AI as a UX Assistant (2023). https://www.nngroup.com/articles/ai-roles-ux/. Accessed 21 Feb 2024
6. Moran, K.: The UX of AI: Lessons from Perplexity. NNGroup (2024). https://www.nngroup.com/articles/perplexity-henry-modisett/. Accessed 21 Feb 2024
7. HAI Standord Univesity: Human-Centered Artificial Intelligence, Stanford University (2023). https://hai.stanford.edu/about. Accessed 11 Oct 2023
8. LeCun, Y., Bengio, Y., Hinton, G.: Deep learning. Nature **521**(7553), 436-444 (2015). https://doi.org/10.1038/nature14539
9. Ciregan, D., Meier, U., Schmidhuber, J.: Multi-column deep neural networks for image classification. In: IEEE Conference on Computer Vision and Pattern Recognition (2012)
10. Martineau, K.: What is generative AI? IBM (2023). https://research.ibm.com/blog/what-is-generative-AI. Accessed 09 Oct 2023
11. Vaswani, A., et al.: Attention is all you need. In: Advances in Neural Information Processing Systems, vol. 30 (2017)
12. Haslam, A., Worthy, P.: Where is the human in human-centered AI? Insights from developer priorities and user experiences. Comput. Hum. Behav. **141**, 107617 (2023). https://doi.org/10.1016/j.chb.2022.107617
13. Baabdullah, A., Alalwan, A., Algharabat, R., Metri, B., Rana, N.: Virtual agents and flow experience: an empirical examination of AI-powered chatbots. Technol. Forecasting Soc. Change **181**, 121772 (2022). https://doi.org/10.1016/j.techfore.2022.121772
14. Gill, S., et al.: Transformative effects of ChatGPT on modern education: emerging Era of AI Chatbots. Internet of Things Cyber-Phys. Syst. **4**, 19–23 (2024)
15. Yao, Y., Duan, J., Xu, K., Cai, Y., Sun, Z., Zhang, Y.: A survey on large language model (LLM) security and privacy: the good, the bad, and the ugly. arXiv preprint arXiv:2312.02003 (2023)
16. Li, Y., et al.: Personal LLM agents: insights and survey about the capability, efficiency and security. arXiv preprint arXiv:2401.05459 (2024)
17. J Wang, W.M., Sun, P., Zhang, M., Nie, J.Y.: Understanding user experience in large language model interactions. arXiv preprint arXiv:2401.08329 (2024)
18. Gerhardsen, J.: Evaluating the user experience of a learning management system:- to improve usability (2023)
19. Corchado, J., López, S., Núñez, J., Garcia, R., Chamoso, P.: Generative artificial intelligence: fundamentals. ADCAIJ **12**(1), e31704 (2023). https://doi.org/10.14201/adcaij.31704
20. Touvron, H., et al.: Llama 2: open foundation and fine-tuned chat models. arXiv preprint arXiv:2307.09288 (2023)

21. Cui, Y., Yang, Z., Yao, X.: Efficient and effective text encoding for Chinese llama and alpaca. arXiv preprint arXiv:2304.08177 (2023)
22. Page, M.J., et al.: The PRISMA 2020 statement: an updated guideline for reporting systematic reviews. Int. J. Surgery 88, 105906 (2021). https://doi.org/10.1016/j.ijsu.2021.105906
23. Liu, Y., Siau, K.: Generative artificial intelligence and metaverse: future of work, future of society, and future of humanity. In: International Conference on AI-generated Content, Singapore (2023)
24. Hassan, A.: Factors affecting the use of ChatGPT in mass communication. In: Emerging Trends and Innovation in Business and Finance, pp. 671–685. Springer Nature Singapore, Singapore (2023). https://doi.org/10.1007/978-981-99-6101-6_49
25. Brandtzaeg, P., You, Y., Wang, X., Lao, Y.: "Good" and "Bad" machine agency in the context of Human-AI communication: the case of ChatGPT. In: International Conference on Human-Computer Interaction, Copenhagen, Denmark (2023)
26. Shin, D., Ahmad, N.: Algorithmic nudge: an approach to designing human-centered generative artificial intelligence. Computer **56**(8), 95–99 (2023). https://doi.org/10.1109/MC.2023.3278156
27. Sison, A.J.G., Daza, M.T., Gozalo-Brizuela, R., Garrido-Merchán, E.C.: ChatGPT: more than a "Weapon of Mass Deception" ethical challenges and responses from the human-centered artificial intelligence (HCAI) perspective. Int. J. Hum.-Comput. Interact. (2023)
28. Weisz, J.D., Muller, M., He, J., Houde, S.: Toward general design principles for generative AI applications (2023). https://arxiv.org/abs/2301.05578. Accessed 17 Feb 2024
29. Shah, C.S., Mathur, S., Vishnoi, S.K.: Continuance intention of ChatGPT use by students. In International Working Conference on Transfer and Diffusion of IT, Nagpur, India (2023)
30. Tlili, A., et al.: What if the devil is my guardian angel: ChatGPT as a case study of using chatbots in education. Smart Learn. Environ. **10**, 15 (2023)
31. York, E.: Evaluating ChatGPT: generative AI in UX design and web development pedagogy. In: Proceedings of the 41st ACM International Conference on Design of Communication, Orlando, Florida. USA (2023)
32. Hernández-Leo, D.: ChatGPT and generative AI in higher education: user-centered perspectives and implications for learning analytics. LASI Spain, Madrid (2023)
33. Mørch, A.I., Andersen, R.: Human-Centred AI in education in the age of generative AI tools. In: Proceedings (2023). http://ceur-ws.org
34. Banh, L., Strobel, G.: Generative artificial intelligence. Electron. Markets **33**(1), 1–17 (2023). https://doi.org/10.5771/1619-2427-2023-17-1-1
35. Peruchini, M., da Silva, G.M., Teixeira, J.M.: Between artificial intelligence and customer experience: a literature review on the intersection. Discover Artif. Intell. **4**(1), 4 (2024)
36. Wang, X., Attal, M.I., Rafiq, U., Hubner-Benz, S.: Turning large language models into AI assistants for startups using prompt patterns. In: International Conference on Agile Software Development, Copenhagen, Denmark (2022)
37. Choi, W., Zhang, Y., Stvilia, B.: Exploring applications and user experience with generative AI tools: a content analysis of reddit posts on ChatGPT. In: Proceedings of the Association for Information Science and Technology (2023)
38. Lo, L.: The art and science of prompt engineering: a new literacy in the information age. Internet Reference Serv. Q., 203–210 (2023)

39. Oniani, D., et al.: Adopting and expanding ethical principles for generative AI from military to healthcare. NPJ Digit. Med. **6**(1), 225 (2023)
40. Guo, M., Zhang, X., Zhuang, Y., Chen, J., Wang, P., Gao, Z.: Exploring the intersection of complex aesthetics and generative AI for promoting cultural creativity in rural china after the post-pandemic era. In: International Conference on AI-generated Content, pp. 313–331. Singapore, Springer (2023). https://doi.org/10.1007/978-981-99-7587-7_27
41. Mao, Y., Rafner, J., Wang, Y., Sherson, J.: A hybrid intelligence approach to training generative design assistants: partnership between human experts and AI enhanced co-creative tools. In: HHAI 2023: Augmenting Human Intellect. IOS Press, pp. 108–123 (2023)
42. Fisher, J.: Centering the Human: digital humanism and the practice of using generative AI in the authoring of interactive digital narratives. In: International Conference on Interactive Digital Storytelling, Kobe, Japan (2023)
43. Qadri, R., Shelby, R., Bennett, C., Denton, E.: AI's Regimes of Representation: a community-centered study of text-to-image models in South Asia. In: ACM Conference on Fairness, Accountability, and Transparency, Chicago, IL, USA (2023)
44. Sun, J., Houde, S., Talamadupula, K., Weisz, J.D.: Investigating explainability of generative AI for code through scenario-based design. In: 7th International Conference on Intelligent User Interfaces, Miami, FL. USA (2022)
45. Kim, P.W.: A framework to overcome the dark side of generative artificial intelligence (GAI) like ChatGPT in social media and education. IEEE Trans. Comput. Soc. Syst. (2023)
46. Huang, Y.: The Future of Generative AI: how GenAI would change human-computer co-creation in the next 10 to 15 years. In: Proceedings of the Annual Symposium on Computer-Human Interaction in Play, Stratford, Canada (2023)
47. Asadi, A.R.: LLMs in design thinking: autoethnographic insights and design implications. In: Proceedings of the 2023 5th World Symposium on Software Engineering, Tokyo, Japan (2023)
48. Yoon, H., Jun, S.: Ethical awareness of UXers in the loop: ethical issues in the Uxer-AI collaboration process from a UX perspective. In: Proceedings of the 25th International Conference on Mobile HCI, Athens, Greece (2023)
49. Zhihan, L.: Generative artificial intelligence in the metaverse era. Cognitive Robot. **3**, 208–217 (2023). https://doi.org/10.1016/j.cogr.2023.06.001
50. Baek, T.H., Kim, M.: Is ChatGPT scary good? How user motivations affect creepiness and trust in generative artificial intelligence. Telematics Inform. **83**, 102030 (2023). https://doi.org/10.1016/j.tele.2023.102030
51. Sarraf, S., Kar, A., Janssen, M.: How do system and user characteristics, along with anthropomorphism, impact cognitive absorption of chatbots-Introducing SUCCAST through a mixed methods study. Decis. Support Syst., 178 (2024)
52. Li, J., Cao, H., Lin, L., Hou, Y., Zhu, R., Ali, A.E.: User Experience Design Professionals' Perceptions of Generative Artificial Intelligence. arXiv preprint arXiv:2309.15237 (2023)
53. Nguyen, T.T., Wilson, C., Dalins, J.: Fine-tuning llama 2 large language models for detecting online sexual predatory chats and abusive texts. arXiv preprint arXiv:2308.14683 (2023)

Bird Detection and Recognition Systems Through Machine Learning and Artificial Intelligent: Advancements and Future Opportunities

Lautaro Rossi Labianca[2], Jaime Álvarez Urueña[1], Javier Curto Hernández[1], and Raúl García Serrada[1](✉)

[1] Air Institute, IoT Digital Innovation Hub (Spain), Deep tech lab. Av. Santiago, Madrigal 39, 37003 Salamanca, Spain
{jalvarez,jcurto,rgarcia}@air-institute.com

[2] BISITE Research Group, University of Salamanca, Edificio Multiusos I+D+I, Calle Espejo 2, 37007 Salamanca, Spain
lross98@usal.es

Abstract. The paper provides an overview of advancements in bird detection and recognition systems using Machine Learning and Artificial Intelligence (AI). It highlights the increasing adoption of wind farms amid rising electricity demand, underscoring their environmental impact on avian species. To address these ecological challenges, the development of bird recognition solutions is crucial. The paper analyzes various techniques, including radar systems, sound recognition, Convolutional Neural Networks (CNNs), electromagnetic detection, YOLOv5, and color segmentation, discussing their features, computational costs, and constraints. It concludes that while deep learning models offer superior results, they need a balance between accuracy and speed, alongside training with large and representative datasets. Ultimately, the paper aims to contribute to efforts aimed at mitigating the adverse effects of wind farms on bird populations through advanced technologies. Additionally, through this paper we intend to shed light over the state of the art on bird detection systems and provide insights that intends to solve some of these drawbacks.

Keywords: Bird Recognition · Bird detection · Machine Learning · Artificial Intelligence · Computer Vision

1 Introduction

Recent technological advancements have spurred an increased demand for electrical energy. Governments are investing in clean energy, including wind farms,

This work is funded from the project "IA4birds: Plataforma de IA Explicable para la predicción y protección de aves en espacios destinados a ser parques eólicos" by Fundación Biodiversidad, FSP (Ministry for Ecological Transition and the Demographic Challenge) within the framework of the order TED/1016/2021, 20th September.

R. Chinthaginjala et al. (Eds.): DCAI 2024, LNNS 1259, pp. 210–219, 2025.
https://doi.org/10.1007/978-3-031-82073-1_21

which globally produce over 976GW with an annual growth rate of 11.4%, aiding in the production of green energy. However, wind farms pose risks to local bird populations and ecosystems, with birds facing potential risks such as death due to collision, habitat loss, and displacement [1].

To minimize the adverse impact of wind farms on birds, integrating flora and fauna recognition and preservation is crucial [2]. The growing awareness of the need to preserve biodiversity has redirected resources and investments towards the development of tools to address this concern. Rapid technological advancements, including GPUs, artificial intelligence, and enhanced computational resources, enable comprehensive analysis and processing of diverse data types such as images, audio, and text, facilitating efforts to mitigate environmental risks [3].

This review paper is the starting one of a set of papers through which will describe the project being taken over at the moment. One of the main project's aim is to safeguard endangered bird's habitats through the deep analysis of their natural living areas while optimizing the installation of wind farms. This review aims to summarize the state of the art of related projects as well as to pinpoint the key of them in order to introduce the development features contained. On the following publications we aim to describe in depth the main pillars of out investigation. These following papers will go through the following modules: (i) Computer Vision Bird Detection through segmentation of on-edge imaging, (ii) Computer Vision on-edge Bird Classification through 2-D Convolutional Neural Networks and (iii) On-edge Bird Classification through sound detection systems coupled with Machine Learning and Deep-Learning methods.

2 Methods

This section presents the objectives within the selection of recognition bird's models, together with the methodology. In this paper, an exploration aimed at understanding and improving the methodologies for recognizing birds, is embarked. Despite the abundance of research in pattern recognition as a whole, the niche of bird recognition presents unique challenges and opportunities that have not been as extensively explored up to 2024.

2.1 Objectives

The objective of this research is to evaluate bird recognition methods by gathering and analysing relevant literature in the field, including most commonly used detection methods. Significant research from various sources, using a range of techniques such as radar, sound recognition and image analysis will be aggregated and classified. The goal is to provide valuable insights and advancements in bird recognition technologies.

2.2 Research Method Followed

There is a significant amount of research in pattern recognition, but our focus is specifically on bird recognition. While there are some research papers and official documents on this topic, there are not as many as we had hoped.

This is the reason why we thoroughly searched for the most interesting and evidence-supported ones. We searched several databases and gathered papers, thesis and publications related to bird recognition, regardless of the year or specific techniques, tools, or technologies. We then conducted research on all the collected papers, taking notes and studying them thoroughly. The groups were classified based on the technique used for recognition: radar, sound and image recognition.

- Recognition through radar.
- Recognition through sounds.
- Recognition through images:
  - YOLOv5.
  - Convolutional neural networks.
  - Segmentation through colours + supervised classification.

2.3 Organisation

In order to organise all the information, we used Mendeley, which is a tool that allows us to make notes on papers, organise and store them. Furthermore, Mendeley counts with an online feature that allows to get the APA citation, paper itself and other information useful for researchers.

2.4 Risk of Bias

This paper addresses bias inherent in research, including individual and search bias. It highlights the inadequacy of relying solely on study type to assess evidence credibility and aims to evaluate factors influencing effect estimates, employing a comprehensive methodology to ensure transparency and trustworthiness in findings.

2.5 Paper Filtering Criteria

The paper filtering criteria were established to ensure the selection of the most relevant research for this review. The criteria included relevance, quality, currency, evidence, and clarity. Papers had to be either related to bird recognition using machine learning and AI or methods related to the goal such as techniques, covering radar, sound, and image recognition methods in order to understand the big picture. Also, Fig. 1 graphs the main findings related to the topic on the main scientific repositories. The research prioritizes computer vision and sound pattern recognition as key areas of focus, with specific emphasis on their main fields. They needed to be peer-reviewed, ensuring expert vetting. While the year of publication was not strict, recent papers were prioritized to capture the latest advancements. Empirical evidence supporting claims was required, though significant theoretical papers were also included. These criteria ensured a comprehensive and high-quality review, with selected papers thoroughly studied and classified, providing valuable insights into the state of the art in bird recognition technologies and ensuring the reliability. Although the related-to-the-topic papers were short, from 39 candidate papers, we selected as relevant 21 publications including papers, congress reports, among others.

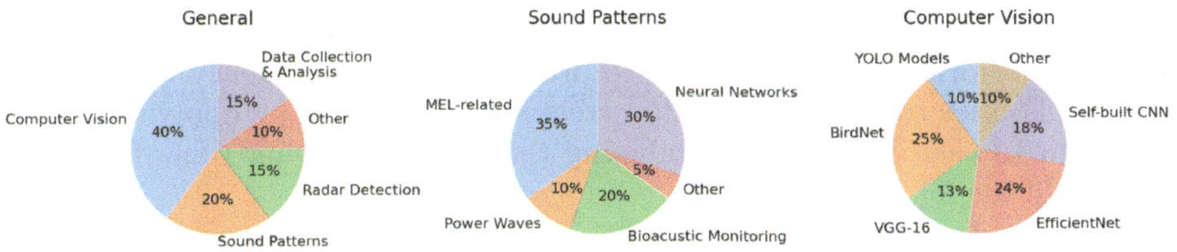

Fig. 1. Distribution of content among the papers reviewed. Computer Vision and Sound recognition were two key elements to understand the topic. Therefore, special attention was put into these categories of research.

3 Results

This section will cover the various techniques used for bird recognition, including their main concepts, advantages, computational cost and main limitations. We will cover three main types of detection: radio waves, audio and images.

3.1 Radar System

This technique differs from others by tracking multiple birds simultaneously rather than individually, using weather and low-flying surveillance radars. Military aircraft, which fly below 2 km where bird detection is most effective, are primarily used for this purpose since commercial planes only briefly operate at this altitude during takeoff and landing [4,5].

Radar Operation. Radar technology, originating in the early 1930 s, operates by transmitting electromagnetic pulses and analyzing the reflected waves to detect objects. Initially aimed at preventing bird strikes at airports, it has since advanced to include bird tracking by classifying objects based on the characteristics of the reflected waves.

3.2 Sounds Recognition

Automated sound recording and analysis techniques for bird surveys and conservation are crucial for efficient avian monitoring, benefiting science and conservation efforts [1]. Sounds, originating from pressure differences in the air, are physical waves with various processing methods, notably Mel spectrograms or computed spectra (GTCC, LPCC, MFCC) for machine learning input. Represented over time on the abscissa (X) axis and amplitude on the ordinate (Y), this representation lacks frequency information, addressed through Fourier transform application.

Sound Recognition Operation. The Fourier transform generates a diagram showing sound frequencies and their produced frequencies. Discrete Fourier transform reveals wave energy variations over time but discards insignificant negative frequencies.

Sound waves, which change frequency over time, can be analyzed using Fourier transform in small intervals to provide valuable frequency information. The fast Fourier transform algorithm creates spectrograms by applying Fourier transform to each period and combining the results, showing time, frequency, and amplitude [6].

Humans demonstrate precise differentiation of low-frequency sounds but struggle with high-frequency distinctions, suggesting a logarithmic scale in human hearing perception. Consequently, spectrograms obtained through Fourier transform application in small time intervals are often passed through the Mel scale, which features a linear segment up to 1000 Hz followed by a logarithmic portion. [7]. By doing so, we would get a Mel spectrogram, and then we can then apply artificial vision techniques (CNN, EfficientNet etc.) to classify the sound.

Spectrums are the result of applying the inverse Fourier transform to the spectrum of the wave on a logarithmic scale. There are several types of spectrums, but the main ones are MFCC (Mel frequency cepstral coefficients) [8], GTCC (Gamma-tone frequency cepstral coefficients) and LPCC (Linear prediction cepstrum coefficients), [9]. These spectrums can be calculated and then fed into a machine learning algorithm. It is widely used in MVS, but other machine learning algorithms can be used, like neural networks or ensembles.

3.3 Convolutional Neural Networks

Convolutional Neural Networks (CNNs), a subset of deep learning, employ convolutional operations ideal for spatially rich image data [10]. These networks also utilize pooling, primarily max pooling, and activation functions [11]. However, challenges include overfitting risks, demanding high-quality and task-specific image datasets for training, necessitating a balance between specificity and generalization [12].

Convolutional Neural Networks Operation. The idea behind this model is to stack convolutional layers, so that many features (the result of applying to an image a convolutional operation) are gathered and fed into the next layer. At the end of all the convolutional layers, typically, a feed-forward neural network is typically stacked to do the predictions and classification.

CNNs, due to their ability to consider spatial information at the pixel level, are well-suited for bird recognition tasks. However, challenges such as the need for extensive training data (mitigated by data augmentation techniques) [13], high computational costs due to the vast number of floating-point operations (FLOPS), and the trade-off between spatial and semantic information capture persist.

To address these issues, various neural network architectures, building upon CNNs, have emerged. Residual networks, for instance, tackle parameter overload and degradation problems by incorporating residual connections [14]. Dense networks, CSP networks, and their combinations further enhance accuracy beyond conventional CNNs [15,16].

Neural Networks Operations to Differentiate Birds from Non-birds Objects. In this context, the Computer Vision module leverages Convolutional Neural Networks (CNNs) to differentiate birds from other objects in the sky. The CNN assigns importance to various aspects of an image and can distinguish between birds and non-bird objects. Through different layers (pooling, fully-connected, filters), the network learns a hierarchy of features from the input images, ranging from simple edges and textures to complex shapes. Ideally, the CNN is trained to recognize bird-specific features such as wings, beaks, and feathers. An innovative feature of this project is the development of a module that can recognize specific bird flying patterns, which will be supported through YOLO5 in future work. CNNs learn spatial hierarchies between features, enabling them to distinguish complex objects like birds, balloons, aircraft, and drones based on shape. Convolution filters identify different features in images, such as the roundness of a balloon or the angularity of a drone. Thorough training with diverse data is crucial for accurate bird feature detection. Pooling operations control overfitting and improve generalization, helping the network distinguish between bird and non-bird images. Fully connected layers process the extracted features, enhancing species classification without losing key information.

3.4 YOLOv5

YOLOv5, an artificial vision model, excels in detecting and classifying objects by encapsulating them within bounding boxes. This model is structured into three main blocks: the backbone, the neck, and the head. The backbone, specifically CSPDarknet53 (also utilized in YOLOv4) [17], is responsible for extracting valuable features from input images using a combination of classical convolutional networks, CSPNets, and residual nets. The neck, employing the PAN model, generates predictions at various levels to capture both semantic and spatial information, thereby addressing the limitations of classical convolutional networks. The head block, identical to the one used in YOLOv4 [17], creates a fixed-size grid for each incoming feature. Each cell within this grid generates three predictions (a number that can be adjusted based on the problem domain). Predefined anchor boxes are then employed to select the best predictions using the IOU measure and the non-max suppression algorithm. It's worth noting that there are typically multiple heads, each specializing in detecting objects of different sizes, thus enabling the detection of objects of various sizes within the same image.

YOLOv5 Operation. Creating a YOLOv5 model from scratch is complex because it involves solving both classification and regression problems for object detection. Developed by Ultralytics [18], YOLOv5 is pre-trained on the COCO [19] dataset but can be fine-tuned with other datasets, such as for bird detection. A key metric in YOLOv5 for selecting the best predictions is the Intersection over Union (IoU) [20], which measures the overlap between predicted and ground truth bounding boxes. Pre-determined anchor boxes, created using K-Means clustering, serve as references for bounding box predictions, varying in size to match the objects in the dataset. YOLOv5 uses multiple prediction heads to handle different object sizes, and each grid cell's predictions are compared to the anchor boxes, selecting the one with the highest IoU as the final prediction.

3.5 Colour Segmentation

The basic idea of image recognition is to train a Machine Learning algorithm (ML) with photos of the birds to be detected and make predictions with pictures that the model has never used before. Deep learning is a subset of ML, so its idea would be the same, feed an algorithm with some data in order to train it and make predictions of other pictures.

This technique produces lots of issues for bird recognition. For instance, the background differs greatly from image to image depending on weather conditions, light etc. Moreover, when there is more than one bird in the picture, modifications must be made to identify them all. A possible solution to these problems lies in segmentation. The core idea is to select all the possible birds in the photo, and then classifiers will identify whether the objects selected are birds and identify its species.

Colour Segmentation Operation. The method involves a challenging step of object extraction. Initially, the background is removed. The image's edge colors are then scanned and ranked by frequency in a histogram. Each pixel is compared, and if its frequency matches the histogram, it's considered background. Otherwise, it's identified as a bird. [21]. Images are depicted using three channels (RGB), resulting in three histograms, one for each color channel. The objects derived from this process are input into a supervised machine learning algorithm. SVM, for instance, provides excellent results relative to computational complexity. It can execute non-linear classification, transforming each input into a higher-dimensional feature space.[3]. Other supervised classifiers like AdaBoost or ensembles classifiers are also possible.

3.6 Comparative Review

Table 1 summarizes the most relevant publications related to this review. In this table a comparison between the main lines of investigation is being displayed. Technological methods for monitoring bird populations provide detailed data and enable comprehensive analysis of avian behavior, migration, and ecology. This integration offers a fine understanding of bird population dynamics and environmental influences, informing conservation strategies and effective management practices to lessen human impacts on these species [2].

Table 1. Comparative study of existing approaches. 'Do' references to **Domain of application**, 'Da' to **Dataset**, 'Ap' to **Approach**, 'Ad' to **Advantages** and 'L' to **Limitations**

#	Features
[1]	**Do:** Sound. **Da:** Various sources. **Ap:** Comparison of techniques to automatically detect and classify avian sounds to species level. **Ad:** Better explanation on how to approach bird-sound exploration and analysis. **L:** The performance of automated classifiers, such as BirdNET, can vary and may be needed to rival manually annotated data.
[3]	**Do:** Images. **Da:** Own dataset using Microsoft's Bing Image Search API v7. 80% used for training and 20% for test (randomised). **Ap:** Create a VGGNet architechture to classify the bird species. **Ad:** Great results of 84.91% in test set regarding the size of the model. **L:** Private dataset not available to the community.
[4]	**Do:** Satellite Imaging. **Da:** Medium-range weather radars (WRs) and short-range radars. **Ap:** tracking algorithms from radar science and assesses their potential. **Ad:** Small scanning radars can track hundreds of birds and their flying patterns over tens of km^2.**L:** Radar accuracy is limited and prone to misinterpretation of biological entities.
[5]	**Do:** Satellite Imaging. **Da:** European Radar surveillance of Animal Movement (ENRAM). **Ap:** To utilize weather radars and their tools to monitor the aerial movement of animals across Europe. **Ad:** examines large-scale aerial movements and behaviors of organisms, abundance and conditions.**L:** Small subsample of individuals are tracked.
[9]	**Do:** Sound. **Da:** 700 curated and segmented audio files from the public database Xeno-Canto. **Ap:** Feature exptraction and comparison of each cepstral's accuracy on a a pre-trained model (SVM). **Ad:** Proving that the GTCC cepstrum-related coefficients are robust and has significant higher accuracy than other related cepstrum analysis. **L:** GTCC analysis is a classic cepstrum analysis and the more groundbreaking cepstrum-related analysis did not outperform GTCC or collaborated complementing GTCC.
[21]	**Do:** Images. **Da:** CUB-200. **Ap:** Eliminate the background of the picture and segment according to colour frequency. **Ad:** Cost-effective method with a segmentation rate of 71% in RGB. **L:** Birds are to be situated at the centre of the image.

4 Conclusions

Renewable energy's rise, notably wind farms, demands solutions to their impact on local bird populations. Bird-recognition systems are crucial for mitigating wind farms' adverse effects on avian species and ecosystems.Renewable energy's rise, notably wind farms, demands solutions to their impact on local bird populations. Bird-recognition systems are crucial for mitigating wind farms' adverse effects on avian species and ecosystems.

Various techniques have been developed for bird recognition, including radar systems, sound recognition, Convolutional Neural Networks (CNNs), YOLOv5, etc. Each of these techniques has its advantages and disadvantages. For instance, while deep learning models and color segmentation provide accurate results, they are computationally expensive. On the other hand, bird recognition through sound can yield good results, but in a noiseless manner is challenging. Radar systems, although effective, are extremely expensive and thus not feasible for widespread use.

In conclusion, while significant strides have been made in the development of bird-recognition systems, there is still room for improvement. Future research should focus on developing cost-effective and efficient bird recognition techniques that can be easily implemented in wind farms. This will not only help in preserving bird species but also ensure the sustainable use of wind energy.

References

1. Scott Brandes, T.: Automated sound recording and analysis techniques for bird surveys and conservation. Bird Conserv. Int. **18**, S163–S173 (2008). https://doi.org/10.1017/S0959270908000415
2. Hasui, E., et al.: Populations across bird species distribution ranges respond differently to habitat loss and fragmentation: implications for conservation strategies, pp. 43–54 (2024). https://doi.org/10.1016/j.pecon.2023.11.003
3. Satyam, R., Saiaditya, G., Sanu, K., Sushila, S.: Image based bird species identification using convolutional neural network. Int. J. Eng. Res. Tech. (IJERT) J. **9**(6) (2020). https://doi.org/10.17577/IJERTV9IS060279
4. Liechti, F., van Gasteren, H.: Current stage of bird radar system. In: 29th Meeting of the International Bird Strike Committee proceedings, pp. 1–3. Australia (2010)
5. Alves, J.A., et al.: Monitoring continent-wide aerial patterns of bird movements using weather radars (2016)
6. Döerfler, M., Bammer, R., Grill, T.: Inside the spectrogram: convolutional neural networks in audio processing. In: 2017 International Conference on Sampling Theory and Applications (SampTA), pp 152–155. IEEE, Tallinn, Estonia (2017). https://doi.org/10.1109/SAMPTA.2017.8024472
7. Umesh, S., Cohen, L., Nelson, D.J.: Fitting the Mel scale. In: 1999 IEEE International Conference on Acoustics, Speech, and Signal Processing. Proceedings. ICASSP99 (Cat. No.99CH36258), pp. 217–220. IEEE, Phoenix, AZ, USA (1999). https://doi.org/10.1109/ICASSP.1999.758101
8. Molau, S., Pitz, M., Schlüter, R., Ney, H.: Computing Mel-frequency cepstral coefficients on the power spectrum. In: 2001 IEEE International Conference on Acoustics. Speech, and Signal Processing Proceedings, pp. 73–76. IEEE Salt Lake City, UT, USA (2001)

9. Ramashini, M., Abas, P. E., Mohanchandra, K., De Silva, L. C.: Robust cepstral feature for bird sound classification. Int. J. Electr. Comput. Eng. (IJECE) **12**(2), 1477–1487 (2022). https://doi.org/10.11591/ijece.v12i2
10. Indolia, S., Goswami, A.K., Mishra, S.P., Asopa, P.: Conceptual understanding of convolutional neural network - a deep learning approach. Procedia Comput. Sci. **132**, 679–688 (2018). https://doi.org/10.1016/j.procs.2018.05.069
11. Dubey, S.R., Singh, S.K., Chaudhuri, B.B.: Activation functions in deep learning: a comprehensive survey and benchmark. Neurocomputing **503**, 92–108 (2022). https://doi.org/10.1016/j.neucom.2022.06.111
12. Aryeni, I., Maulidiah, H.M., Toar, H., Wicaksono, M.J.W., Gunawan, I.: Application of computer vision for real-time detection of fruit color and size in fruit sorter. J. Appl. Electr. Eng. (2023). https://doi.org/10.30871/jaee.v7i2.6740
13. Shorten, C., Khoshgoftaar, T.M.: A survey on image data augmentation for deep learning. J. Big Data **6**(1), 1–48 (2019). https://doi.org/10.1186/s40537-019-0197-0
14. He, K., Zhang, X., Ren, S., Sun, J.: Deep residual learning for image recognition. In: 2016 IEEE Conference on Computer Vision and Pattern Recognition (CVPR), pp. 770–778. IEEE, Las Vegas, NV, USA (2016). https://doi.org/10.1109/CVPR.2016.90
15. Huang, G., Liu, Z., van der Maaten, L., Weinberger, K.Q.: Densely connected convolutional networks. In: 2017 IEEE Conference on Computer Vision and Pattern Recognition (CVPR), pp. 2261–2269. IEEE, Honolulu, HI, USA (2017). https://doi.org/10.1109/CVPR.2017.243
16. Wang, C.-Y., Liao, H.-Y.M., Yeh, I.-H., Wu, Y.-H., Chen, P.-Y., Hsieh, J.-W.: CSPNet: a new backbone that can enhance learning capability of CNN. 2020 IEEE/CVF Conference on Computer Vision and Pattern Recognition Workshops (CVPRW), pp. 1581–1583. IEEE, Seattle, WA, USA (2020). https://doi.org/10.1109/CVPRW50498.2020.00203
17. Ning, M., Lu, Y., Hou, W., Matskin, M.: YOLOv4-object: an efficient model and method for object discovery. In: 2021 IEEE 45th Annual Computers, Software, and Applications Conference (COMPSAC), pp. 31-36. IEEE, Madrid, Spain (2021). https://doi.org/10.1109/COMPSAC51774.2021.00016
18. Ultralytics YOLOv5 code in github Homepage. https://github.com/ultralytics/yolov5. Accessed 20 Feb 2024
19. COCO Homepage. https://cocodataset.org. Accessed 20 Feb 2024
20. Rezatofighi, H., Tsoi, N., Gwak, J., Sadeghian, A., Reid, I., Savarese, S.: Generalized intersection over union: a metric and a loss for bounding box regression. In: 2019 IEEE/CVF Conference on Computer Vision and Pattern Recognition (CVPR), pp. 658–666. IEEE, Long Beach, CA, USA (2019). https://doi.org/10.1109/CVPR.2019.00075
21. Marini A,. Facon J,. Koerich A.L.: Bird species classification based on color features. In: 2013 IEEE International Conference on Systems, Man, and Cybernetics, pp. 4336–4341. IEEE, Manchester, UK (2013). https://doi.org/10.1109/SMC.2013.740

Dynamic FOG Computing for the Efficient Management of IoT Infrastructures

Lucía Arnau Muñoz[1], José Vicente Berná Martínez[1](✉), Carlos Calatayud Asensi[2], and Iren Lorenzo Fonseca[1]

[1] University of Alicante, Carretera San Vicente del Raspeig, 03690 Alicante, Spain
{lucia.arnau,jvberna,iren.fonseca}@ua.es

[2] Aguas de Valencia S.A., Avda. Marqués del Turia, 46005 València, Spain
ccalatayud@globalomnium.com

Abstract. Nowadays, the use of IoT devices is widely standardised in different areas of our lives, from work, home, educational institutions, etc. Such amount of sensing data needs to be controlled in an efficient way, and FOG computing is one of the most used paradigms in this kind of environments, due to its ability to solve the computational needs of data generating devices, reducing the traffic between them and decentralising the data processing. Due to the constant growth of IoT environments, the FOG layer must be able to configure and scale as the infrastructure grows, adapting to the new computational requirements of the environment. These operations are still performed manually by system administrators, resulting in an inefficient and error-prone practice. This work proposes the design and development of a FOG layer capable of dynamically controlling the configuration of the logical components of the layer, providing the infrastructure with new nodes, a coordinator and a balancer, alongside the existing FOG processor nodes. The coordinator node balances the load distribution for the processor nodes, while the balancer will perform the load distribution under the coordinator's guidelines. This proposal automates the operation of IoT infrastructures, provisioning with the minimum resources to cover the computational needs. To validate the proposal, the model has been deployed on the campus of the University of Alicante as a FOG layer of the sensorisation system.

Keywords: FOG · Internet of Things · dynamic management · efficient architectures · load balancing

1 Introduction

Nowadays, where millions of sensors are integrated every day in cities, industries or educational centres, the ability to collect and process data in real time has become an invaluable asset for a wide range of applications and services. On the campus of the University of Alicante, a smart sensorisation project is being developed, where the necessary IoT infrastructure [1] is being deployed to implement any type of sensorisation, as well as being able to process and manage all the information generated by the sensors, to offer a multitude of valuable services to the community. Implementing sensorisation systems

R. Chinthaginjala et al. (Eds.): DCAI 2024, LNNS 1259, pp. 220–229, 2025.
https://doi.org/10.1007/978-3-031-82073-1_22

within a university campus, comparable in infrastructure to a city, is very beneficial for the people in it, as many useful services can be deployed for the users in it. One of the main uses in the university is for efficient resource management, as sensorisation can help monitor and optimise the use of resources such as energy, water and physical space on the university campus. This can lead to significant cost reductions and greater environmental sustainability. On the other hand, security-related services can be created, as sensor systems help to maintain security in their institutions by detecting and responding quickly to emergency situations, such as fires or improper access to certain locations [2], thus protecting staff and students within the institution. For example, CO2, presence and temperature sensors can provide early warning of a serious situation in progress. Not surprisingly, it is also used for services related to campus activity such as research, as the use of sensorisation at the university serves as a testing and experimentation environment for research projects in areas such as artificial intelligence. The data collected can be used to analyse patterns, identify trends and develop innovative solutions.

Thanks to the extensive and heterogeneous sensing system that has been implemented on campus, we can collect millions of different data in real time, generating a constant flow of information that we can later exploit from other services and applications [3]. However, the raw data captured requires meticulous processing before it can be use, as we do not get much information from the output value alone. To convert this data into user-readable data, it is necessary to carry out a processing that includes enriching it with meta-information stored in databases, filtering and cleaning, and so on. This analysis and transformation process is essential to generate meaningful services from the collected data.

To address this processing need, we have opted for the FOG Computing paradigm, applied to IoT infrastructures [4]. FOG Computing is a paradigm that allows adapting different types of architectures by using multiple layers of computational infrastructure, combining resources such as network edge nodes close to the final information store [5]. The use of this paradigm provides features such as: data and applications are at the edge of the network, storage and computation are distributed, geographical distribution of many nodes, heterogeneity of devices, data and applications, and finally, it facilitates real-time communications [6]. This paradigm is very suitable when it comes to building systems that perform real-time processing and analysis of IoT-generated data [7] and therefore we find that it is an option to solve our needs.

The union of IoT and FOG Computing generates an environment in which countless types of IoT devices, from sensors at the Edge [8], to their controllers in intermediate layers such as gateways, generate large amounts of information in real time. However, a new challenge arises, the variability of the workload of the elements belonging to the FOG layer, since the amount of packets that are processed at the FOG layer is not constant [9], but varies according to factors such as the time of day, the day of the week or the season of the year at that time. This variability creates the need to guarantee the minimum computational capacity necessary for the operation of the system even at critical times, without increasing costs excessively, and without manually reconfiguring computational infrastructures, as this is inefficient.

In this work we propose as a solution a FOG layer with dynamic and automatic scaling of infrastructures. Instead of allocating fixed resources that may be inefficient at times of low load or insufficient at times of high demand, our FOG layer will be equipped with components that will automatically manage resources, ensuring the necessary computing capacity at all times.

The rest of the article is organised in the following chapters. In Sect. 2, a description of the current situation of the campus is given, together with the problems this presents. In Sect. 3, the design of the FOG architecture that solves the problem is presented. Section 4 explains the communication protocol used to develop the solution. Section 5 describes the new components in the architecture, along with the existing ones, and their functionalities. Section 6 describes the implementation of the architecture on the campus of the University of Alicante, specifying the tests carried out. Section 7 draws the main conclusions of the article and the lines of future work.

2 Background and Problem Definition

The architecture implemented at the University of Alicante consists of a series of static nodes, each of which processes a certain type of data, from climatology to environmental measurements, such as CO2. Since each node processes information in a fixed way, without taking into account the ideal load distribution between them, this results in an inefficient solution, due to the imbalance of work involved. This FOG layer scheme has a serious efficiency problem because the data messages to be processed vary according to the time of day (night vs. day), the day of the week (Monday vs. Sunday) and the time of year (holiday vs. work). This results in nodes not processing the same amount of messages and therefore wasting a lot of computational power. In the Fig. 1 shows a graph with the number of messages processed per second on three nodes on a weekday. The yellow line shows the total number of messages reaching the FOG layer, and then the rest of the lines show how many messages each node processes. This is because the messages processed can be of 5 types: WIFI, Weather, Energy, Cars and Miscellaneous. The WIFI, Miscellaneous and Cars data sources are processed by node 1, Climatology by node 2, and Energy by node 3.

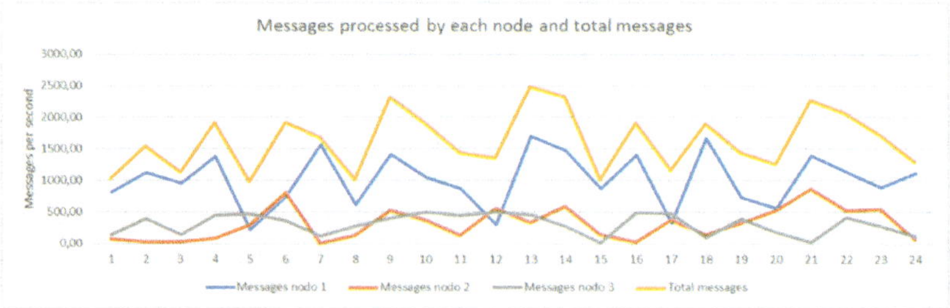

Fig. 1. Messages per minute processed per node (blue, orange and grey) and total messages arriving.

As can be seen, the amount of messages processed varies and nodes process high amounts, while, at the same time, other nodes hardly process any data. In the Fig. 2 shows the state of power consumed by each node and at each instant, when messages are processed, together with the amount of messages received at that moment, being remarkable the imbalance of work between nodes.

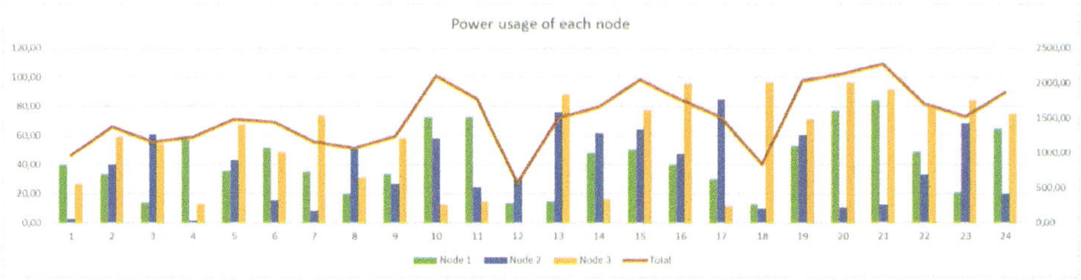

Fig. 2. Power in use at each node when processing messages.

This is why it has been decided to address the problem by generating a dynamic and auto-scalable FOG layer, so that only the necessary resources come into operation when they are required, and making balanced and equal use of the resources that are running. And of course, always ensuring sufficient computational capacity to process the data messages received.

3 Overall Design of the Solution

A new architecture is proposed, as shown in Fig. 3, where the FOG Layer evolves and incorporates new elements, where two levels are differentiated, a control level, which will be a new level added to the existing architecture, and the processing level, where the existing processing nodes are located. The control layer will consist of a new coordinator element and a balancer element. The processing layer contains the previously existing processing nodes, but new functionalities have been added to each node to support automation and control operations from the control layer.

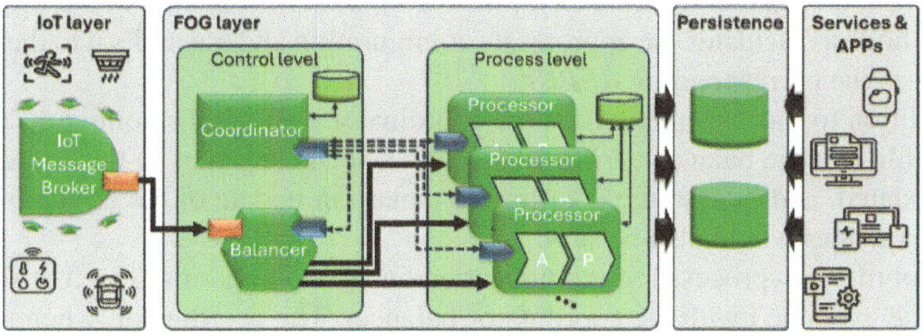

Fig. 3. Dynamic FOG architecture and scaling automation.

The FOG layer contained in our architecture will have three types of components: a balancer, capable of obtaining the messages from the different sensorisations from the Gateways and application gateways, and forwarding them to their corresponding processor node; a controller node that will be in charge of monitoring the status of the balancer and the nodes, so that it can organise the workload that will reach each node; and finally the processor nodes where the data acquisition and processing operations are carried out. Communication between the components for their control and management will be carried out by means of a new protocol defined in this architecture, which we call *Simple FOG Management Protocol - SFMP.*

4 Simple FOG Management Protocol - SFMP

This protocol has been developed exclusively to solve the communication between the components of this new architecture, which is carried out by means of a connection-oriented communication protocol [10], in which each request message has an associated response message. The internal structure of each message consists of the following elements: UID, the message identifier; Type, the type of message, request or response; Operation, action to be carried out by the receiving element of the message. In response messages, this parameter will contain a NULL, while in request messages, there are three types of operations; Data, objects of information to be sent. The operations supported by *Operation* are: GETSTATUS, the coordinator node requests the balancer node for its status; INFOSTATUS, status of the balancer node to the controller; SETCONFIG, new configuration for the balancer node; REGISTER, the processor node reports its availability; and UNREGISTER, the processor node notifies the end of its availability.

Communication between coordinator and balancer will occur for each change in the system that is registered. A change in the system occurs when a new processing node becomes part of the infrastructure, or when a processing node is no longer accessible (discussed in the following subsection on communication between coordinator and processor). When the coordinator deems it appropriate (a change has occurred, a certain period has passed), it requests the balancer to know its status; the balancer responds with a *REPLY* to this request. If a *REPLY* to a *REQUEST* does not occur, the sender of the *REQUEST* repeats the sending of the request. Then the balancer sends its status to the controller and receives a confirmation *REPLY.* The coordinator with the information received from the state of the balancer, plus the internal information about the available processor nodes, calculates the appropriate configuration and transmits it to the balancer, and receives the corresponding *REPLY.*

In addition to this scheme, where the coordinator initiates the communication, it is also possible that the balancer, from time to time, sends its status to the coordinator, to keep it updated, and in case the coordinator makes an update of the configuration, the relevant messages would also be sent.

The coordinator-processor communication has three scenarios. The first scenario is like the one explained with the coordinator-balancer. The coordinator, when necessary, asks the processors about their status (in case there is any change, or any drop), and updates its internal information to later configure the balancer if necessary. The other two communication scenarios are the operations: REGISTER and UNREGISTER. In both operations, the processor node informs the coordinator node of its intentions, i.e. to be available for use in the infrastructure, or to disconnect from it. Both messages will send their internal information, both their name or node ID, and the message load they tolerate. The main difference between them is the order of execution of the *REPLY* messages from the coordinator. In the case of *REGISTER*, the coordinator node notifies the processor node with a response message, *REPLY*, and carries out the control of the rest of the elements in case it is necessary to redistribute the load. For the *UNREGISTER*, when the processor notifies the coordinator, it does not get an immediate response, since a reconfiguration of the message load distribution of the rest of the elements is carried out first, without taking this node into account, so the infrastructure continues to operate without this element, without losing messages during the process of disconnecting the node. Once the new distribution is finished, the response message is sent to disconnect, eliminating the node from the list of available resources.

5 Components of the Architecture

Once the communication protocol of the FOG components has been defined, we will define the components themselves.

The balancer will be the element of the architecture in charge of receiving all traffic coming from the physical sensor layer, through gateways and application gateways, and redirecting this traffic to the corresponding node that will process the data packets. The balancer will send messages according to the load distribution dictated by the coordinator node. The balancer measures the traffic flow and sends this value to the coordinator node from time to time, which will be responsible for calculating how to divide the message flow between the available processing nodes in the system. The balancer periodically receives from the coordinator, the list of active processor nodes that are part of the FOG layer, and the weighting of traffic distribution between these nodes to ensure a balanced use of resources.

This coordinator node shall be responsible for calculating the most efficient traffic distribution from the balancer node to the processor nodes. When appropriate, the controller node will issue a new configuration message to the balancer node. The coordinator has a database or local information, in which it keeps a list of the processor nodes that belong to its environment, that is, the resources available to it, together with their corresponding data such as IP addresses, processing capacity, identifier, specific characteristics such as unavailability times, etc., so that it is aware of the resources that make up the infrastructure. As mentioned in the SFMP description, the coordinator uses *REGISTER*, *UNREGISTER* and *INFOSTATUS* messages to keep itself informed of the status of the infrastructure.

The calculation algorithm for distributing the workload among the processing nodes shall use a node processing capacity (*NPC*) indicator that is communicated during the *REGISTER* operation. This *NPC* shall be estimated for each node and may be different for each node. This is because nodes do not have the same physical characteristics, varying for example CPU, RAM, network gateways, GPU, etc. To calculate the *NPC*, a stress test of the node will be performed with a prepared message dataset, thus obtaining a capacity indicator. With the *NPCs* of all registered nodes, the per-node message load distribution ($\%m$) on each node i is calculated using the following equation.

$$\%m_i = \frac{NPC_i}{\sum_{j=1}^{na} NPC_j} \tag{1}$$

where *na* in this case is the number of active nodes in the controller's list.

In order to perform the calculations, it is necessary to know *na*, i.e. how many and which nodes should be active in the system. For this purpose, the information of each node is used. Let $N = \{n_1, n_2, n_3, \ldots N_{nr}\}$ be the list of nodes registered in the controller node, where *nr* is the total number of registered nodes. Each n_i is defined by a tuple called $n_i = \{NPC, IP, name\}$, where *NPC* has already been defined, *IP* is the physical *IP* address of the node (to be able to communicate with it) and *name* is its unique identifier. The controller node, as mentioned above, receives the number of messages handled in the balancer, *rm* (rm received messages). Therefore, it is a matter of calculating:

$$Minimum\, na\, such\, that \sum_{j=1}^{na} NPC_j \geq rm \tag{2}$$

NPC_j being the *NPC* value of a node of *N*, and *na* a value between 1 and *nr*.

A node may be disconnected from the system, in which case it will issue a *UNREGISTER* message to the controlling node. This starts the message chain again to redistribute the load among the remaining registered nodes *nr*, and will recalculate *na* and the $\%m_i$.

On the other hand, once it is known how the processing nodes are going to be activated, we go on to define the functions they will fulfil in the architecture. Each processor node retains the acquisition and processing functionalities described in the background, functions through which it acquires and processes the data it receives. But the nodes have been modified, so that instead of having a publish-subscribe based connection as in the original scheme (Fig. 1), they have been implemented with a reactive interface that can be invoked by the balancer to send it a data packet. In this way, it is now the balancer that sends the data to be processed to the processors, whereas before the processors were listening on the message channel. In addition to this new interface, they also implement the SFMP protocol described above.

6 Implementation at the University of Alicante

The feasibility test of the proposed FOG architecture has been implemented in the infrastructures of the University of Alicante, for the management of the sensorisation of different areas with which they work. The San Vicente campus of the University of Alicante has an extension of more than one million square metres, more than 70 buildings, roads, a desalination plant, a lighting system, video surveillance, large landscaped green areas and many other services. As mentioned above, it is comparable to 95% of small and medium-sized cities with a population of 40,000 inhabitants or less.

The number of sensorisations carried out on the campus currently exceeds 2 million records per hour, as it is made up of several sensorisation subsystems, some of which are made up of thousands of sensors, while others are just a few dozen. Some of these sensorisations are carried out every 15 min, others every 30 s, depending on the variation that the sensorised value may undergo in each measurement. There are periodic sensorisations, at specific times of the day at specific times, for example, measuring water consumption, electricity consumption or connections to the wifi network. Other sensorisations are triggered by events, for example, access to a room, measuring the state of a light fitting, taking the temperature of a room or the humidity of the garden. This means that our FOG layer receives messages, an indeterminate, variable number, which can be huge or tiny.

Our FOG layer infrastructure will consist of three physical machines, one of high processing capacity (Node 1), one of medium capacity (Node 2), and finally an embedded device (Node 3) of limited capabilities. The sensing has been deployed using LORA technology, where all sensing will be captured through access points (AP). The IoT layer uses TheThingStack and MQTT integration with the balancer.

The coordination and balancing nodes are created on virtual machines on the servers of the university's data processing centre. The rest of the processing nodes will consist of the physical machines mentioned above. The list of machines available to the coordinator node will be ordered by the processing capacity they can tolerate, i.e. node 1 will go first, then node 2 and the last node to be used will be node 3. In this way, the coordinator will activate processing nodes, from the most powerful to the last, the Raspberry PI. The use of the last machine, Raspberry Pi, will be used as an additional element of control of the infrastructure, since when this machine starts to operate it means that the workload of the entire infrastructure is abundant, so more resources will have to be added to the infrastructure. A routine has been programmed in this processing node so that when it starts up, it alerts the system administrator of this situation. When this happens and new resources have been added to the list, this node 3 will go to the end of the list of available resources of the coordinator node, being the last resource to be used and allowing again to monitor the workload of the FOG layer.

For the *NPC* calculation of each node, a dataset of 100,000 messages was obtained, equivalent to one hour of nightly data reception in the infrastructure. With this dataset, a stress test was performed for each node, injecting the entire list of values and measuring the number of messages they were able to handle per second. The calculated NPCs were: Node 1 $NPC = 1350$; Node 2 $NPC = 600$; Node 3 $NPC = 20$. To avoid possible saturations of the infrastructure, the controller node will not allow any node to exceed 90% of its processing, i.e. when the number of messages to be processed exceeds 90% of the messages processed by the active nodes, the next in the list will be activated. Similarly, when the number of messages to be processed falls below 90%, the node will be deactivated. The following figure shows the results of the tests carried out during 24h.

As can be seen in Fig. 4, there is now only one node active by default (the dotted red line indicates the number of active nodes), node 1, while the message processing demand does not reach a certain level (90% of the maximum 1350 messages it can process). When this limit is exceeded, node 2 is activated, and if 90% of both is exceeded, node 3 is activated.

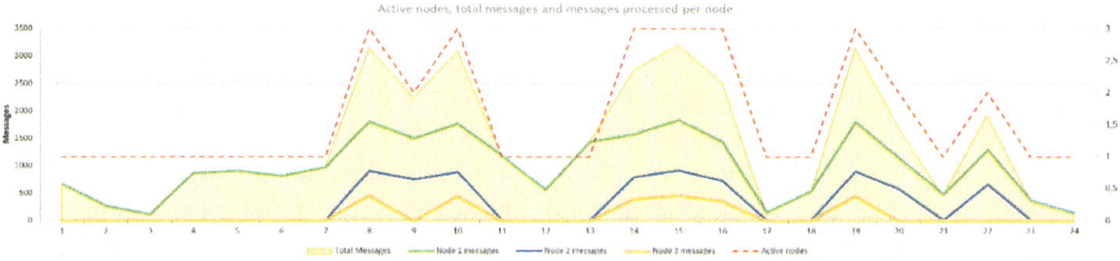

Fig. 4. Active nodes, total messages received and messages processed by each node.

Another interesting aspect is if we look at the power used by each node while they are processing, Fig. 5. In contrast to what was previously observed on the generic FOG layer, now the nodes are balanced, taking advantage of the computational power. This implies a drastic decrease in computational waste.

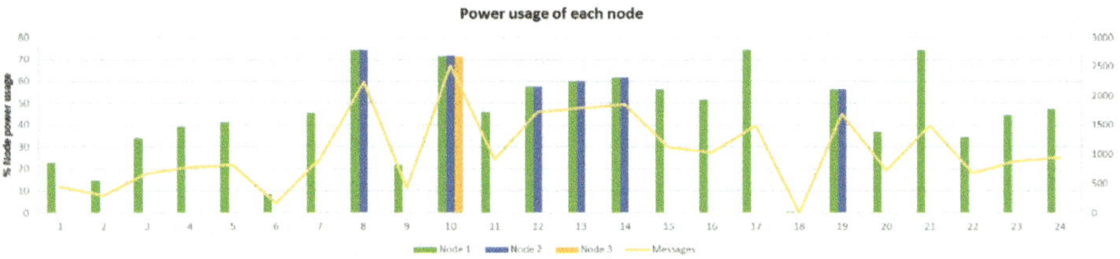

Fig. 5. Power measurement used for processing.

7 Conclusions

This work has proposed the design and development of an advanced dynamic and self-scalable FOG layer. The objective of this proposal is to maximise the performance of the infrastructures through the distribution of load on its nodes, but maintaining a commitment to efficiency, using only the necessary resources. This allows parallel computing of the workload, but with minimal costs.

Several important conclusions can be drawn from the results obtained, the first of which is that the architecture is capable of providing the developed FOG layer with the necessary flexibility to adapt to the IoT environments for which it is implemented, since the use of different types of heterogeneous IoT devices is a constantly changing scenario. Then, which can be automated, using this autonomously configurable Control Level, the processing needs of the architecture can be configured by the computational capacity in use by the node, so that the infrastructure is at all times covered with the minimum needs to ensure the service of data capture and processing. The fact that the message load to be handled by each node is adapted to its characteristics, i.e. the computational load it tolerates, means that a node is not saturated from the moment it is activated, but is adapted to its limits, ensuring the availability of data at all times by prioritising the correct operation of the nodes and, therefore, of the infrastructure. On the other hand, it was necessary to be able to parallelise the different computational processes, as this makes it easier to provide the infrastructure at the FOG layer. This way we can keep adding physical resources, increase the number of nodes, and the layer will automatically adjust its configuration to the needs and available resources. Finally, the decision to split the FOG layer into two different levels again facilitates auto-scaling and data management. The new control layer with the coordination and balancing nodes, in addition to the processing layer with the task execution nodes already existing in the infrastructure, will allow for better capabilities in the future. This tiering provides a modular architecture, facilitating further development and scalability of the system.

As a future line of work, we are currently making progress, on the one hand, in improving the efficiency of the FOG layer. Now that we have continuous monitoring of the workload of the FOG layer, we can introduce predictive algorithms that prepare the infrastructure for the early activation of nodes. This can facilitate shutting down machines completely at times when it is known they will not be needed, or moving them to a dormant state when they are likely to be needed.

Acknowledgments. This work has received funding from the UAIND22-01B project "Adaptive control of urban supply systems" of the Vice-Rectorate for Research of the University of Alicante, and the CENID2024/01 project - "Protection of infrastructures through digital intelligence. Detection of anomalies and preventive maintenance".

Disclosure of Interests. The authors have no competing interests to declare that are relevant to the content of this article.

References

1. Khan, J.Y., Yuce, M.R., (eds.): Internet of Things (IoT): Systems and Applications. CRC Press (2019)
2. Macia Pérez, F., Berna Martínez, J.V., Lorenzo Fonseca, I.: Modelling and implementing smart universities: an it conceptual framework. Sustainability **13**(6), 3397 (2021)
3. Patel, K.K., Patel, S.M., Scholar, P.: Internet of things-IOT: definition, characteristics, architecture, enabling technologies, application & future challenges. Int. J. Eng. Sci. Comput. **6**(5) (2016)
4. Soumyalatha, S.G.H.: Study of IoT: understanding IoT architecture, applications, issues and challenges. In: 1st International Conference on Innovations in Computing & Net-working (ICICN16), CSE, RRCE. I. J. of Advanced Networking & Applications, vol. 478 (2016)
5. Atlam, H.F., Walters, R.J., Wills, G.B.: Fog computing and the Internet of Things: a review. Big Data Cogn. Comput. **2**(2), 10 (2018)
6. Das, R., Inuwa, M.M.: A review on fog computing: issues, characteristics, challenges, and potential applications. Telematics Inform. Reports, 100049 (2023)
7. Yacchirema, D.C., Sarabia-Jácome, D., Palau, C.E., Esteve, M.: A smart system for sleep monitoring by integrating IoT with big data analytics. IEEE Access **6** (2018)
8. Towards the Distributed Edge - An IoT Review (2018, 1 December). IEEE Conference Publication|IEEE Xplore. https://ieeexplore.ieee.org/abstract/document/8603663
9. Al-Joboury, I.M., Al-Hemiary, E.H.: Performance analysis of Internet of Things protocols based fog/cloud over high traffic. J. Fund. Appl. Sci. **10**(6S), 176–181 (2018)
10. Mondal, A., Bhattacharjee, S.: A reliable, multi-path, connection oriented and independent transport protocol for IoT networks. In: 2017 9th International Conference on Communication Systems and Networks (COMSNETS), pp. 590–591. IEEE (2017)

ChatGPT and Semiotic Representation Theory: Innovating the Measurement of Cognitive Demand in Mathematical Tasks

Eduardo Puraivan[1,2,3], Connie Cofré-Morales[7(✉)], Miguel Rodríguez[3], Tamara Lasnibat-Godoy[4], Juan Tapia[5], Carlos Hervás-Gómez[6], and María Dolores Díaz-Noguera[6]

[1] Escuela de Ciencias, Universidad Viña del Mar, Valparaíso, Chile
epuraivan@uvm.cl, eduardo.puraivan@postgrado.uv.cl, eduardo.puraivan@upla.cl
[2] Escuela de Ingeniería Informática, Universidad de Valparaíso, Valparaíso, Chile
[3] Facultad de Ciencias de la Educación, Universidad de Playa Ancha, Valparaíso, Chile
mrodriguez@upla.cl
[4] Instituto de Matemática, Pontificia Universidad Católica de Valparaíso, Valparaíso, Chile
tamara.lasnibat.g@mail.pucv.cl
[5] Fakultät für Mathematik, Univertät Bielefeld, Bielefeld, Germany
juan.tapia@uni-bielefeld.de
[6] Facultad de Ciencias de la Educación, Universidad de Sevilla, Andalucía, Spain
{hervas,noguera}@us.es
[7] VRA, Universidad Viña del Mar, Valparaíso, Chile
connie.cofre@docente.uvm.cl

Abstract. Determining the level of cognitive demand in school tasks is a significant challenge, marked by the lack of consensus among evaluators when assigning cognitive complexity in different areas of knowledge. In view of this problem, we propose a methodology based on the Theory of Registers of Semiotic Representation to guide the assignment of cognitive demand in mathematical tasks, specifically focused on the quadratic function. We have used ChatGPT to implement this methodology, applying it to mathematical tasks extracted from school textbooks. The results show that the analysis methodology proposed for the classification of these tasks is simple to implement with ChatGPT and with the experts. Moreover, on the classification of the 3 tasks analyzed, the experts evaluated ChatGPT's performance well with respect to 2 of them, and on a third one there were discrepancies.

Keywords: ChatGPT · Level of cognitive demand · Mathematics tasks

R. Chinthaginjala et al. (Eds.): DCAI 2024, LNNS 1259, pp. 230–240, 2025.
https://doi.org/10.1007/978-3-031-82073-1_23

1 Introduction

In the area of education, the design of didactic and/or evaluative tasks implicitly entails a level of cognitive demand [10]. Normally this level of cognitive demand is established before designing the task and is subject to the learning objectives to be achieved [16]. However, the level of cognitive demand can be analyzed by knowing the task, as is the case with the tasks proposed in school textbooks. The most plausible way to do this is to use taxonomies or models that make explicit the actions that a learner must perform in response to the cognitive transformations demanded by a given task. One of the best-known taxonomies most commonly used by teachers to structure the didactic tasks they design is Bloom's taxonomy [11]. Said taxonomy makes explicit 6 levels in the dimension of cognitive processes which are: Remember, Understand, Apply, Analyze, Evaluate and Create. These levels advance in complexity but should not be interpreted as linear, since a student can work at one of the higher levels without necessarily having gone through all of the previous levels [7]. On the other hand, in the specific area of mathematics there is also a model that considers 4 levels of cognitive demand that a student may face when developing a task in that subject. This model was proposed by Stein and Smith [12], who suggest that the use of this model provides insight into the type of thinking required by the tasks to the students. The description of the 4 levels contemplated by this model can be found in Sect. 2 of this work.

Identifying the level of cognitive demand of a predesigned task, for example those presented in textbooks, is a complicated work that requires a slow analysis that in some cases can even be subjective [10], which also becomes a problem when you want to assign these levels on a large number of tasks due to the time and resources that this demands [14]. In addition, it has been reported that in many cases teachers experience difficulties in developing learning objectives that are appropriate to the levels of Bloom's Taxonomy [16]. This difficulty can be extrapolated to the design and review of didactic tasks with respect to the levels of cognitive demand they imply. About the difficulty that subjectivity in the classification of tasks can mean, the authors [12] also mention that in their work with teachers it is observed that they do not always agree among themselves or with the experts on the classification of tasks, which is why they designed their model to classify the levels of cognitive demand of tasks specifically in the subject of mathematics. On the other hand, discordance was also detected among expert judgment in the study conducted by Cofré-Morales et al. [2] who classified History tasks using ChatGPT based on the levels of Bloom's taxonomy and a model specifically applied to History tasks. This experience also considered a classification made by experts, where it was observed a marked discordance. In turn, the tasks was classified and argued by ChatGPT, a work that received a high valuation by the experts, both in the classification and in the argumentation.

The problems mentioned above regarding the classification of tasks according to their levels of cognitive demand allows us to identify the need to work on a tool that supports this labour. A tool that can classify tasks would allow teachers and educational entities to verify whether the tasks that are analyzed reach the level of cognitive demand expected. To address this problem, we used ChatGPT to perform the classification of the levels of cognitive demand of didactic tasks of Mathematics, specifically tasks on the content of the quadratic function. To analyze and classify the level of cognitive demand, we used as a framework the Theory of Registers of Semiotic Representation (TRSR) [5] and model of [12]. Thus, with the help of specialists in Didactics of Mathematics (DM) we have operationalized that theory with the levels of Stein's model [12] in order to classify and argue the selected tasks.

In the following, we propose the theoretical foundations that support this work in Sect. 2; in the Sect. 3 we detail the methodology used to carry out the experiments. Then in Sect. 4 we present the results derived from the experiments performed. Finally, in Sect. 5 we develop the conclusions of this work.

2 Related Work

2.1 Fundamentals from the Didactics of Mathematics

In the Theory of Registers of Semiotic Representation (TRSR), mathematical activity is mainly centered on semiotic representations [5]. The activity that is linked to the production of a representation is called semiosis, while the conceptual apprehension of mathematical objects is called noesis. In this theory, there will be no noesis without semiosis [6]. On the other hand, a register of semiotic representation must allow the three cognitive activities linked to semiosis. First, the formation of an identifiable representation is recognized, i.e. a type of format associated with the object. Then, the treatment of a register that corresponds to the transformations that can be performed on this format. Finally, the conversion between two or more registers of semiotic representation that account for the transit between one and the other. It should be pointed out that conversion is a highly complex cognitive task, since there are no specific rules to guide it [6]. According to [4], an integral understanding of a mathematical concept is based on the coordination of at least two of its representation registers, and this is evidenced in the fluency and spontaneity with which the cognitive activity of conversion is carried out.

2.2 Cognitive Demand in Mathematics

In order to understand what will be understood by cognitive demand in this research, in the challenge of selecting a mathematical task, it is important to clarify what we mean by task. In this sense, in [3] it is stated that a task is a requirement that emerges from some area of knowledge and that a learner must solve. On the other hand, a task involves three aspects that characterize it: (a) the products that a learner will obtain, (b) the operations that must be performed, and (c) the resources that arise from it to solve the task. In [13] a task is defined as a problem or set of problems that deal with a concept or a mathematical content related to that concept. Thus, we will understand the cognitive demand of a task as the level of complexity of the different actions that are necessary for a learner to be able to respond to the task [3,13]. Therefore, in the following, we present criteria for the classification of a task, based on the level of cognitive demand [15] and the use of semiotic representation registers.

Lower-Level Cognitive Demands. The lower-level demands category is divided into two main types of activities: memorization and procedures without connections. These activities are essential in the learning process, although they do not require a high level of cognitive processing. `Memorization` involves both the reproduction of data, rules, formules or definitions previously learned. The time required for resolution is short, as there is no procedure [12]. They are not ambiguous and have no connections. They are mathematically related activities, i.e. facts, definitions or terms are remembered or recognized [15]. The *didactic approach* from the TRSR is Without treatment, or conversion of semiotic representation registers. The `Procedures without connections` are algorithms, they use previously learned procedures. Have no connection to concepts or meanings underlying the procedures used. Focus on obtaining a correct answer rather than developing understanding. Require explanations that focus only on describing the process used [12]. Related Mathematical Activities: Solves a routine problem that requires several steps. Interprets information in a graphical manner [15]. The *didactic approach* from the TRSR is Requires a treatment in the given register, to recognize if there is the presence of significant symbolic units.

Higher-Level Cognitive Demands. High cognitive demand activities foster the development of a deep understanding of mathematical concepts, encouraging students to apply their knowledge in a more integrated and reflective manner. Thus, the `Procedures with connections`, considers the use of procedures intended to develop deeper levels of understanding of mathematical concepts and ideas. There are connections close to underlying conceptual ideas, unlike limited algorithms. It is usually represented in multiple ways. Related mathematical activities: Interpret information from a complex graphical form. Establishes and/or justifies a conjecture. Develops logical arguments for a concept. [12]. These are mathematical activities related so that they all have the same format. Performs a multi-step procedure that requires decision making [15]. The *didactic approach* from the TRSR is Treatments and conversions (of mathematics). On the other hand, `Doing mathematics` requires complex, non-algorithmic thinking. They lead students to explore and understand the nature of mathematical concepts, procedures, or relationships. They also lead students to access relevant knowledge and experiences. They demand considerable cognitive effort and may involve some level of anxiety for the student, due to the unpredictable nature of the solution process required [12]. Related mathematical activities: Relates mathematical concepts to other content areas. Relates mathematical concepts to real-world applications. Applies a mathematical model to a situation or problem [15]. The *didactic approach* from the TRSR is Treatments and conversions (given in context).

2.3 Cognitive Demand and Artificial Intelligence (AI)

With respect to the identification of the levels of cognitive demand of didactic tasks, several *software* have been developed. In some cases, AI is used to perform large-scale classification to alleviate the difficulties in manual processing, and in other cases other computer techniques are employed. In [14] they designed a software using weakly supervised Machine Learning (ML) to assign binary labels and perform the classification of the cognitive demand of didactic tasks taking as reference the levels of Bloom's Taxonomy but in binary language, i.e., "lower-order" and "higher-order" cognitive labels. Other studies have developed tools that automatically classify tasks by analyzing the semantics and syntax of the tasks to determine the corresponding level [10]. These tools assign a level of cognitive demand based on the words used in the question's wording.

Regarding the use of AI, ChatGPT has been used in experiments to classify didactic tasks. In [8] ChatGPT was used to design an evaluative script, which included a series of evaluative tasks and inputs such as the grading guideline. Although this was not the main focus of the study, ChatGPT was asked to assign a level of cognitive demand based on Bloom's taxonomy to the questions it generated. It was observed that ChatGPT is capable not only of generating questions but also of classifying didactic tasks by assigning a cognitive demand level. The study by Cofré-Morales et al. [2] also used ChatGPT to classify predesigned didactic tasks extracted from textbooks, focusing on validating ChatGPT's classifications through expert judgment. Despite some disagreement among the judges regarding the assigned levels, at least one of the two judges evaluating each task agreed with ChatGPT's classification. This agreement is considered an indicator that ChatGPT can assign levels accurately and in line with expert evaluations. Therefore, it can be concluded that ChatGPT is a valuable support tool for addressing the challenges of assigning cognitive demand levels to a set of tasks.

3 Methodology

In the development of this study, we have focused on the exploration of school mathematics tasks, specifically related to the quadratic function. The selection of these tasks has been made based on the analysis of the Mathematics text for the second year of secondary education, which is used in Chile from 2020 to the present time [1]. This didactic material serves as the basis for our study due to its relevance and widespread use in the Chilean educational context. In order to carry out our research, in a first phase, we have extracted all the tasks on quadratic function from the previously mentioned text. Then we have focused on 5 specific tasks, these tasks were subjected to an analysis using TRSR (See Sect. 2.1), a framework that allows us to examine in detail the cognitive demands involved in mathematical tasks. To ensure the objectivity and accuracy of the analysis, each task was evaluated by a team of experts in mathematics education, using a set of established criteria based on the TRSR. Thus, the experts, in order to operationalize the general criteria presented in the conceptual framework, detailed in Sect. 2.1 and Sect. 2.2, have prepared a matrix with the criteria for classifying according to the cognitive demand a task on the quadratic function theme, presented in Table 1. This matrix is then used to train ChatGPT to analyze new tasks.

Table 1. Didactic criteria from the Theory of Registers of Semiotic Representation to classify tasks according to cognitive demand in the context of quadratic functions.

Cognitive demand	Description
Lower cognitive demand. Memorization	The analytical expression, table, or graph of a quadratic function is recognizable at a glance. This is reinforced by the heading of the task which specifies the quadratic function. There is no need for treatment or conversion.
Lower cognitive demand. Procedures without connection	The analytical expressions, tables, or graphs associated with a quadratic function are not recognizable at a glance. There is a need to perform a treatment on the given register in order to respond to the task.
Higher cognitive demand. Procedures with connection	The analytical expressions, tables, or graphs are presented without specifying a quadratic function. There is a connection with other mathematical notions in different domains of mathematics, and invariants associated with the quadratic function are verified. It requires the treatment and conversion between at least two registers of semiotic representation.
Higher cognitive demand. Doing Mathematics	It is required to discover invariants, formulate hypotheses to connect them with the knowledge of a quadratic function. The statements make a direct connection with mathematical objects or phenomena of everyday life. It requires the formation, treatment, and conversion of registers.

4 Experimentation and Results

In the experiments, we provided the framework described in Table 1 to ChatGPT-4 for its evaluation of the tasks. Our study focused on three tasks (extracted from the previously mentioned Mathematics text and different from the five used by experts to construct Table 1). Below, we present an example of the results obtained for one of the tasks.

The analysis performed by ChatGPT-4 for each question of one of the tasks is as follows:

Question a : Build a table for each of the functions. Identify at least 4 points that belong to the function.

- *Cognitive demand*: Lower (Memorization)
- *TRSR*: Direct identification of points in the graphic register without processing or conversion to other registers.
- *Justification*: The task involves observing and transcribing points from a graph (graphic record) to a table (tabular record), which does not require complex treatments or conversions.

Question b : At what points do the X axis intersect? And the Y axis?

- *Cognitive demand*: Lower (Procedures without connections)
- *TRSR*: Operation within the graphic register, possibly with minimal conversion to the algebraic register when interpreting the intersection points.
- *Justification*: The task is to identify specific points of intersection with the axes, which is done within the graphic register with a basic understanding of the coordinates.

Question c : Replace in the equations the value x = 0 and obtain the points (0, f(0)), (0, g(0)) and (0,h(0)). What do they correspond to graphically?

- *Cognitive demand*: Lower (Procedures without connections)
- *TRSR*: Treatment within the algebraic register with a direct interpretation in the graphic register.
- *Justification*: It involves substituting one value and calculating another within the same algebraic equation, with an immediate graphical interpretation without complex conversion to other registers.

Question d : Obtain the discriminants of the equations f(x)= 0, g(x)= 0 and h(x)= 0. How are their discriminants related to the number of points at which the functions intersect the X axis?

- *Cognitive demand*: Higher (Procedures with connections)
- *TRSR*: It requires conversions between the algebraic and graphical register by relating the discriminant to the number of graphical roots/intersections.
- *Justification*: The task requires a conceptual analysis of the quadratic function and its discriminant to determine the number and nature of the roots, which implies a deep understanding of the registers involved and their interrelation.

4.1 Expert Analysis

To evaluate the use of our methodology (Table 2) implemented with ChatGPT-4, we asked two experts in mathematical didactics to rate the work performed by the chat. The following key questions were posed to guide the initial evaluation by dimensions: Does ChatGPT adequately assign the level of cognitive demand to the mathematical tasks? (D_1), Does ChatGPT correctly apply the Theory of Registers of Semiotic Representation (TRSR) in the analysis of the tasks? (D_2), Are the justifications provided by ChatGPT with respect to the proposed methodology clear and coherent? (D_3). The scale used for the assessment considers the following 5 options: Strongly disagree (1), Disagree (2), Neither agree nor disagree (3), Agree (4), Strongly agree (5).

Table 2. Experts' assessment of ChatGPT's work. (D1:Cognitive demand, D2: TRSR, D3: Justification).

Task	Dimension	Expert 1	Expert 2	Average
1	D_1	5	5	5
	D_2	5	5	5
	D_3	5	5	5
2	D_1	3	4	3.5
	D_2	2	2	2
	D_3	2	3	2.5
3	D_1	3	4	3.5
	D_2	4	5	4.5
	D_3	4	4	4

Based on the results of the Table 2 in Tasks 1 and 3 ChatGPT performed well. However, Task 2 exhibits the largest discrepancies (task shown as an example previously). In this regard, we provide the experts' critical analysis of the Task to identify opportunities for improvement. In this task, to address question a), it is necessary to start by examining the graph to identify points on it. Then, these points must be expressed in a tabular record, i.e., a conversion must be performed. Although conversion is commonly associated with 'procedures with connection', when considering the level at which the task is applied, the level corresponds to 'memorization', in line with the categorization made by ChatGPT. However, there is discrepancy regarding the application of the TRSR criteria and justifications, as ChatGPT indicates that no conversion is performed.

For question b), the points of intersection of the graph with the axes must be identified, which is a straightforward process and requires neither treatment nor conversion, thus falling into a 'memorization' level. This contrasts with ChatGPT's categorization, which classifies it as a 'procedure without connection'. In addition, there is discrepancy in the application of the TRSR criteria, as it suggests the possibility of conversion, but this is not the case as it always remains in a graphic record.

In relation to question c), it is necessary to determine the point at which the preimage is 0 for each of the functions. Then, it must be recognized that this point represents the intersection between the graph and the y axis. This process entails a treatment and a conversion, so its level is 'procedures with connection'. This classification contrasts with the level suggested by ChatGPT. Furthermore, discrepancies arise in the application of TRSR criteria and in justification, as ChatGPT suggests that the interpretation of the point found is done directly, which may not be obvious to a student learning about quadratic functions.

As for question d), there is consensus on all criteria used and justifications.

In summary, when analyzing this case, it appears necessary to furnish ChatGPT with a more detailed explanation of the concept of conversion. This is concluded due to the misinterpretations by the artificial intelligence observed in three out of the four questions in the task.

5 Conclusions

To achieve a high standard of quality in education, ensuring the achievement of learning objectives is essential. One way to achieve this is to perform a review of the cognitive demand of the tasks designed to verify that these are at a level in line with the objectives. This review can be performed using as a reference a taxonomy such as Bloom's [14], or another model that is specifically focused on the subject being worked on. In this case, for the cognitive demand of mathematics tasks we have chosen the levels of [12], to which we have added elements of the TRSR [4–6] to obtain a framework with a clear and precise description for tasks on the quadratic function.

An interesting result is that the proposed methodology (shown in the Table 1) is easy to implement both with ChatGPT and with the experts who participated in the evaluation stage of the work performed by ChatGPT. On the other hand, of the 3 tasks analyzed, only one of them shows serious discrepancies with the experts' judgment. This situation was previously discussed (see Sect. 4.1). On the other hand, the best evaluation was obtained for Task 1, which only implied a low cognitive demand.

In short, our work proposes a methodology for the analysis of the cognitive demand of mathematical tasks, but specifically adding the Theory of Registers of Semiotic Representation applied to the study of the quadratic function. This allows us to prove that to perform task classification according to the cognitive demand, it is possible to use computer tools based on AI, which allows to speed up the work, or to contribute to the decision making process regarding the most appropriate level for the task. These results complement the findings of Puraivan et al. [9], as the experiments in this paper were conducted with a much more specific framework with respect to cognitive demand. However, experimentation with a low number of tasks is a limitation that we hope to overcome in future work.

Acknowledgements. E. Puraivan has been partially funded by the Escuela de Ingeniería Informática, Universidad de Valparaíso, Chile, through grant No. 01.016/2020 and Beca de Doctorado Nacional ANID 21232242. T. Lasnibat has been funded by Beca de Doctorado Nacional ANID 21240828.

References

1. Barrera Yáñez, R., Manuel, R.H., Morales Valderrama, K., Ortiz Solis, N., Norambuena Morales, P., Díaz Valenzuela, E.: Texto del estudiante Matemática 2° medio. Editorial SM, Chile (2020)
2. Cofré-Morales, C., Puraivan, E., Huencho-Iturra, K., Astudillo-Vásquez, M., Herváz-Gómez, C., Díaz-Noguera, M.: Using artificial intelligence to assess the level of cognitive complexity involved in didactic tasks. a case study on historical thinking. In: Díaz-Noguera, M.D., Herváz-Gómez, C., Sánchez-Vera, F., de los Á. Romero Aguirre, G. (eds.) Artificial Intelligence and Education: Enhancing Human Capabilities, Protecting Rights, and Fostering Effective Collaboration Between Humans and Machines in Life, Learning, and Work. McGraw Hill, México, 1st edn. (2024)
3. Doyle, W.: Academic work. Rev. Educ. Res. **53**(2), 159–199 (1983). https://doi.org/10.3102/00346543053002159
4. Duval, R.: Registros de representación semiótica y funcionamiento cognitivo del pensamiento, pp. 173–201. Grupo Editorial Iberoamérica (1998)
5. Duval, R.: Semiosis y pensamiento humano: Registros semióticos y aprendizajes intelectuales. Universidad del Valle/Peterlang, Cali, Colombia (1999)
6. Duval, R.: Un tema crucial en la educación matemática: La habilidad para cambiar el registro de representación. La Gaceta de la RSME **9**(1), 143–168 (2006)
7. Förster, C., Rojas-Barahona, C.: Aprendizaje y evaluación: lo que no se evalúa, no se aprende. In: Förster, C. (ed.) El poder de la evaluación en el aula. Mejores decisiones para promover aprendizajes, pp. 43–74. Ediciones Universidad Católica de Chile (2017)
8. Kwan, C.C.L.: Exploring ChatGPT-Generated Assessment Scripts of Probability and Engineering Statistics from Bloom's Taxonomy, pp. 275–286. Springer Nature Singapore (2023). https://doi.org/10.1007/978-981-99-8255-4_24
9. Puraivan, E., Cofré-Morales, C., Lasnibat, T., Hervás-Gómez, C., Díaz-Noguera, M.D., Rodríguez, M.: Empirical evidence on the use of chatgpt to determine the level of cognitive demand in didactic tasks. In: Proceedings of the 2024 19th Iberian Conference on Information Systems and Technologies (CISTI). IEEE (2024), accepted
10. Rawat, A., Kumar, S., Singh Samant, S.: A systematic review of question classification techniques based on bloom's taxonomy. In: 2023 14th International Conference on Computing Communication and Networking Technologies (ICCCNT). IEEE (2023). https://doi.org/10.1109/icccnt56998.2023.10308403
11. Stanny, C.: Reevaluating bloom's taxonomy: what measurable verbs can and cannot say about student learning. Educ. Sci. **6**(4), 37 (2016). https://doi.org/10.3390/educsci6040037
12. Stein, M., Smith, M.: Mathematical tasks as a framework for reflection: from research to practice. Math. Teach. Middle School **3**(4), 268–275 (1998)
13. Stein, M.K., Lane, S.: Instructional tasks and the development of student capacity to think and reason: an analysis of the relationship between teaching and learning in a reform mathematical project. Educ. Res. Eval. **2**(1), 50–80 (1996). https://doi.org/10.1080/1380361960020103
14. Wang, Z., Manning, K., Mallick, D.B., Baraniuk, R.G.: Towards Blooms Taxonomy Classification Without Labels, pp. 433–445. Springer International Publishing (2021). https://doi.org/10.1007/978-3-030-78292-4_35

15. Webb, N.: Criteria for alignment of expectations and assessments in mathematics and science education. Natl. Inst. Sci. Educ. (1997)
16. Li, Y., Rakovic, M., Poh, B.X., Gasevic, D., Chen, G.: Automatic classification of learning objectives based on bloom's taxonomy (2022). https://doi.org/10.5281/ZENODO.6853191

Dynamic Hybrid Model for Biopesticides Production Using *Bacillus Thuringiensis* Strains

Sergio Figueroa-Cardona, Carlos E. Robles-Rodríguez, Rim El-Jeni, Luc Fillaudeau, and César A. Aceves-Lara(✉)

TBI, Université de Toulouse, CNRS, INRAE, INSA, Toulouse, France
aceves@insa-toulouse.fr

Abstract. Commercial biopesticides are primarily based on the bacterium *Bacillus Thuringiensis* which produces insecticidal proteins. Dynamic models have been proposed to describe the production of these proteins, but their approximation can still be improved. Dynamic hybrid modelling combines kinetic models with data-driven algorithms which could be promising to improve the existing dynamic models. This work presents a dynamic hybrid model using Support Vector Machine (SVM) to predict the specific production of protein and spore by different strains of *B. thuringiensis*. The dynamic hybrid model was trained and validated with independent datasets of batch fermentations using the production rates of biomass, protein and spores computed with a kinetic model and the strain type as predictors. Additionally, Shapley values were calculated to determine the pertinence of each predictor. The Normalized root mean squared errors (NRMSE) revealed an improvement of the dynamic hybrid model of 12% for proteins and 7% for spores over the dynamic model. Although good, the model could be further improved by training the model with a larger quantity of data.

Keywords: *B.thuringiensis* · Support Vector Machine · hybrid modelling · optimization · fermentation

1 Introduction

Bacillus thuringiensis is a facultative anaerobic Gram-positive sporulating bacterium, commonly used in the production of biopesticides [1]. Most commercial biopesticides are of microbial origin and are primarily based on this microorganism [2]. *B. thuringiensis* has been shown to be toxic to various organisms, such as lepidopterans, coleopterans, dipterans, or nematodes, but is considered safe for mammals. Thus, the products based on this bacterium provide effective and environmentally benign control of several insects in agricultural, forestry, and disease-vector applications [3].

Its insecticidal effect is mainly due to the production of crystalline inclusions consisting of multiples insecticidal proteins, known as δ-endotoxins or Cry proteins, which are related to the spore formation of the bacteria [1]. *B. thuringiensis* must experience certain modifications of its physiology to produce these δ-endotoxins, which makes its culture a challenging labor. One possibility to optimize the production of these δ-endotoxins

R. Chinthaginjala et al. (Eds.): DCAI 2024, LNNS 1259, pp. 241–249, 2025.
https://doi.org/10.1007/978-3-031-82073-1_24

is to use a model-based approach in which several simulations of mathematical models are assessed. However, only a few models are available that can correctly emulate the dynamics of the growth and production phases of *B. thuringiensis* culture [4].

Dynamic hybrid modelling integrates both data-driven and mechanistic approaches and could offer a cost-effective solution for modelling complex biochemical processes where the underlying mechanisms are not entirely known. Although more efficient in utilizing data compared to purely data-driven methods, the construction of dynamic hybrid models still require the collection of large sets of experimental data through time-consuming experiments for new processes [5].

Hybrid modelling has been applied in fermentation-related bioprocesses, to compare the performance of kinetic models and its hybrid counterpart, to describe the production of a compound of interest, as the case study of yeast astaxanthin production under uncertainty using Gaussian processes [6]. Other algorithms used before are artificial neural network (ANN) and response surface methodology (RSM), to optimize the biohydrogen production by dark fermentation [7].

In this paper, Support Vector Machine (SVM) was chosen based on its versality [8], the type of data available and its ability to work well with small datasets [9]. This method is described in Sect. 2.

The aim of this paper is to propose a dynamic hybrid model to represent the dynamics of the fermentation of *B. thuringiensis* with a special focus on the products: the protein and the spore. The dynamic hybrid model is described in Sect. 2, and the results obtained are detailed in Sect. 3.

2 Materials and Methods

2.1 Organism and Culture Media

Three strains of *B. thuringiensis* were studied: BLB1, HD1 and Lip, a Lebanese strain [10]. Luria broth (LB) was used for inoculum production, whereas a semi-synthetic medium (SSM). For the SSM, concentrated glucose (Sol 2) and all salts solutions (Sol 3, 4, 5) were prepared and sterilized separately and added before inoculation to the rest of medium (Sol 1) previously sterilized.

2.2 Experimental Setup

The model was calibrated with datasets collected from batch experiments performed at 30 °C in a 3 L Biostat B plus fermenter (Sartorius; Germany) containing 1.8 L of the SSM medium. pH was regulated continuously at 6.8 using solutions of 1 M H_2SO_4 and 3 M NaOH. Dissolved oxygen was continuously monitored by an optical oxygen sensor and maintained at 25% of pO2-saturation with constant aeration rate (VVM = 10 with Qair = 0.18 min·L − 1) and variable stirring (from 250 to 1200 rpm).

2.3 Experimental Data

A dataset of 9 batches was obtained based on the experimental setup: 3 batches per tested strain. Each batch contains the measurements of 4 variables: biomass, substrate, protein and spore concentration. Table 1 presents the list of the available data by batch and strain.

Table 1. Description of the experimental data.

Strain 1 (BLB1)		Strain 2 (HD1)		Strain 3 (Lip)	
Batch	Data per variable	Batch	Data per variable	Batch	Data per variable
1	10	4	10	7	26
2	10	5	7	8	8
3	11	6	26	9	10

As the hybrid model is based on a dynamic model, additional data was generated based on these experimental measurements to train the hybrid model. In this case, 100 data were generated per each training batch. The distinction between training and validation batches is detailed in Sect. 3.1.

2.4 Dynamic Model

The dynamic model proposed by Monroy et al. [11] has been used as basis of this study. The model represents the mass balances for *B. Thuringiensis* in a batch reactor as described in Eqs. (1), (2), (3) and (4)

$$\frac{dX}{dt} = (\mu - k_d) \cdot X = r_X \tag{1}$$

$$\frac{dS}{dt} = -\frac{\mu \cdot X}{Y_{XS}} = r_S \tag{2}$$

$$\frac{dPro}{dt} = X \cdot k_{pro} = r_{pro} \tag{3}$$

$$\frac{dSpo}{dt} = X \cdot k_{spo} = r_{spo} \tag{4}$$

The variables described in these equations are the concentrations of biomass (X), substrate (S), protein (Pro) and spore (Spo), expressed in g.L^{-1}. The parameter Y_{XS} denotes the yield coefficient between biomass and substrate in gX.gS^{-1}. The biomass growth rate is represented by μ and its decay constant as k_d, both in h^{-1}. In this particular case, the Contois expression was used to calculate the biomass growth rate, as given in Eq. (5).

$$\mu = \frac{\mu_{max} \cdot S}{(K_c \cdot X) + S} \tag{5}$$

The term μ_{max} is the maximal growth rate and K_c is the Contois specific constant .

The parameter k_{pro} is the specific kinetic constant for protein (gPro.gX^{-1}.h^{-1}) and k_{spo} is the specific kinetic constant for spores (CFU$\times 10^{-5}$.gX^{-1}.h^{-1}). The mass balances in Eqs. (3) and (4) correspond to the production rates for protein (r_{pro}) and spore (r_{spo}).

2.5 Dynamic Hybrid Model

In this work, protein and spores were modelled by a data-driven approach. Therefore, two new equations were included in the model as follows,

$$\frac{dPro}{dt} = r^*_{pro} \tag{6}$$

$$\frac{dSpo}{dt} = r^*_{spo} \tag{7}$$

where r^*_{pro} and r^*_{spo} are calculated as the derivative of a sigmoidal function that describes the proteins and spore concentrations, respectively.

These output variables were modelled using a Support Vector Machine (SVM) approach. SVM is a non-parametric and non-linear technique, which has attracted great attention in recent years due to its stability, robustness, and generality, especially for the cases involving high-dimensional regression or classification analysis [12].

In this work, SVM is used to predict accurately the desired output y through Eq. (8).

$$y = w^T \varphi(x) + b \tag{8}$$

where φ(x) is the nonlinear mapping of the input x into a high dimensional feature space. The determination of this equation is based on an optimization problem, which aims to find parameters that improve the efficiency of the y estimation [13]. This is achieved by using a Kernel function, which can be linear, quadratic, cubic and Gaussian, according to the mapping of the data.

SVM has been used to design soft sensors to produce protein on the same process by *B. thuringiensis* where variables as pO_2, agitation and strain number were categorized as the most important features/variables to describe protein production [14].

In this case, a model for r^*_{pro} was defined based on five predictors: *B. thuringiensis* strain type used in the experiment (*strain*), the biomass growth rate (μ) from Eq. (5), and the consumption/production rates. This last three were calculated from the dynamic model, Eqs. (2) to (4) (r_S, r_{pro}, r_{spo}). The SVM model can be expressed as a function of the form:

$$r^*_{pro} = f(strain, \mu, r_S, r_{pro}, r_{spo}) \tag{9}$$

For the r^*_{spo} model, r^*_{pro} is included in the predictors to increase accuracy.

$$r^*_{spo} = f(strain, \mu, , r_S, r_{pro}, r_{spo}, r^*_{pro}) \tag{10}$$

The variable *strain* was defined as an integer from 1 to 3, assigned as 1: BLB1, 2: HD1 and 3: Lip respectively.

Experimental data were filtered and smoothed so that the numerical derivative of protein and spore of the experimental data could have a similar behavior to the production rates calculated with the dynamic model. Using this filtered data, the model could be trained to follow the expected evolution of the products. For instance, to avoid decreases in product concentration and any experimental disturbances. The filtered data is presented in the Results section.

3 Results

3.1 Training Tests

The model was trained using experimental data from nine batch essays described in Sect. 2.3. Six out of these datasets were used for model training (two per each strain) and the remaining three datasets (Batch No. 3, 6 and 9) were used for model validation (one for each strain of *B. thuringiensis*). The batches for training were selected based on the values obtained for r^*_{pro} and r^*_{spo} in order to have the validation sets inside the range of the training sets.

SVM modelling was performed using the Regression Learner tool from Matlab2020a. The hybrid model was integrated and solved in Matlab2020a.

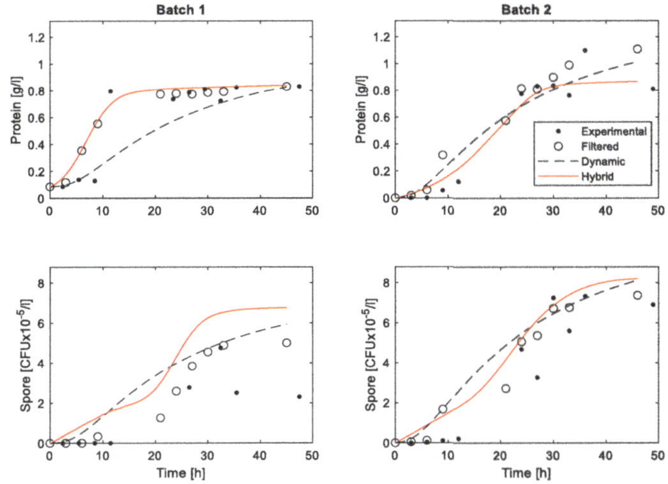

Fig. 1. Training tests for protein and spore concentration with strain 1.

The results presented in Fig. 1 depict the performance of the dynamic and hybrid model for protein and spore production in comparison to the experimental and filtered data from 3 batches using strain 1. The dynamic hybrid model presents a similar behavior than the dynamic model for protein concentration, while there is an improvement in the spore concentration, where the hybrid model fits better the experimental data.

Figure 2 shows that the dynamic hybrid model is better at describing protein production than the dynamic model, but the opposite is observed for the spore in batch 4. These variations can be explained by the wide range of values that a product concentration can take at the same conditions for the same strain and the fact that there is a trade-off between the simulation of both batches.

Figure 3 shows that both models follow the dynamics for proteins. The hybrid model fits better for spores when experimental data has higher values and starts in a low concentration point (Fig. 3 Batch 8).

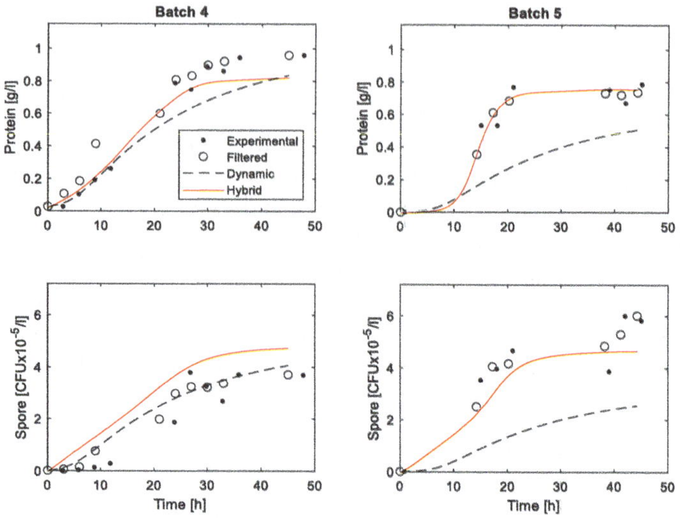

Fig. 2. Training tests for protein and spore concentration with strain 2.

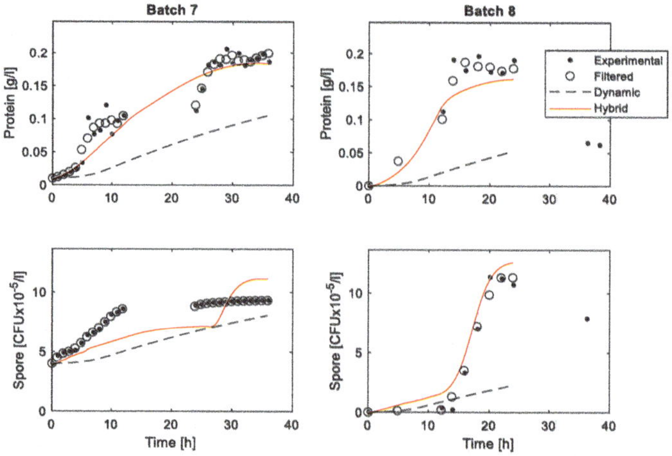

Fig. 3. Training tests for protein and spore concentration with Strain 3.

3.2 Validation Tests

The batches with higher uncertainty on protein and spore concentrations were used for model validation. For this reason, significant differences are observed between models and data in Fig. 4.

There are some improvements in the prediction of protein production for the case of strain 1, depicted in Fig. 4 Batch 3. In the case of the spores, the differences are notorious, but the hybrid model is more likely to consider drastic changes in spore concentration because it takes information from its production rate, so it has an important advantage over the dynamic model.

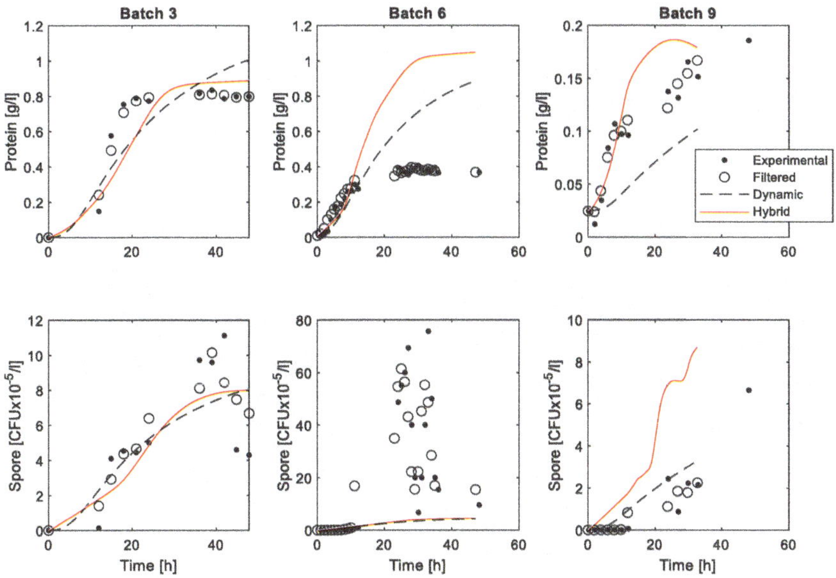

Fig. 4. Validation tests for Protein production. (Batch 3: strain 1, Batch 6: strain 2, Batch 9: strain 3).

The Normalized Root Mean Square Error (NRMSE) was calculated for each test to assess the overall performance of the models in both the training and validation stages. A distinction is made regarding the evaluation on the experimental and the filtered data. The values of NRMSE are presented in Table 2.

Table 2. NRSME comparison.

Variable	Data classification	Experimental data	Dynamic model	Hybrid model
Protein	Training	Original	0.4231	0.0691
		Filtered	0.4156	0.0540
	Validation	Original	0.4637	0.4049
		Filtered	0.4713	0.4144
Spore	Training	Original	0.4373	0.1143
		Filtered	0.4340	0.1065
	Validation	Original	0.3067	0.2879
		Filtered	0.3514	0.3262

The NRMSE shows that the dynamic hybrid model has better accuracy than the dynamic model in all tests, but more importantly, in the validation tests.

3.3 Shapley Values

To consider the influence of each predictor in the models, a Shapley values evaluation was performed over both models. This indicator represents the deviation of the Shapley values, taking the mean as a reference. Therefore, the predictors with a higher deviation will be more determinants to the model.

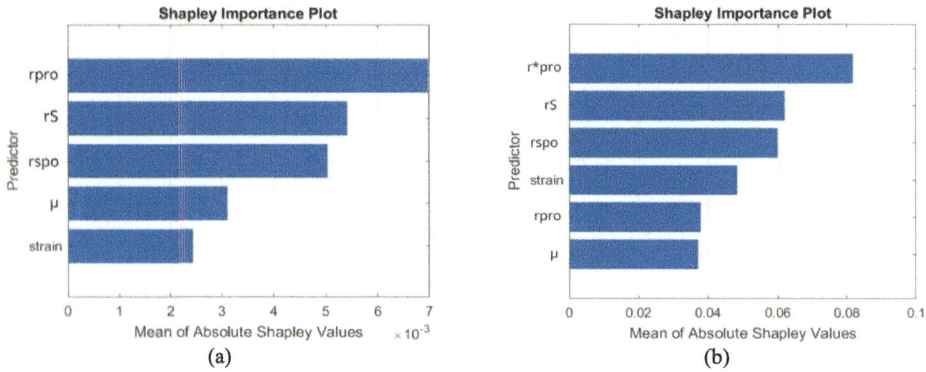

Fig. 5. Shapley importance plots for protein production rate (a) and spore production rate (b) model.

In Fig. 5.a, it is observed that the influence of the dynamic model predictors is very high, being the protein production rate the main predictor in the protein model, as expected. Nevertheless, in Fig. 5.b the output of the protein model has a significant impact over the accuracy of the spore rate model. The spore rate from the dynamic model is in third place, but its deviation score is not considerably high compared to the previous places. As the deviations from the mean value are not significant, no predictor can be neglected in the computing of both models.

4 Conclusion

A dynamic hybrid model was proposed to describe biopesticides production produced by three different strains of *B. Thuringiensis*. This model was able to fit protein and spore production. The dynamic hybrid model was compared against a dynamic kinetic model. The dynamic hybrid model has shown better performance than the kinetic dynamic model for the proteins production but for spores there is a gap in some data sets. The NRMSE values for the dynamic hybrid model present an improvement of 12% for proteins and 7% for spores over the dynamic model. The analysis of Shapley values shows all predictors are necessary for both models. This approach will be applied with different kinds of substrates as raw material in future works.

Acknowledgments. This work was supported by the 'Alternative Biopesticides For Safe Integrated Pest And Water Management Around Mediterranean' (SAFWA) project, which has received funding from the European Union's Horizon 2020 research and innovation program under the grant agreement No2022/Section 2.

References

1. Schnepf, E., et al.: Bacillus thuringiensis and its pesticidal crystal proteins. Microbiol Mol Biol Rev **62**(3), 775–806 (1998). https://doi.org/10.1128/mmbr.62.3.775-806.1998
2. Jallouli, W., Driss, F., Fillaudeau, L., Rouis, S.: Review on biopesticide production by Bacillus thuringiensis subsp. kurstaki since 1990: focus on bioprocess parameters. Process Biochem. **98**, 224–232 (2020). https://doi.org/10.1016/j.procbio.2020.07.023
3. Rowe, G.E., Margaritis, A.: Bioprocess design and economic analysis for the commercial production of environmentally friendly bioinsecticides from Bacillus thuringiensis HD-1 kurstaki. Biotechnol. Bioeng. **86**(4), 377–388 (2004). https://doi.org/10.1002/bit.20146
4. Navarro-Mtz, A.K., Pérez-Guevara, F.: Construction of a biodynamic model for cry protein production studies. AMB Express **4**(1), 79 (2014). https://doi.org/10.1186/s13568-014-0079-y
5. Kay, S., Kay, H., Rogers, A.W., Zhang, D.: Integrating hybrid modelling and transfer learning for new bioprocess predictive modelling. In: Kokossis, A.C., Georgiadis, M.C., Pistikopoulos, E., (eds.) Computer Aided Chemical Engineering, vol. 52, in 33 European Symposium on Computer Aided Process Engineering, vol. 52, pp. 2595–2600. Elsevier (2023). https://doi.org/10.1016/B978-0-443-15274-0.50412-1
6. Vega-Ramon, F., Zhu, X., Savage, T.R., Petsagkourakis, P., Jing, K., Zhang, D.: Kinetic and hybrid modeling for yeast astaxanthin production under uncertainty. Biotechnol. Bioeng. **118**(12), 4854–4866 (2021). https://doi.org/10.1002/bit.27950
7. Wang, Y., et al.: Optimization of dark fermentation for biohydrogen production using a hybrid artificial neural network (ANN) and response surface methodology (RSM) approach. Environ. Prog. Sustain. Energy **40**(1), e13485 (2021). https://doi.org/10.1002/ep.13485
8. Vapnik, V., Golowich, S., Smola, A.: Support vector method for function approximation, regression estimation and signal processing. Adv. Neural Inf. Process. Syst. MIT Press (1996). https://proceedings.neurips.cc/paper_files/paper/1996/hash/4f284803bd0966cc24fa8683a34afc6e-Abstract.html Accessed 21 May 2024
9. Li, D.-C., Liu, C.-W.: A class possibility based kernel to increase classification accuracy for small data sets using support vector machines. Expert Syst. Appl. **37**(4), 3104–3110 (2010). https://doi.org/10.1016/j.eswa.2009.09.019
10. El Khoury, M., Azzouz, H., Chavanieu, A., Abdelmalak, N., Chopineau, J., Awad, M.K.: Isolation and characterization of a new Bacillus thuringiensis strain lip harboring a new cry1Aa gene highly toxic to Ephestia kuehniella (Lepidoptera: Pyralidae) larvae. Arch. Microbiol. **196**(6), 435–444 (2014). https://doi.org/10.1007/s00203-014-0981-3
11. Monroy, T.S., et al.: Dynamic model for biomass and proteins production by three bacillus thuringiensis ssp kurstaki strains. Processes **9**(12), 2147 (2021). https://doi.org/10.3390/pr9122147
12. Hamedi, H., Mohammadzadeh, O., Rasouli, S., Zendehboudi, S.: A critical review of biomass kinetics and membrane filtration models for membrane bioreactor systems. J. Environ. Chem. Eng. **9**(6), 106406 (2021). https://doi.org/10.1016/j.jece.2021.106406
13. Wang, J., Yu, T., Jin, C.: Online estimation of biomass in fermentation process using support vector machine1. Chin. J. Chem. Eng. **14**(3), 383–388 (2006). https://doi.org/10.1016/S1004-9541(06)60087-6
14. Robles Rodriguez, C.E., et al.: Soft-sensors for monitoring B. Thuringiensis Bioproduction. In: Matsui, K., Omatu, S., Yigitcanlar, T., González, S.R. (eds.) Distributed Computing and Artificial Intelligence, Volume 1: 18th International Conference. DCAI 2021. Lecture Notes in Networks and Systems, vol. 327, pp. 129–136. Springer, Cham (2022). https://doi.org/10.1007/978-3-030-86261-9_13

An Analysis of Different Text Representation Schemes for an Immune Clustering Algorithm

Matheus A. Ferraria[1(✉)], Pedro P. Balbi[1], and Leandro N. de Castro[2,3]

[1] FCI - PPGEEC - Programa de Pós-Graduação em Engenharia Elétrica e Computação, Universidade Presbiteriana Mackenzie, São Paulo, SP, Brazil
72208635@mackenzista.com.br, pedrob@mackenzie.br

[2] Graduate Program in Technology, School of Technology, State University of Campinas (Unicamp), Limeira, São Paulo, SP, Brazil
ldecastrosilva@fgcu.edu

[3] Human-Centered Artificial Intelligence and Natural Computing Research Group, Florida Gulf Coast University, Fort Myers, FL 33965, USA

Abstract. This research investigates the challenges and effectiveness of various text representation methods (standard vector, grammar-based, and distributed), when applied to clustering short texts. The study explores Bag-of-Words for standard vector, Linguistic Inquiry and Word Count (LIWC), Part-of-Speech Tagging (POS-Tagging), and the Medical Research Council Psycholinguistic Database (MRC) for grammar-based, and Word2Vec, fastText, Doc2Vec, and SentenceBERT for distributed representations. Utilizing the aiNet bio-inspired clustering algorithm, the results reveal surprising findings, with grammar-based representations demonstrating competitive performance despite their simplicity, while standard vectors exhibit known challenges like high dimensionality. The study contributes insights into the properties of different text representations, providing a foundation for optimizing their application in clustering tasks with short and informal texts.

Keywords: Clustering · Information retrieval · Natural Computing · NLP · Text Mining

1 Introduction

The advent of social media platforms promoted a transformation in social interactions, eliminating geographical constraints and facilitating faster and easier communication among users. The widespread popularity of these platforms has resulted in an unprecedented surge in data generation, holding great potential

Supported by CNPq for the research grant PQ 303356/2022-7; CAPES for the projects STIC-AmSud (CAMA) No. 88881.694458/2022-01; Mackenzie-PrInt No. 88887.310281/2018-00; and FAPESP for grant 2021/11905-0.

R. Chinthaginjala et al. (Eds.): DCAI 2024, LNNS 1259, pp. 250–260, 2025.
https://doi.org/10.1007/978-3-031-82073-1_25

for extracting valuable insights not readily attainable from other data sources [9].

The process of structuring textual data is an extremely necessary step towards being fully able to leverage this massive data that is constantly being generated. Although necessary, this step is very flexible and it allows different strategies to be applied, each one leveraging different linguistical aspects (e.g. grammatical, semantical, statistical) that can lead to very different representations. The existence of many different textual representations makes choosing one a much more challenging task, as representations can lead to different results depending on the task and type of text they are being used.

This study aim at investigating and assessing some text representation categories, namely: standard vector (Bag-of-Words), grammar-based (Linguistic Inquiry and Word Count; Part-of-Speech Tagging; Medical Research Council Psycholinguistic Database (MRC)), and distributed (Word2Vec; fastText; Doc2Vec; SentenceBERT). The experiments are conducted using the aiNet clustering algorithm [4], inspired by immune network theory, and subsequently evaluated using a combination of performance metrics. To ensure a comprehensive analysis, two short-text social media datasets are employed for experimentation. The goal of this study is to have obtain a better understanding of how each representation may affect the quality of the obtained clusters.

While similar studies exist [2,6,14], they mainly focus on more popular representations such as n-gram and word2vec; and do no take into consideration representations that leverage grammatical or syntactical information which could provide valuable assessments and discoveries for this area.

2 The aiNet Clustering Algorithm

The *immune network theory*, introduced in the 1970 s, challenged the traditional lock-and-key model of immune recognition, proposing a dynamic network where antibodies and immune cells interact dynamically [4]. Antibodies possess receptors for self-recognition and recognizing foreign antigens, forming an internal network in equilibrium. Invading antigens disturb this network, whose receptors are referred to as *internal images* [3,4,7].

In aiNet, antigens (Ag) are the input data (objects), while antibodies (Ab) are the prototypes learned from the input data, which correspond to the immune network internal representations of the antigens. The Ag-Ab interaction is the network's main mechanism to reinforce and improve the antibodies present in the system. This interaction is based on the complementarity of shapes between Ag and Ab. For the algorithm, antigens and antibodies are represented by N-dimensional vectors and the affinity between Ag and Ab can be measured either with similarity or dissimilarity measures [5].

Figure (1) presents a pseudocode of the aiNet learning algorithm and its main steps. The algorithm works as follows. An initial set of antibodies Ab and a matrix of memory cells M are created, serving the purpose of maintaining sets of prototypes that will represent clusters of data in the available datasets.

After initialization, an iterative search process starts until a certain number of iterations has been run. Within this loop, a number of steps occur. First, a partial random population is created and added to Ab. Then, for each input object (Ag), its affinity with all prototypes (Abs) is calculated and the N_b highest affinity ones are selected and cloned proportionally to affinity (the higher the affinity, the larger the clone size) and mutated inversely proportionally to affinity (the lower the affinity, the higher the mutation rate). A number $\zeta\%$ of the highest affinity mutated clones are selected, and the redundant ones, based on a similarity threshold ϵ, are eliminated. Those whose affinity with the antigen are smaller than a threshold σ_s are eliminated. After all these steps are performed for each input object, the remaining prototypes are added to the set of M memory cells, and another suppression step removes redundancy within M.

```
initialize the antibody set Ab
initialize the memory matrix M
while stopping criterion is not met do
  initialize a random population with size m and concatenate it with Ab
  for each input object do
    calculate its affinity with every member of Ab
    select the N_b highest affinity antibodies
    clone the N_b antibodies proportional to their affinity
    mutate the clone inversely proportional to their affinity
    select the ζ% highest affinity clones
    for each clone in the selected clone set do
      if aff > σ_p then
        remove (prune) it from the set of selected clones
      end if
      if aff < σ_s then
        remove (suppression) it from the set of selected clones
      end if
    end for
    store the remaining clones in M
  end for
  calculate the affinity between all objects in M and suppress those with affinity
  smaller than σ_s
  replace Ab with M
end while
```

Fig. 1. The aiNet learning algorithm.

The main parameters necessary to run the algorithm are max_{it} (maximal number of iterations), m (number of antibodies to be generated during the random population initialization), N_b (number of antibodies to be selected for cloning), N_c (multiple of the number of clones to be generated), $\zeta\%$ (percentage of antibodies to be selected), σ_p (pruning threshold), and σ_s (suppression threshold used to control redundancy).

At the end of aiNet training, the memory matrix M generated in the last iteration will contain the prototypes found based on the learning from the input data. From this matrix, it is possible to calculate the affinity among its antibodies and find groups of prototypes representing groups of data.

3 Text Representations

This research focuses on investigating different strategies and assess how they can impact clustering algorithms. To provide a more segmented discussion and presentation, three categories are proposed based on the strategy employed by each representation: standard vector, grammar-based, and distributed representations.

3.1 Standard Vector

Standard vectors employ statistical information from the input to produce outputs containing word or sequence frequencies [6]. For this category only the N-gram model is studied.

The N-gram model is a straightforward approach in NLP, representing textual data by treating every sequence of n tokens as a distinct feature [10]. Unigrams, or Bag-of-Words (BoW) for sequences of size one, are commonly used in this model. $TF-IDF$ assesses word importance in a document based on its relative frequency in that document and the inverse of the number of documents containing the word [2]:

$$TF(t,d) = \frac{f_{t,d}}{\sum_{t' \in d} f_{t',d}} \tag{1}$$

$$IDF(t,D) = \log \frac{N}{\mid \{d \in D : t \in d\} \mid} \tag{2}$$

$$TF-IDF(t,d,D) = TF(t,d) \cdot IDF(t,D) \tag{3}$$

where $TF(t,d)$ is the relative frequency of term t within document d, and $f_{t,d}$ is the number of times the term t occurs in document d

3.2 Grammar-Based

Grammar-based representations utilize grammatical rules, lexical information, and external dictionaries to categorize words and generate predefined outputs [11].

Linguistic Inquiry and Word Count (LIWC) is a textual analysis tool that uses an external dictionary to extract meaningful information by structuring the text into a predefined set of categories provided by a dictionary. This structuring process is responsible for assigning one or multiple categories to each word, which are then used to generate the final representation.

Part-Of-Speech (POS) Tagging was originally written by [13] as a tool to assign a tag based on its grammatical function, such as noun, verb, and adjective, to each word of the text. When structuring a document via POS-Tagging, a count is made of the number of words in each tag. Thus, at the end of the structuring process, there is a matrix in which each attribute refers to a tag [13].

Medical Research Council Psycholinguistic Database (MRC) is a machine usable dictionary of psycholinguistic information containing 150,837 words, where each word is composed of 26 linguistic and psycholinguistic attributes. The attributes are obtained from publicly available sources and structured in a single dictionary [15]. The MRC representation consists of mapping words contained in the dictionary into a vector representing the 26 attributes mentioned before.

3.3 Distributed

Distributed representations, on the other hand, leverage pre-trained models or neural networks to learn language intricacies, producing representations rich in semantic information [9].

Word2Vec is a word representation model that converts words into numerical vectors. This model is much more complex than the previous representations, as it requires a neural network to be trained before actually converting each word. It was created with the goal of overcoming recurring limitations of previous representations, such as sparsity, high dimensional vectors, and the inability of extracting contextual and syntactic information [9].

FastText is a representation that learns the meaning of each word by analyzing its semantics and occurrences, similarly to Word2Vec. However, its main difference arises from the use of character-level information and morphological properties [1]. Due to its similarities many consider fastText to be an extension of Word2Vec [6].

Doc2Vec is a text representation that aims at generating numerical vectors that provide a general understanding of variable sequences of text such as sentences, paragraphs and even documents. The main difference between Doc2Vec and Word2Vec is that Doc2Vec uses a single vector to represent variable sequences of text. This single fixed size vector is called *paragraph vector* and is unique to each sequence [8].

SentenceBERT is a text representation built on top of Bidirectional Encoder Representation from Transformers (BERT). This representation was created with the goal of extending existing BERT models to derive sentence embeddings that encompass semantical, syntactical and contextual information [12].

4 Performance Assessment

The goal of this research is to provide an assessment of how each text representation studied here behaves when paired with short texts, a challenging type of text with scarce contextual information, and the aiNet clustering algorithm.

The aiNet algorithm was merely chosen here as a tool to obtain the necessary results for this study, it is important to notice that other algorithms with different strategies (e.g. density based, centroid based) might lead to different results. However such assessment would take the focus off the representations themselves.

4.1 Experimental Methodology

As the paper focus is on the text representations, to avoid an exponential number of experiments caused by multiple variations of both the aiNet and each text representation hyperparameters, a fixed set of hyperparameters was chosen for aiNet and used for all representation schemes.

The experiments were divided into two phases, a preliminary phase used to find the most suitable hyperparameters for the aiNet, and a final phase to explore the text representation hyperparameters and generate the final results. Both phases use the following steps:

1. The dataset Ds undergoes some pre-processing steps to make the data easier to manipulate. After these steps a cleansed dataset Ds_c is generated;
2. Ds_c is then used as input to one of the methods previously described, generating text representation T;
3. The aiNet algorithm is then trained by using T as the input data. The training step here refers to the process of generating the aiNet antibodies.
4. After the training step, a Minimal Spanning Tree (MST) algorithm is used to segment the network prototypes (antibodies) into different clusters.
5. Performance evaluation is calculated by applying the selected performance measures.

To assess the results obtained, two types of clustering measures were chosen: internal and external. These evaluation measures are detailed in a future section.

4.2 Datasets

Given that there are different types of text (e.g. News, Reviews, Abstracts), each one with its own intricacies and characteristics that can exert influence on the final results, we decided to use a specific type of text across all experiments to ensure consistency among the results.

For this research only short texts are used, a type commonly found in social media and brief messages. It often possesses length restrictions that lead to a more informal nature containing *slangs*, abbreviations and *emojis*. This informal nature combined with its length restriction provides a challenging scenario for text representation as information and context are more scarce.

Based on these characteristics two datasets were chosen, each dataset comes from a different platform. Despite their different sources they wonderfully represent the restrictions and challenges of working with this type of text.

1. **CIC Truth Seeker (2023)**: Dataset composed of 134,198 *tweets* extracted to provide insights regarding fake news propagation and social media interactions. This data is divided into three classes, however only two of them are referenced in the dataset (*Agree* and *Disagree*). To reduce the computation required by the experiments a stratified sample of 6,705 objects is going to be used. This dataset is available at https://www.kaggle.com/datasets/sudishbasnet/truthseekertwitterdataset2023.
2. **SMS Spam Collection**: A collection of 5,574 text messages gathered with the goal of improving spam detection. From the original dataset, 415 objects needed to be removed for containing empty texts and duplicated information, leaving a total of 5,159 objects to be used in the experiments. This dataset is divided into two classes: ham and spam. The original dataset is available at https://archive.ics.uci.edu/dataset/228/sms+spam+collection.

4.3 Pre-processing Steps

The following pre-processing techniques are applied: lowercase and apostrophe normalization; accent, special character and additional space removal; quotation, number (Bag-of-Words only), punctuation (all except grammar-based) and stopword (all except grammar-based) removal; contraction; separation and lemmatization (all except grammar-based). During the accent removal step the text is converted into UTF-8, due to this conversion unicode emojis are removed. However simpler versions (e.g. :O, ;D) are kept and the way they are used varies between the studied representations.

4.4 aiNet's Hyperparameters Exploration

After a set of preliminary experiments the following aiNet hyperparameters were found to better fit the requirements: $Nmax_{it} = 25$; $m = 500$; $N_b = 15$; $N_c = 10$; $\zeta\% = 10$; $\sigma_p = 0.5$; $\sigma_s = 0.10$.

4.5 Evaluation Measures

To provide a comprehensive assessment of aiNet-generated clusters, this paper employs a combination of internal and external measures. To maintain consistency across calculations, the cosine measure, which aligns with the aiNet algorithm, is chosen for internal measures dependent on similarity and dissimilarity:

1. **Dunn Index (DU)**: Dunn Index combines inter- and intra-cluster distances to provide a cluster quality measure. It ranges from $[0, \infty]$ with higher values indicating more cohesive and separated clusters.

2. **Davies-Boudin Index (DB)**: The Davies-Boudin Index uses inter- and intra-cluster distances from a centroid perspective to determine cluster quality. It varies in $[0, \infty]$, with lower values indicating better quality.
3. **Silhouette Score**: The Silhouette Score evaluates the cohesiveness of the cluster. It ranges in $[-1, 1]$, where higher values indicate higher quality clusters.
4. **Clustering Accuracy** (Cl_{ACC}): Clustering accuracy measures similarity between obtained and original clusters, ranging in [0,1].
5. **Normalized Mutual Information (NMI)**: NMI quantifies agreement between clusterings, ranging between [0,1].

5 Results and Discussions

Table 1 summarizes the results of the evaluation process, presenting a detailed overview of eight text representations in the context of both the CIC Truth Seeker Dataset and the SMS Spam Collection. This analysis takes into account the five selected measures presented, with superior results highlighted for each measure. Despite the intentional selection of short-text datasets, some disparities in characteristics become apparent.

Table 1. Results obtained for both CIC Truth Seeker (2023) and SMS Spam Collection dataset.

		Internal			External	
Datasets	**Representations**	**DI**	**DB**	**Sil**	**NMI$_{sqrt}$**	**Cl$_{ACC}$**
CIC	Bag-of-Words	-	-	-	-	-
	LIWC	0.546	1.170	0.290	0.001	95.41%
	POS-Tagging	0.489	1.767	0.407	0.001	95.41%
	MRC	**1.000**	1.051	**0.901**	0.001	95.41%
	Word2Vec	0.299	1.619	0.355	0.002	95.41%
	fastText	0.309	**0.613**	0.553	0.002	95.41%
	Doc2Vec	0.358	1.385	0.446	0.001	95.41%
	SentenceBERT	0.496	0.958	-0.038	**0.020**	**95.42%**
SMS	Bag-of-Words	**0.995**	3.010	-0.001	0.002	87.56%
	LIWC	0.660	0.550	0.255	0.001	87.56%
	POS-Tagging	0.747	0.471	0.419	0.023	87.67%
	MRC	0.722	0.595	0.793	0.008	87.56%
	Word2Vec	0.411	0.172	0.776	**0.532**	**94.20%**
	fastText	0.407	1.340	0.549	0.020	87.65%
	Doc2Vec	0.875	**0.016**	**0.969**	0.002	87.56%
	SentenceBERT	0.707	3.254	0.004	0.044	88.12%

The obtained results reinforce the intricacies of clustering evaluation measures, revealing a lack of consensus regarding the most effective representation across different metrics.

External measures, in particular, exhibit inconsistency, suggesting varying degrees of divergence from the ground truth labels. The Cl_{ACC} results, closely mirroring the original dataset class distribution, imply a significant influence of the mapping process on shaping the final outcomes.

The results of the CIC Truth Seeker (2023) dataset reveals important observations. Bag-of-Words highlights inherent challenges related to dimensionality, resulting in a single cluster. Surprisingly, the MRC, despite relying on a dictionary from 1999 that lacks consideration for contemporary slangs and abbreviations, outperforms newer alternatives like LIWC and POS-Tagging. For the distributed representations, fastText's morphological capabilities proved advantageous, outpacing Word2Vec. Equally surprising is the resilience of Doc2Vec, originally designed for larger texts, which surpassed Word2Vec in many metrics. However, SentenceBERT, despite its advanced linguistic capabilities, exhibits suboptimal cluster cohesion, as evidenced by its silhouette score.

For the SMS Spam Collection dataset, the overall higher results for both DB and Silhoutte hint at its increased complexity in representation. While Bag-of-Words achieves some success, its limitations become apparent in elevated DB and negative Silhouette values. Grammar-based representations, especially the MRC, consistently yield superior clusters, with other representations demonstrating competitive performance based on internal measures. In the domain of distributed representations, this dataset proved more advantageous with word2Vec outperforming fastText, Doc2Vec maintaining its strong performance, while SentenceBERT continues to lag behind other representations, unable to effectively leverage its sophisticated language knowledge.

6 Conclusions and Future Trends

This research proposed an investigation into the effectiveness of different text representations for clustering tasks, employing two distinct short-text datasets: the CIC Truth Seeker Dataset; and the SMS Spam Collection. The study evaluated eight text representations, including Bag-of-Words, grammar-based representations (MRC, LIWC, POS-Tagging), and distributed representations (Word2Vec, fastText, Doc2Vec, SentenceBERT). The evaluation involved five selected measures, and the results showed differences in performance across representations.

In general, it is possible to notice that despite sharing some similarities the datasets chosen for this study produce very different results, the results show that tweets are fairly more complicated to deal across most representations, specially those that do not use neural networks or language models. Such complications are related to its more informal natural, scarse context and higher usage of *slangs*.

The results obtained further reinforce the idea that clustering evaluation measures can provide somewhat conflicting results. Based on our results it is not possible to determine which representation would be more adequate for this type of text. The conflicting results obtained from both datasets highlight the need for more experiments with a larger number of datasets to confidently draw a conclusion regarding which method obtains the best clusters.

The experiments performed brought insights regarding each text representation studied here. However, it also highlighted a couple of points that could be improved to extend this study. When it comes to clustering algorithms, many strategies can be adopted such hierarchical clustering and others based on density, like DBSCAN. Another point highlighted by our results is the impact that a dataset can have on the end results. For future researches it would be interesting to include a wider variety of short text datasets to allow for a more grounded analysis. Finally, although this research focused on short texts, the use of other types of texts (e.g. news, reviews) would be interesting, as it could allow for a greater knowledge regarding the text representations influence on clustering.

References

1. Bojanowski, P., Grave, E., Joulin, A., Mikolov, T.: Enriching word vectors with subword information. Trans. Assoc. Comput. Linguist. **5**, 135–146 (2017)
2. Cahyani, D.E., Patasik, I.: Performance comparison of TF-IDF and word2vec models for emotion text classification. Bull. Electr. Eng. Inform. **10**(5), 2780–2788 (2021)
3. Dasgupta, D., Yu, S., Nino, F.: Recent advances in artificial immune systems: models and applications. Appl. Soft Comput. **11**(2), 1574–1587 (2011)
4. de Castro, L.N., Timmis, J.: Artificial immune systems: a new computational intelligence approach. Springer Science & Business Media (2002)
5. de Castro, L.N., Von Zuben, F.J.: aiNet: an artificial immune network for data analysis. In: Data Mining: a Heuristic Approach, pp. 231–260. IGI Global (2002)
6. Hang, S.: Clustering short texts: categorizing initial utterances from customer service dialogue agents (2021)
7. Jerne, N.K.: Towards a network theory of the immune system. Ann. Immunol. **125C**(1–2), 373–389 (1974)
8. Le, Q., Mikolov, T.: Distributed representations of sentences and documents. In: International Conference on Machine Learning, pp. 1188–1196. PMLR (2014)
9. Naseem, U., Razzak, I., Khan, S.K., Prasad, M.: A comprehensive survey on word representation models: from classical to state-of-the-art word representation language models. Trans. Asian Low-Resource Lang. Inf. Process. **20**(5), 1–35 (2021)
10. Pibiri, G.E., Venturini, R.: Handling massive n-gram datasets efficiently. ACM Trans. Inf. Syst. (TOIS) **37**(2), 1–41 (2019)
11. Puigcerver, J., Toselli, A.H., Vidal, E.: Querying out-of-vocabulary words in lexicon-based keyword spotting. Neural Comput. Appl. **28**, 2373–2382 (2017)
12. Reimers, N., Gurevych, I.: Sentence-BERT: sentence embeddings using Siamese BERT-networks. arXiv preprint arXiv:1908.10084 (2019)
13. Toutanova, K., Klein, D., Manning, C.D., Singer, Y.: Feature-rich part-of-speech tagging with a cyclic dependency network. In: Proceedings of the 2003 Human Language Technology Conference of the North American Chapter of the Association for Computational Linguistics, pp. 252–259 (2003)

14. Wang, P., Xu, B., Xu, J., Tian, G., Liu, C.L., Hao, H.: Semantic expansion using word embedding clustering and convolutional neural network for improving short text classification. Neurocomputing **174**, 806–814 (2016)
15. Wilson, M.: MRC psycholinguistic database: machine-usable dictionary, version 2.00. Behavior Res. Methods Instrum. Comput. **20**(1), 6–10 (1988)

AI-Powered Crop Monitoring for Precision Agriculture

Saifeldin Hassan, Zainab Khan, Abhilasha Singh, Jinane Mounsef(✉), and Omar Abdul Latif

Department of Electrical Engineering and Computing Sciences, Rochester Institute of Technology, Dubai, UAE
jmbcad@rit.edu

Abstract. This study explores the integration of machine learning (ML) technologies in precision agriculture, with a specific focus on the application of drones equipped with advanced sensors. The core of this research lies in the deployment of Artificial Intelligence (AI) algorithms to facilitate real-time identification and classification of key agricultural factors, particularly crop health. In this work, YOLOv8 is trained on annotated datasets comprising both RGB and Normalized Difference Vegetation Index (NDVI) images and its performance was compared with Support Vector Machine (SVM) algorithm. The study reveals that YOLOv8 exhibits superior performance compared to SVM for both types of images, achieving an average accuracy rate of 86.5%. This highlights the pivotal role of integrating drone technology and sensory vision in revolutionizing precision agriculture, offering valuable insights into the future of smart cities and sustainable farming, especially in areas with limited water resources.

Keywords: Artificial Intelligence · Multispectral sensors · Precision agriculture · Crop health monitoring · Irrigation optimization · Drone technology · Smart cities

1 Introduction

Agriculture is a vital field for the maintenance of food stability and security, but it is also one of the largest freshwater-consuming human activities, accounting for 70% of freshwater resources worldwide [3]. Technological advancements in the past 20 years have made human lives easier and have increased the efficiency of resources. This has led to unprecedented population growth, creating a need to maximise food production while reducing resource consumption, as the world population continues to grow and is projected to reach 9.7 billion by 2050 [2]. Aridity and soil erosion pose significant threats to global arable land, compromising the agricultural process and challenging conventional methods used to boost crop yields. These issues impact around 40% and 20% of the world's arable land, respectively [7]. Precision agriculture aims to optimize resource efficiency through the use of data-driven technologies, including drones, sensors,

R. Chinthaginjala et al. (Eds.): DCAI 2024, LNNS 1259, pp. 261–270, 2025.
https://doi.org/10.1007/978-3-031-82073-1_26

and cloud computing to maximize productivity while minimizing resource waste and reducing costs [5].

Recent years have seen significant research and developments in the use of drones and deep learning in the agricultural sector. Zhang *et al.* examined the use of unmanned aerial vehicle low-altitude remote sensing (UAV-LARS) in agricultural monitoring in China. They discussed the advancements, applications, and impact of UAV technology in enhancing agricultural practices and monitoring within the Chinese context [11]. Sharma *et al.* provided a systematic review of ML applications in agriculture, focusing on the prediction of soil parameters like organic carbon and moisture content. They also addressed crop yield prediction, as well as the detection of diseases and weeds in crops, and species identification [9]. Reda *et al.* explored the use of ML algorithms for estimating soil organic carbon and total nitrogen from soil samples collected from four agricultural sites in Morocco. The authors showed that ensemble AI methods produced better results compared to Back Propagation Neural Networks [8]. Zhu *et al.* successfully used a hybrid model combining particle swarm optimization (PSO) and Extreme Learning Machine (ELM) to predict daily evapotranspiration in Northwest China, aiming to optimize water requirements for crops in agricultural practices [12]. Kamir *et al.* used ML models like Random Forest (RF), XGBoost, Cubist, Multilayer Perceptron (MLP), Support Vector Regression (SVR), Gaussian Process, k-NN, and Multivariate Adaptive Regression Splines (MARS) to identify yield gap hotspots in wheat production. The authors evaluated the performance of NDVI time-series data across Australia using the MOD13Q1 dataset and found that the SVR model had superior performance [4].

This study focuses on optimizing water use in agriculture by leveraging drones equipped with RGB cameras as well as multispectral sensors to collect data from crop fields. This data is further processed by ML algorithms to distinguish between healthy and unhealthy crops, enabling precise irrigation practices.

2 Objectives

This research aims to integrate advanced AI models into drone technology, enhancing real-time detection capabilities of stressed crops within agricultural landscapes. The core of this study involves the application of YOLOv8 and SVM models, trained on RGB and multispectral imagery of potato plants [10], to accurately assess crop health. This approach seeks to refine the precision of these models and extend their applicability to various agricultural conditions. A significant objective is to design a system that is both scalable and adaptable, accommodating diverse field sizes and crop types, thus ensuring its efficacy in a broad spectrum of agricultural settings.

3 Methodology

This section provides a detailed explanation of the methodology, including the proposed system, the dataset used, the hardware setup, and the ML models used for crop monitoring. The architecture of the proposed system is divided into four main layers: sensors, network/communication, application, and client layers, as illustrated in Fig. 1.

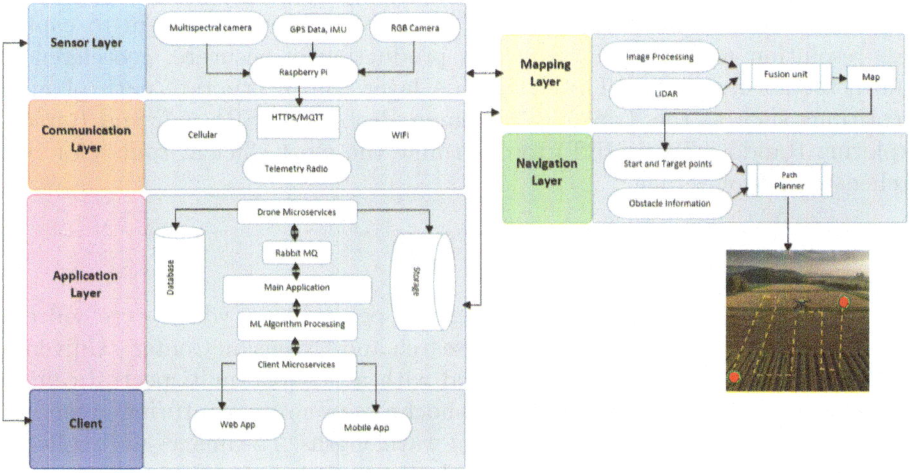

Fig. 1. A detailed illustration of the proposed drone-based crop monitoring system architecture, highlighting the various layers involved, from the sensors and drone for data collection to the client for crop health assessment.

3.1 Proposed System

The proposed crop monitoring system includes four main layers. The initial layer is the sensor layer, tasked with gathering data on plant moisture levels. This layer employs RGB and multispectral cameras to capture data across the visible spectrum related to water stress measurements. The subsequent layer is the network/communication layer. This layer is responsible for accurately labeling and transmitting the collected data to either a ground station or other predetermined destinations. This ensures that the data can be thoroughly analyzed and subsequently applied to inform decisions related to crop management, as referenced in [1]. The data processing primarily takes place in the application layer. Within this layer, one of the focuses is on the NDVI index. Any variation in this metric serves as an indicator of potential issues in the plant's photosynthetic processes, often manifested as fluctuations in the red wavelength of light [6]. The application layer, which is the primary focus of this paper, processes the video feed received from the communication layer, performing pre-processing tasks such as resizing and modifying brightness and contrast. Subsequently, these images are analyzed to determine crop health, employing AI algorithms such as YOLOv8 and SVM. The final layer is the client layer designed as the system's front-end, accessible to end-users via a web application that supports cross-device compatibility and has the potential for mobile application adaptation. Due to restrictions on drone usage in outdoor environments, this paper primarily focuses on the application layer for monitoring crop health.

A future development of the proposed system will involve the integration of two additional layers: the mapping and navigation layers. Within the mapping layer, drones will use data from the client and sensor layers to capture high-resolution images, facilitating the production of accurate, georeferenced, and distortion-free maps. The navigation layer will incorporate path planning algorithms such as the Traveling Salesman Problem (TSP), A*, and Rapidly-exploring Random Tree (RRT) to determine the most efficient routes for comprehensive area coverage.

3.2 Dataset Used

The dataset in [10] contains aerial images of potato crop, which were collected from a field located at the Aberdeen Research and Extension Center - University of Idaho, using a Parrot drone equipped with a Sequoia multispectral camera. This camera features RGB and four monochrome sensors, capturing green, red, red-edge, and near-infrared bands of light wavelengths. The images were collected by flying the drone over the potato field at a low altitude of 3 m. The dataset includes RGB images sized at 750 $\times$ 750 pixels and spectral images sized at 416 $\times$ 416 pixels. The image patches were extracted from the aerial high-resolution images by using cropping, rotating, and resizing operations. Each image is associated with ground-truth annotations, related to the regions with healthy and stressed plants, outlined in the images with rectangular bounding boxes. We augmented the dataset from 360 to 823 RGB images by applying rotations of -5 and $+5°$. We then subdivided the new dataset into three subsets: 716 images for training, 72 for validation, and 35 for testing. Additionally, NDVI images were generated using the formula:

$$NDVI = (NIR - Red)/(NIR + Red) \tag{1}$$

where "NIR" is near-infrared and "Red" denotes red spectral images. These NDVI images are vital for assessing vegetation health and are a common measure in remote sensing research.

3.3 Hardware Setup

The PX4 V2 500 drone is built with carefully selected components to create a complete and reliable UAV system. Its structure, the Holybro X500V2 Frame Kit, is crafted from strong carbon fiber, providing durability and agility for aerial operations. For propulsion, it uses Holybro 2216 KV880 Motors and Holybro BLHeli ESC 20A controllers, ensuring stable flight dynamics. Combined with 1045 Propellers, this setup delivers effective thrust and responsive mobility. The Pixhawk5x flight controller, along with the M8N GPS Module, enables precise navigation and real-time data processing. Additionally, it features a dedicated camera attachment for enhanced imaging capabilities. Power distribution and management are efficiently handled by specialist boards and modules. The detailed hardware and datasheet of components used is described in Table 1 and Fig. 2, respectively.

Table 1. Description of the hardware components used to build the UAV.

Components	Model	Estimated Weight (g)
Multispectral Camera	Micasense RedEdge-M	150–200
Motors (each x 4)	Holybro 2216 KV880	200–280
LiPo 5200mAh Battery	DUPU 5200MAH 4S 25C	400–500
Raspberry Pi	Model 3B+	45
Frame Kit	X500V2	610
Propellers	-	10–15
Power Module with Voltage Regulator	LM7805	10–20
GPS Module	M8N	10–20
Telemetry Radio	FrSky X8r	5–10
ESCs (each x 4)	-	20–60

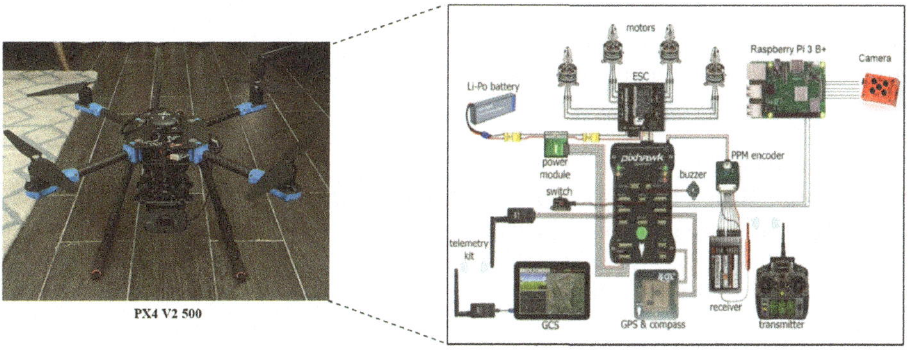

Fig. 2. Component breakdown of the PX4 V2 500, illustrating the connectivity of sensors to the Pixhawk controller.

3.4 ML Models

This section focuses on the YOLOv8 and SVM models for detecting and classifying stress condition of potato crops due to premature plant senescence.

YOLOv8 model is trained on a dataset of RGB and NDVI images [10], which include both healthy and stressed plant conditions. YOLOv8 leverages its convolutional neural network (CNN) architecture, integrating features like Feature Pyramid Networks (FPN) and a bottom-up pathway for enhanced feature extraction at multiple scales. Hence, YOLO can perform real-time object detection in a single pass through the network, making it suitable for applications with strict latency requirements. The training process iteratively adjusts the weights to minimize loss, thereby enhancing localization accuracy and reducing the risk of false positives and negatives. By correlating specific spectral signatures and morphological characteristics with plant health conditions, YOLOv8 learns to identify and classify the health status of crops accurately, thus facilitating the early detection of stress symptoms in plants, as shown in Fig. 3.

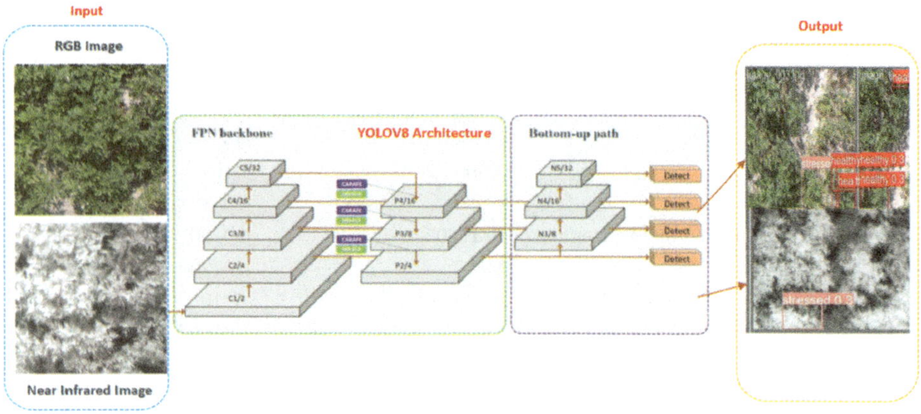

Fig. 3. YOLOv8 architecture for crop health detection, with RGB and spectral images processed through the FPN backbone and bottom-up path during training.

Alternatively, SVM serves as a powerful tool in crop health monitoring by classifying crops based on spectral features, providing valuable insights for improved crop management. In this context, we used SVM with a linear kernel to distinguish healthy from stressed crops using features collected from spectral reflectance and RBG values. During the training process, these features are used to establish a hyperplane or decision boundary that optimally differentiates the data classes, as shown in Fig. 4. During the training phase, the hinge loss function, serving as the cost function, is systematically reduced to identify the model's optimal parameters that best describe the characteristics of the dataset.

4 Results and Discussion

YOLOv8 and SVM were trained on a NVIDIA GeForce RTX 3070 Laptop GPU running Windows 10, equipped with an AMD64 processor and Python 3.11.

YOLOv8 training involved RGB and NDVI images from the augmented dataset over 100 epochs, with a learning rate set to 0.01, a momentum coefficient of 0.937, and using the Adam optimizer for the training process.

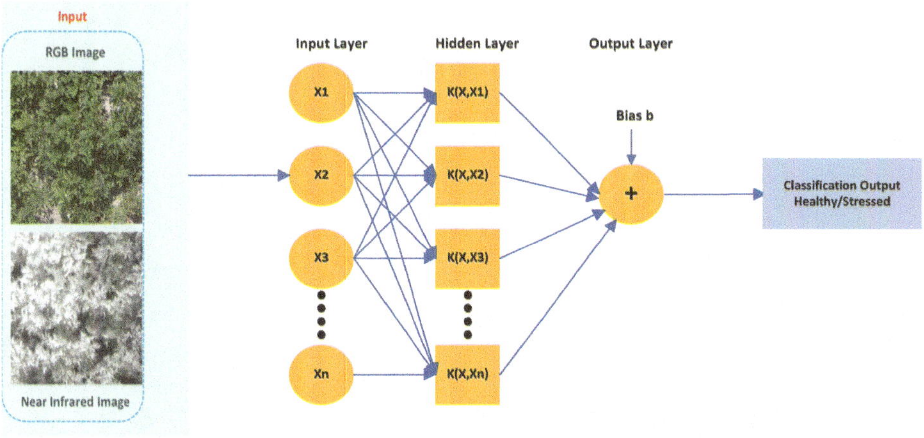

Fig. 4. SVM architecture for crop health detection.

The confusion matrices for RGB and NDVI images are presented in Fig. 5(a) and (b). The YOLO model's performance on the RGB dataset showed a precision of 0.89, a recall of 0.93, and an F1 score of 0.91. Further for the NDVI dataset, the precision was 0.80, recall was 0.82, and the F1 score was 0.80. Precision-recall curves for both RGB and NDVI images are depicted in Fig. 6(a) and (b).

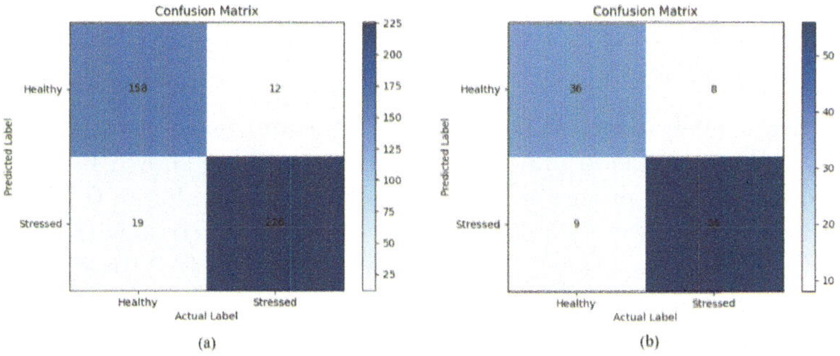

Fig. 5. Confusion matrices of YOLOv8. (a) RGB images, (b) NDVI images.

Balancing precision and recall is vital because high precision ensures that most of the predicted stressed health conditions are correct, whereas high recall means that a large percentage of actual stressed conditions are identified. YOLOv8's detection results, as shown in Fig. 7(a) and (b), reveal that the model effectively identified stressed and healthy crops in RGB images.

Further these images were fed to SVM and its performance on classifying crops using RGB and NDVI images revealed significant misclassification, often

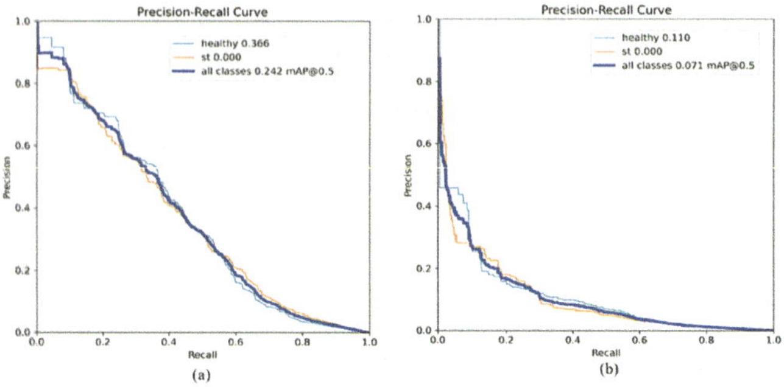

Fig. 6. PR curve of YOLOv8. (a) RGB images, (b) NDVI images.

Fig. 7. YOLOv8 detection results for crop health classification. (a) RGB images, (b) NDVI images.

misidentifying NDVI images as stressed and RGB images as healthy, as shown in Fig. 8(a)-(b). The confusion matrix for SVM classifier is shown in Fig. 9(a)-(b). For RGB datasets, the precision was 0.38 with a recall value of 0.39, and an F1 score of 0.38. Similarly for NDVI images, the precision was 0.36 with a recall value of 0.27, and an F1 score of 0.31. The low performance of SVM is technically due to linear decision boundaries, which are less effective for the complex, non-linear patterns present in the images but YOLOv8 excels in handling such complexities.

The performance evaluation of both algorithms, based on accuracy and inference time, is summarized in Table 2. It can be seen that YOLOv8 outperformed SVM, achieving higher validation average accuracies of 86.5% for RGB (including stressed and healthy) and 68.5% for NDVI images (including stressed and healthy), with a quicker average inference time of 0.09 s for 35 test RGB and NDVI images. Conversely, SVM showed lower average accuracies of 52.12% for RGB (including stressed and healthy) and 35.49% for NDVI images (including stressed and healthy), with a significantly longer inference time of 420.237 s.

Table 2. Performance comparison for different approaches.

Method	Images	Performance Metrics		
		Accuracy Healthy (in %)	Accuracy Stressed (in %)	Inference Time (s)
YOLOv8	RGB	87	86	0.078
	NDVI	76	61	0.102
SVM	RGB	38.04	66.2	346.780
	NDVI	35.85	35.13	493.695

Fig. 8. SVM results for crop health classification. (a) RGB images, (b) NDVI images.

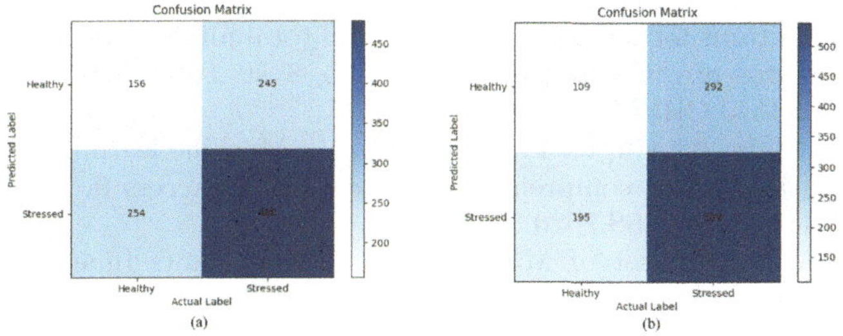

Fig. 9. Confusion matrices of SVM classifier. (a) RGB images, (b) NDVI images.

5 Conclusion

This work successfully transformed RGB and multispectral data into accessible metrics to optimize farming operations and improve crop yields. We demonstrated YOLOv8's superior performance in crop health classification over SVM in terms of both accuracy and inference time. Future work will focus on testing with real-time captured datasets and designing a drone navigation and mapping layer. Additionally, a mobile app will be developed to provide timely notifications of crop health to farmers, aiming to make precision agriculture more effective and sustainable.

References

1. Damigos, G., Lindgren, T., Sandberg, S., Nikolakopoulos, G.: Performance of Sensor Data Process Offloading on 5G-Enabled UAVs. 23 (2023). https://doi.org/10.3390/s23020864
2. Fukase, E., Martin, W.: Economic growth, convergence, and world food demand and supply. World Dev. **132**, 104954 (2017). https://doi.org/10.1016/J.WORLDDEV.2020.104954
3. Ghermani, N., Yin, L., Tao, F., Chen, Y., Wang, Y.: Reducing agriculture irrigation water consumption through reshaping cropping systems across China. Agric. Forest Meteorol. **312**, 108707 (2022). https://doi.org/10.1016/J.AGRFORMET.2021.108707
4. Kamir, E., Kamir, E., Waldner, F., Hochman, Z.: Estimating wheat yields in Australia using climate records, satellite image time series and machine learning methods. ISPRS J. Photogram. Remote Sens. **160**, 124–135 (2020). https://doi.org/10.1016/J.ISPRSJPRS.2019.11.008
5. Monteiro, A., Santos, S., Goncalves, P.: Precision agriculture for crop and livestock farming-brief review. Animals **11**(8), 2345 (2021). https://doi.org/10.3390/ANI11082345
6. Neupane, K., Baysal-Gurel, F.: Automatic identification and monitoring of plant diseases using unmanned aerial vehicles: a review. Remote Sens. **13**(19) (2021). https://doi.org/10.3390/RS13193841
7. Pravalie, R., et al.: Arable lands under the pressure of multiple land degradation processes. A global perspective. Environ. Res. **194**, 110697 (2021). https://doi.org/10.1016/J.ENVRES.2020.110697
8. Reda, R., et al.: A comparative study between a new method and other machine learning algorithms for soil organic carbon and total nitrogen prediction using near infrared spectroscopy. Chemometr. Intell. Lab. Syst. **195**, 103873 (2019). https://doi.org/10.1016/J.CHEMOLAB.2019.103873
9. Sharma, A., Jain, A., Gupta, P., Chowdary, V.: Machine learning applications for precision agriculture: a comprehensive review. IEEE Access **9**, 1 (2021). https://doi.org/10.1109/ACCESS.2020.3048415
10. Vakanski, A.: A Dataset of Multispectral Potato Plants Images. https://www.webpages.uidaho.edu/vakanski/Multispectral_Images_Dataset.html. Accessed 17 May 2024
11. Zhang, H., Wang, L., Tian, T., Yin, J.: A review of unmanned aerial vehicle low-altitude remote sensing (UAV-LARS) use in agricultural monitoring in China. Remote Sens. **13**(6), 1221 (2021). https://doi.org/10.3390/RS13061221
12. Zhu, B., Yu, F., Daozhi, G., Jiang, S., Lu, Z., Ningbo, C.: Hybrid particle swarm optimization with extreme learning machine for daily reference evapotranspiration prediction from limited climatic data. Comput. Electron. Agric. **173**, 105430 (2020). https://doi.org/10.1016/J.COMPAG.2020.105430

A Soft Sensing Approach for Efficient Monitoring of Nanobody-Based Scorpion Antivenom Production

Juan C. Acosta-Pavas[1], David Camilo Corrales[2], Irene Martínez-Menéndez[1], Susana María Alonso Villela[1], Balkiss Bouhaouala-Zahar[3,4], Georgios K. Georgakilas[5], Konstantinos Mexis[6], Stefanos Xenios[6], Theodore Dalamagas[5], Antonis Kokosis[6], Michael O'donohue[1], Luc Fillaudeau[1], and César A. Aceves-Lara[1](✉)

[1] TBI, Université de Toulouse, CNRS, INRAE, INSA, Toulouse, France
{acostapa,aceves}@insa-toulouse.fr
[2] INRAE, UMS (1337) TWB, 135 Avenue de Rangueil, 31077 Toulouse, France
[3] Laboratoire des Biomolécules, Venins et Applications Théranostiques (LBVAT), Institut Pasteur de Tunis, 13 Place Pasteur, BP-74, 1002 le Belvédère, Tunis, Tunisia
[4] Faculté de Médecine de Tunis, Université Tunis El Manar, 15 rue Djebel Lakhdhar, 10007 Bab Saadoun Tunis, Tunisia
[5] Athena Research Center, Marousi, Greece
[6] School of Chemical Engineering, National Technical University of Athens, Iroon Polytechneiou 9, 15780 Zografou, Athens, Greece

Abstract. The growing demand for recombinant-based products has increased rapidly in recent years. One of the current market scopes is the production of recombinant scorpion antivenoms in *E. coli*. However, there are limitations in real-time quantification of Critical Quality Attributes (CQAs), such as recombinant proteins or biomass concentrations. Soft sensors merge as an alternative to solve this problem. In this work, six data-driven soft sensors were developed for online monitoring of the recombinant protein concentration. First, feature selection methods were tested to determine the relevant online variables to train the soft sensors. Then, the protein experimental dataset was augmented using a hybrid model. The K-Nearest Neighbors (KNN), and the Radial Basis Function (RBF) Support Vector Machine (SVM) obtained the best results with R^2 values higher than 0.996, *RMSE* and *MAE* values less than 0.058 and 0.051. However, the KNN soft sensor was computed in less time than the RBF SVM soft sensor. These results show the feasibility of KNN and RBF soft sensors for real-time monitoring of recombinant scorpion antivenoms.

Keywords: Process Monitoring · *E. Coli* · Antivenom Production · Machine learning · Data-Driven Soft Sensor · Feature Selection

R. Chinthaginjala et al. (Eds.): DCAI 2024, LNNS 1259, pp. 271–281, 2025.
https://doi.org/10.1007/978-3-031-82073-1_27

1 Introduction

The growing demand for recombinant-based products has increased rapidly in recent years [1]. One of the current market scopes is the production of recombinant scorpion antivenoms in *E. coli*, given the health problems related to scorpion stings in tropical and sub-tropical countries [2]. Usually, serotherapy uses fragments of equine antibodies (F(ab)'2) to target the neurotoxins in scorpion venom. Through protein engineering, toxin-specific nanobodies (VHH), which are fragments of camelid antibodies, have been found to have better tissue penetration and lower molecular weight (15 kDa vs 100 kDa) than their equine counterparts [3]. *E. coli* is commonly used as a host in the pharmaceutical industry to enhance recombinant protein production [4, 5].

However, this bioprocess's measurement, monitoring, control, and optimization represent a challenge due to the complexity, cost, or limitation in physical sensors to real-time quantification of Critical Quality Attributes (CQAs) such as recombinant proteins. Soft Sensors have emerged as an alternative solution to this problem [6].

Soft Sensors offer a fast and low-cost approach to monitoring and controlling bioprocesses by integrating real-time screening and control of critical process variables [7]. Soft Sensors are models that estimate a hard-to-measure property using relatively easy measurements [8].

A few studies developed in the context of soft sensors applied in protein recombinant production. Gundinger et al. [9] implement a soft sensor for controlling the feeding strategy during the fed-batch cultivations for the production of a recombinant model antigen-binding fragment (Fab) in the periplasm of *E. coli* via phosphate (PO4)-sensitive phoA expression system (pAT). The soft sensor was trained using four features: CO_2 evolution rate, O_2 uptake rate (measured by Off-gas analysis), the substrate feed rate, and the reactor broth mass. The prediction of the soft sensor was the biomass concentration.

Vaitkus et al. [10] studied the use of soft sensors in protein recombinant production using a cell strain of *E. coli*. Three types of soft sensors were analyzed: Support vector, relevance vector, and random forest. The authors proposed two types of datasets; the first dataset was generated by a first-principle mathematical model developed for the recombinant protein production process. The second dataset was collected during real cultivation of the cell strain of *E. coli* BL21 (DE3) pET21-IFN-alfa-5 in a 7L bioreactor. The features of the soft sensors were the glucose and lactose feed rates, oxygen uptake rate, time after induction, carbon dioxide production rate (CPR), and feeding flow rate. The outputs were optical density (OD), acetate quantity, protein, and wet biomass. The random forest model obtained accurate and appropriate results based on the Root Mean Squared Error.

Recently, Medl et al. [1] developed an artificial neural-network-based soft sensor (ANN-based OD soft sensor) for the real-time estimation of the OD in a microbioreactor system in the production of the single-chain variable fragment (scFv) from *E. Coli*. The ANN-based OD soft sensor was trained using the volume-specific cumulative ammonia and volume-specific carbon feed additions as features. The soft sensor accurately predicted the OD ($>95\%$).

This paper compares six different data-driven soft sensors to efficiently (in terms of computation time) and effectively (in terms of accuracy) monitor the production of nanobody-based scorpion antivenom. The contributions of this work are summarized as follows:

- It was designed six machine learning models to drive six soft sensors to monitor the recombinant protein concentration online. Feature selection was performed to determine the relevant online variables for model training
- To train the machine learning models, the experimental protein datasets were augmented using simulation data from a dynamic hybrid model tailored to the nanobody-based antivenom production process
- The machine learning models were tested for their accuracy in monitoring the recombinant protein concentration and for their efficiency in computation time

2 Development of Soft Sensors

2.1 Experimental Data

Two strains, *E. coli* CH10–12 and *E. coli* NbF12–10, were used. Seven experiments were conducted in a working volume of 2L at a temperature of 37 °C for the batch and fed-batch modes. Dissolved oxygen (pO_2) was kept over 15%, and the pH was regulated at 7 for the entire experiment. Induction was carried out in each reactor at the temperature displayed in Table 1 [2].

Table 1. Protein concentration measurements were taken at each experiment. *DP: Data Points

Experiment	*E. coli* Strain	Induction temperature (°C)	Protein concentration experimental DP*	MCFS experimental DP*
1	CH10–12	28	17	6468
2		30	15	6570
3		33	21	4208
4		37	21	4109
5		29	28	12998
6		32	27	11504
7	NbF12–10	29	25	11091

Protein content was monitored over time for each experiment, with a total between 15 and 28 experimental measurements (Table 1). The bioreactor was equipped with a Multifunctional Controller Software BioPAT® MFCS (Sartorius) software, capable of measuring every 5 s the 13 variables listed in Table 2.

Table 2. Online measurements are considered in the algorithm training. *ACIT and BAST are the control variables of acid and base pump that regulate the pH.

On-line measurements	
pH	Glucose Added (g)
Dissolved oxygen (%)	Glucose feed setup (g/h)
Pressure $(mbar)$	Substrate weight (g)
Temperature $(°C)$	Base weight (g)
Volume (L)	ACIT*
Air Flow (NL/min)	BAST*
Agitation (RPM)	

2.2 Dynamic Hybrid Model

The nanobody-based antivenom production process was simulated using the dynamic hybrid model proposed and validated in previous work [11]. Four mass balances for biomass, glucose, proteins, and volume are given by Eq. (1) – (4).

$$\frac{dX}{dt} = r_X - X\frac{F_{in}}{V} \tag{1}$$

$$\frac{dS}{dt} = \frac{F_{in}}{V}(S_{in} - S) - r_S - P\frac{F_{in}}{V} \tag{2}$$

$$\frac{dP}{dt} = r_P - P\frac{F_{in}}{V} \tag{3}$$

$$\frac{dV}{dt} = F_{in} \tag{4}$$

X (g/L) and $S(g/L)$ are the concentrations of biomass and glucose; $P(mg/L)$ is the concentration of nanobody proteins; $V(L)$ is the volume of the liquid phase in the bioreactor; $F_{in}(L/h)$ is the glucose flowrate; $S_{in}(g/L)$ is the concentration of glucose in the feed, $r_X(g/h)$ and $r_S(g/h)$ the production rate of biomass and glucose, respectively.

The production rate of nanobody proteins $r_P(mg/h)$ is defined as follows:

$$r_P = \begin{cases} 0\,if\ there\ is\ not\ induction \\ q_P X if\ there\ is\ an\ induction \end{cases} \tag{5}$$

$q_P\left(mg_{protein}/g_{CDW}/h\right)$ intended for the specific productivity of the nanobody production, which was modeled using a Support Vector Machine (SVM) learner. The SVM was trained with experimental data provided by [2], using a Radial Basis Function (RBF) Kernel with a 5-fold cross-validation [11].

2.3 Experimental Data Augmentation

This work aims to build soft sensors capable of real-time monitoring of the protein concentration. Nevertheless, given the small number of experimental instances for the protein concentration, between 15 and 28 (see Table 1).

Offline measurements, such as protein concentration are difficult and time-consuming to obtain, so samples are taken only sporadically. In this work, there are only between 15 and 28 offline measurements that were generated during experiments and a large number of online data with the MFCS software, between 4000 and 12000. It was decided to increase the protein data points by using the dynamic hybrid model. The augmented dataset increases the training set volume and ensures a comprehensive representation of the protein concentration dynamics. In this sense, the soft sensors can learn the relationships between online measurements and accurately forecast protein concentration over time. The simulated data were generated at the time corresponding to the MFCS datasets.

2.4 Training of Machine Learning Algorithms

2.4.1 Feature Selection

A feature selection was developed using a SHAP (SHapley Additive explanations) technique, which is a game theoretic approach to explain the output of any machine learning model. By utilizing the traditional Shapley values from game theory and their related extension, it establishes a connection between the best credit allocation and local explanations [12]. Algorithms such as CART Decision Tree (DT), Random Forest (RF), K-Nearest Neighbors (KNN), and RBF SVM algorithms were used to explore the SHAP values.

The number of instances was decreased to reduce computational complexity. It was summarized data at a higher level of granularity, grouping instances and utilizing the mean of each group as the summary statistic. This approach facilitates the summarization of the dataset and allows the application of the SHAP values technique without losing important information. The results of each algorithm were compared, and the features with the greatest effect on the variable to be predicted were selected. The parameters simulations for each algorithm were the same described in Sect. 2.4.2.

2.4.2 Training and Testing

Six different machine learning algorithms were explored to build the soft sensors: CART DT, RF, KNN, Quadratic SVM, Cubic SVM, and RBF SVM. All the algorithms were trained using the Scikit-learn Python library.

The simulation conditions of each algorithm are listened bellow:

- CART DT: The maximum number of branches in the tree was set to 5.
- RF: The maximum number of trees and their maximum number of branches was set to 5
- KNN: The number of neighbors was set at 2

- For the Quadratic, Cubic, and RBF SVM the regularization parameter is 10, 10, and 100, respectively. The margin of tolerance in the optimization process is 0.1 for all the SVM. For the Quadratic and Cubic SVM, the independent term in the kernel function corresponds to 1.0

The dataset was generated using the seven experiments in Table 1 and the variables in Table 2, ensuring an equilibrated data proportion at each experiment. The initial size of the dataset was (14007 × 13). However, after applying the feature selection, it was reduced to (14007 × 6). The dataset was performed, to ensure an equilibrated data proportion at each experiment. Therefore, a random 80/20 split was performed; 80% (11205 × 6) of the dataset was used for training and 20% (2802 × 6) for testing. A fivefold cross-validation with three repetitions was performed to guarantee the correct distribution of the instances and maintain a balanced representation across folds. It is important to highlight that the dataset was normalized to train the SVM models.

2.4.3 Statistical Analysis

The performance of soft sensors for protein production was analyzed using the Mean Absolute Error (*MAE*), Mean Squared Error (*MSE*), Root Mean Squared Error (*RMSE*), and Coefficient of Determination $\left(R^2\right)$, Eq. (6) – (8).

$$MAE = \frac{1}{n}\sum_{i=1}^{n}\left|Y - \widehat{Y}\right| \tag{6}$$

$$MSE = \frac{1}{n}\sum_{i=1}^{n}\left(Y - \widehat{Y}\right)^2 \tag{7}$$

$$R^2 = 1 - \frac{\sum\left(Y_{exp} - \widehat{Y}\right)^2}{\sum\left(Yexp - \overline{Y}\right)^2} \tag{8}$$

With $RMSE = \sqrt{MSE}$, n the number of instances, Y the testing label set for the protein concentration, $\widehat{Y}$ the predicted value of the protein concentration using the soft sensors, and $\overline{Y}$ the mean value of Y.

3 Results and Discussion

3.1 Feature Selection

Table 3 displays the ranking of SHAP values for each algorithm. It shows that glucose added, substrate weight, and temperature are one of the most important variables in protein production. Especially the temperature, since the experiments were designed to consider the temperature changes applied during the induction.

Variables such as the BAST and the base weight are directly related to the system's pH control. The pressure is related to gases generated by biomass growth and the agitation with the mass transfer process. As the ranking could change depending on the algorithm used, both the ranking and the expertise in the process were used to select the variables. Therefore, six variables were selected to train the algorithms: glucose added, temperature, base weight, substrate weight, pressure, and BAST.

Table 3. Feature selection rank using different algorithms.

Ranking	DT	RF	KNN	RBF SVM
1	Glucose Added	Glucose Added	BAST	Glucose Added
2	Temperature	Temperature	ACIT	Volume
3	Agitation	Agitation	Base weight	Base weight
4	Base weight	Base weight	Substrate weight	Substrate weight
5	Pressure	Pressure	Glucose Added	Pressure
6	pH	BAST	Agitation	Temperature
7	Dissolved oxygen	ACIT	Glucose feed setup	ACIT
8	Volume	Volume	Pressure	Dissolved oxygen
9	Substrate weight	Dissolved oxygen	Dissolved oxygen	BAST
10	BAST	Substrate weight	Volume	Agitation
11	ACIT	Air Flow	Temperature	Glucose feed setup
12	Glucose feed setup	pH	Air Flow	Air Flow
13	Air Flow	Glucose feed setup	pH	pH

3.2 Training and Test of Machine Learning Algorithms

Table 4 shows the statistical analysis of the different soft sensors for protein production monitoring using the six variables selected in the feature selection. The six trained algorithms present a good performance $\left(R^2 > 0.92\right)$. However, the KNN and the RBF SVM show the best results with R^2 values higher than 0.997. Values of 1.023×10^{-5} and 0.003 were obtained in the *MSE*, while values of 3.198×10^{-3} and 0.058 were obtained in the *RMSE* for the KNN and RBF SVM, respectively, indicating that those models present less variance between the experimental protein concentration and the predicted with the model. Similarly, values of 7.768×10^{-4} and 0.050 were obtained in the *MAE*, which means that KNN and RBF SVM could be the more consistent and precise models.

Table 4. R^2, *MSE*, *RMSE*, *MAE*, and simulation time in the training and test of different machine learning algorithms.

Metrics /Model	CART DT	RF	KNN	Quadratic SVM	Cubic SVM	RBF SVM
Training						
R^2	0.970	0.972	0.999	0.926	0.993	0.997
MSE	0.020	0.017	1.023×10^{-5}	0.074	6.928×10^{-3}	0.003

(continued)

Table 4. (*continued*)

Metrics /Model	CART DT	RF	KNN	Quadratic SVM	Cubic SVM	RBF SVM
RMSE	0.140	0.132	3.198×10^{-3}	0.272	0.083	0.058
MAE	0.078	0.073	7.768×10^{-4}	0.192	0.068	0.050
Test						
R^2	0.970	0.972	0.999	0.921	0.992	0.996
MSE	0.020	0.016	3.583×10^{-5}	0.072	6.942×10^{-3}	0.003
RMSE	0.140	0.126	5.986×10^{-3}	0.268	0.083	0.058
MAE	0.079	0.070	1.398×10^{-3}	0.189	0.068	0.051
Simulation Time (s)	0.378	1.279	0.406	353.092	594.018	14.083

In the test, the KNN and the RBF SVM present the best performance, with values higher than 0.996 in the R^2, and values less than 0.058 and 0.051 in the *RMSE*, and *MAE*. Although both models show good results in terms of R^2, *RMSE*, and *MAE*, it is necessary to consider the computational cost of each one. Table 4 shows that the computational time for training and testing is approximately 35 times less implementing KNN concerning RBF SVM, making it a suitable candidate for use as a soft sensor for protein monitoring.

Figure 1 displays the scatter plot for the six-trained models using the test dataset. It is confirmed that KNN and RBF SVM present less dispersion between the experimental data and the predictions performed for each model.

Regardless of the results obtained with other soft sensors $\left(R^2 > 0.92\right)$, there is evidence of a dispersion between the model estimations and the experimental data in the test, particularly for the CART DT, RF, and quadratic SVM. Whereas the last one presented a greater dispersion. This could be associated with the nature of the kernel, which does not allow for adequate capture of interactions between different features. The CART DT and RF dispersion is based on how the DTs are developed, which consists of recursively partitioning the feature space. i.e., the trees divide the feature space into regions and assign a constant value to each region at the end of the branches.

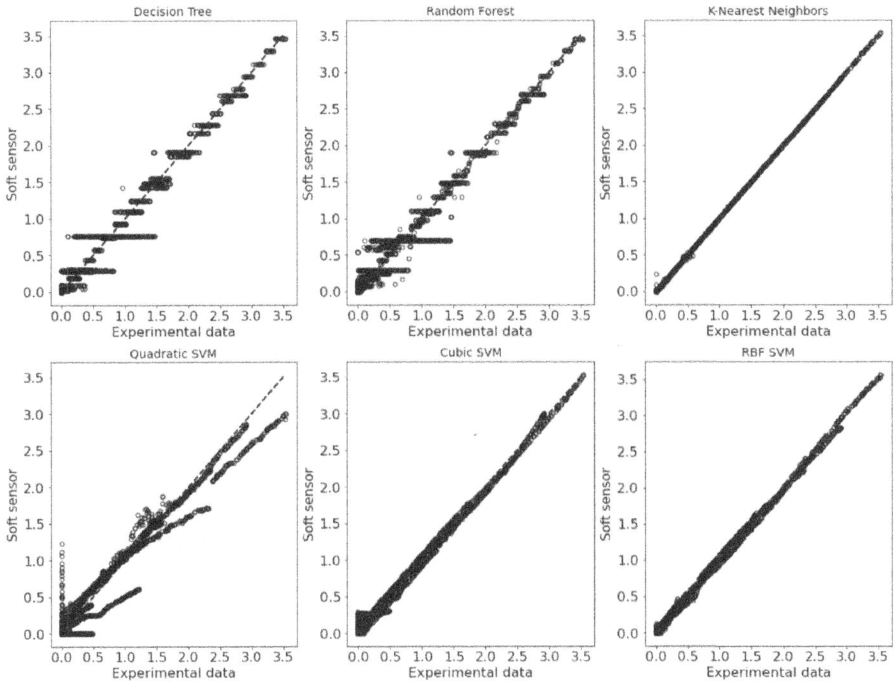

Fig. 1. Scatter plot: evaluation of the six-trained models using the test dataset.

3.3 Soft Sensor Monitoring Application

This work focuses on the formulation of soft sensors for recombinant protein monitoring. Figure 2 displays the behavior of four experiments with the KNN soft sensor for protein monitoring. Notably, the KNN soft sensor can predict the protein concentration for different induction temperatures (Fig. 2a-c) and different strains of microorganisms (Fig. 2d). The time that the KNN soft sensor takes to predict the protein concentration for a single point is on the order of 1.006×10^{-3} s. This makes soft sensors an ideal alternative for application in processes where physical sensors cannot quantify variables of interest through variables that can be measured online.

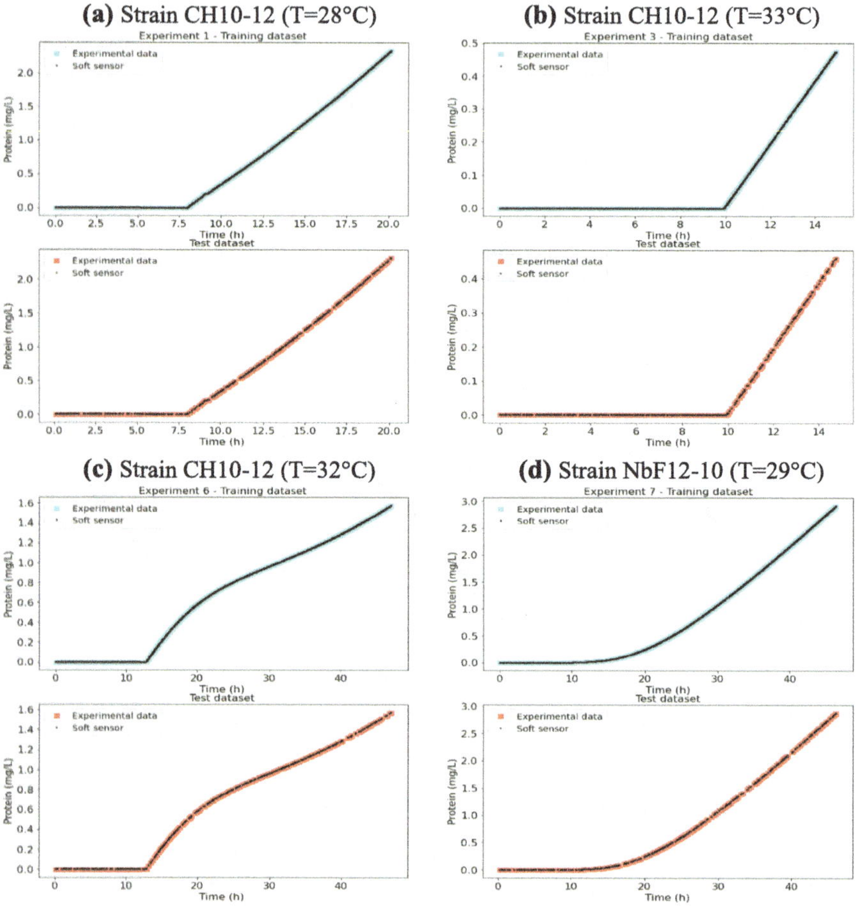

Fig. 2. Protein prediction using KNN soft sensor for each experiment.

4 Conclusions

Six machine learning algorithms were trained and utilized as soft sensors for real-time monitoring of protein production. SHAP values can serve as an alternative to support feature selection for training different algorithms. Nevertheless, this selection should also be guided by process knowledge. In the training and testing process, the KNN and RBF SVM soft sensors exhibited the best performance, with R^2 values higher than 0.996, *RMSE* and *MAE* values less than 0.058 and 0.051, respectively. Regarding computational time, the KNN soft sensor had a computation time of 35 times less than the RBF SVM soft sensor. These results show the capability of the KNN and RBF SVM soft sensors to accurately predict the recombinant protein concentration, making this type of soft sensor a suitable alternative for real-time monitoring (in order of 1×10^{-3} s in the case of recombinant proteins) of variables that cannot be easily measured. In future work, deep learning could be explored to predict the future behavior of the concentration of recombinant proteins or biomass growth rate.

Acknowledgments. This work was funded (or co-funded) by the European Union under the Horizon Europe project Bioindustry 4.0, grant n. 101094287.

Compliance with Ethical Standards. This article does not contain any studies with human participants or animals performed by any of the authors.

References

1. Medl, M., Rajamanickam, V., Striedner, G., Newton, J.: Development and validation of an artificial neural-network-based optical density soft sensor for a high-throughput fermentation system. Processes **11**, 297 (2023). https://doi.org/10.3390/pr11010297
2. Alonso Villela, S.M., Ghezal-Kraïem, H., Bouhaouala-Zahar, B., Bideaux, C., Aceves Lara, C.A., Fillaudeau, L.: Effect of temperature on the production of a recombinant antivenom in fed-batch mode. Appl. Microbiol. Biotechnol. **105**, 1017–1030 (2021). https://doi.org/10.1007/s00253-021-11093-5
3. Bouhaouala-Zahar, B., Ben Abderrazek, R., Hmila, I., Abidi, N., Muyldermans, S., El Ayeb, M.: Immunological aspects of scorpion toxins: current status and perspectives. Inflamm. Allergy - Drug Targets. **10**, 358–368 (2011). https://doi.org/10.2174/187152811797200713
4. Geethalakshmi, S., Narendran, S., Pappa, N., Ramalingam, S.: Development of a hybrid neural network model to predict feeding method in fed-batch cultivation for enhanced recombinant streptokinase productivity in Escherichia coli. J. Chem. Technol. Biotechnol. **87**, 280–285 (2012). https://doi.org/10.1002/jctb.2712
5. Alonso Villela, S.M., Kraïem-Ghezal, H., Bouhaouala-Zahar, B., Bideaux, C., Aceves Lara, C.A., Fillaudeau, L.: Production of recombinant scorpion antivenoms in E. coli: current state and perspectives. Appl. Microbiol. Biotechnol. **107**, 4133–4152 (2023). https://doi.org/10.1007/s00253-023-12578-1
6. Brunner, V., Siegl, M., Geier, D., Becker, T.: Challenges in the development of soft sensors for bioprocesses: a critical review. Front. Bioeng. Biotechnol. **9**, 722202 (2021). https://doi.org/10.3389/fbioe.2021.722202
7. Iglesias, C.F., Ristovski, M., Bolic, M., Cuperlovic-Culf, M.: RAAV manufacturing: the challenges of soft sensing during upstream processing. Bioengineering **10**, 229 (2023). https://doi.org/10.3390/bioengineering10020229
8. Kadlec, P., Gabrys, B., Strandt, S.: Data-driven soft sensors in the process industry. Comput. Chem. Eng. **33**, 795–814 (2009). https://doi.org/10.1016/j.compchemeng.2008.12.012
9. Gundinger, T., Kittler, S., Kubicek, S., Kopp, J., Spadiut, O.: Recombinant protein production in e. coli using the phoA expression system. Fermentation **8**, 181 (2022). https://doi.org/10.3390/fermentation8040181
10. Vaitkus, V., Brazauskas, K., Repšytė, J.: Soft-sensors based on black-box models for bioreactors monitoring and state estimation. In: Proceedings of the 2020 12th International Conference on Bioinformatics and Biomedical Technology, pp. 43–47. ACM, Xi'an China (2020). https://doi.org/10.1145/3405758.3405780
11. Corrales, D.C., Villela, S.M.A., Cescut, J., Daboussi, F., Fillaudeau, L., Aceves-Lara, C.A.: Dynamic Hybrid Model for Nanobody-based Antivenom Production (scorpion antivenon) with E. coli CH10–12 and E. coli NbF12–10. (2024)
12. Štrumbelj, E., Kononenko, I.: Explaining prediction models and individual predictions with feature contributions. Knowl. Inf. Syst.. Inf. Syst. **41**, 647–665 (2014). https://doi.org/10.1007/s10115-013-0679-x

System for the Efficient Management of Drinking Water Infrastructures Using Artificial Intelligence-Based Optimisation Algorithms

Carlos Calatayud Asensi[1], José Vicente Berná Martínez[2](✉), Lucia Arnau Muñoz[2], and Iren Lorenzo Fonseca[2]

[1] Aguas de Valencia S.A., Avda. Marqués del Turia, 46005 València, Spain
ccalatayud@globalomnium.com

[2] University of Alicante, Carretera San Vicente del Raspeig, 03690 Alicante, Spain
{jvberna,lucia.arnau,iren.fonseca}@ua.es

Abstract. Energy consumption has become the main concern of any company or organisation. However, there are services where energy consumption takes a back seat to meet the requirements. This is the case of drinking water services. In order to ensure the supply to the population, until now, the strategy followed consisted of maximising storage, at any cost. This way of operating infrastructures generates energy inefficiency, as well as undermining sustainability and risking greater water losses in the event of breakdowns. Nowadays, thanks to the historical information available, it is possible to generate algorithms capable of minimising consumption while optimising water storage and ensuring water supply. In this work we propose a system through which, using prediction algorithms and a heuristic search for the optimal solution, the optimal behaviour of a drinking water production system is calculated. The system is divided into two blocks, the prediction and the control generator, so that they are able to evaluate the current state of the system and obtain a behaviour adequate to the needs with the minimum consumption. This proposal has been instantiated on a real drinking water infrastructure of a city of 5000 inhabitants in the southeast of Spain, demonstrating the viability of the proposal and the improvement compared to traditional behaviour.

Keywords: planning and scheduling · real time system · decision-making support system · drinking water infrastructures

1 Introduction

Supplying drinking water to the population is currently one of the most pressing problems in many cities affected by drought, and it also depends on many different factors [1]. Although climate change [2] is not one of the most important determinants of water scarcity, there is also strong competition for natural resources, an increase in the cost of maintaining drinking water for human consumption and a rise in the cost of energy to

R. Chinthaginjala et al. (Eds.): DCAI 2024, LNNS 1259, pp. 282–291, 2025.
https://doi.org/10.1007/978-3-031-82073-1_28

produce it [3]. In addition, old water infrastructures need to be updated or completely renovated, and the mobility patterns of people have changed, with some cities emptying in summer while others multiply their population by a factor of 10 or more [4]. All of these factors affect the water supply model. Traditionally, a city uses large water storage reservoirs that are kept full to the maximum to ensure supply. This refilling action is attempted during the hours of the day when energy costs are cheap [5], although this is not always possible and, increasingly, the cost of energy need not be a determining factor. Moreover, this mode of operation is contraindicated for several reasons:

- There is a risk of large losses of drinking water in the event of incidents and breaks in the infrastructure. When a tank ruptures and needs to be emptied in order to take action, if there is a lot of stored water, emergency emptying will be necessary and therefore water will be lost. If infrastructure elements such as pumps, pipes or valves break down, more water will also be lost. By storing more water, it also keeps the pressures of the infrastructure higher, which facilitates the occurrence of breakdowns.
- It implies the need to maintain large quantities of controlled potable water in storage [6]. Water storage has an added cost, the chemicals used to make the water drinkable.
- There is also an indirect negative environmental effect. In many cases, water is extracted from underground, from natural aqueducts, and although water consumption will be the same whether the water is in the reservoir or underground, the reality is that aquifers suffer greater stress, and it is completely unnecessary to move water from its natural reservoir (the aquifer) to an artificial reservoir when it is not yet going to be consumed.
- Keeping a lot of water in storage leads to a lack of water sheet renewal and possible formation of by-products from the decomposition of chlorine (THM, chlorites and chlorates…) used for water disinfection [7, 8]. Stagnation zones occur within water reservoirs, where water does not move unless the water table is lowered sufficiently for the water to move.
- Finally, and one of the main reasons, keeping the water level at a minimum has another energy purpose. This means that the booster pumps must pump water with sufficient pressure to overcome the column of water in the tank, which means that if there is 2 m of water in the tank instead of 1 m, the pumps will consume more electrical energy to pump the same amount of water. This situation generates energy waste [9].

To avoid these negative effects, drinking water supply needs to be managed rationally, and this involves maintaining a delicate balance between objectives that are sometimes completely at odds with each other:

- Objective 1. Minimise stored drinking water.
- Objective 2. Ensure water supply for the population.
- Objective 3. To generate water at the lowest possible cost.

To generate an efficient control, capable of meeting the objectives and doing so in a context subject to so many external circumstances, sophisticated artificial intelligence algorithms [10] are needed to generate the optimal behaviour of the drive pumps in current situations, but also predicting future ones. In other words, an optimisation system with prediction and continuous monitoring capabilities is needed.

This article deals with the proposal of an intelligent predictive control system for water infrastructures, validated through a real use case of a city of 5000 inhabitants in the southeast of Spain, which can ensure the objectives described. To this end, the rest of the article is structured as follows. Section 2 describes the proposed model divided into two modules, prediction, and control generation. Sections 3 and 4 go into the prediction and control modules in more detail, describing the algorithms that form them. Section 5 instantiates the model on a real infrastructure and demonstrates its usefulness with the calculation using data extracted from the city. Finally, Sect. 6 presents the conclusions of the present work and defines the future lines of our development.

2 Predictive Optimised Control Model

The control system will be composed of two modules: a module for predicting the city's water needs, the expected consumption; and a module capable of generating optimised behaviour based on the current state of the infrastructures and the consumption calculated by the first module. The first module will make a prediction of expected consumption at a time from a prediction start time to an end time. With this prediction and the state of the infrastructures (stored water and state of the pumps), the second module will calculate the optimal sequence of pump starts and stops to supply the city and meet the objectives. Figure 1 shows a diagram of the proposed model and the modules that make it up.

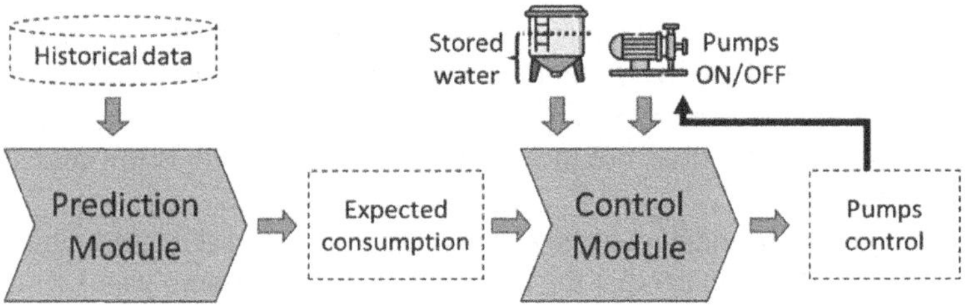

Fig. 1. General model of the Predictive Control System.

3 Prediction Module

The prediction will use information from the historical consumption data of the city or place of study (in our examples we will use the years 2021, 2022 and 2023 as the model will be instantiated to these years when testing). Through the study of the actual data, it is possible to observe how the consumption pattern remains similar over time when comparing similar dates. Consumption is conditioned by the time of the year, i.e. the week of the year, the day of the week and the time of the day when consumption occurs. Figure 2 shows an example with two similar weeks, week 6 (approximately the first week of February), of the year 2023 (blue line) and 2022 (orange line), plotting consumption over the days.

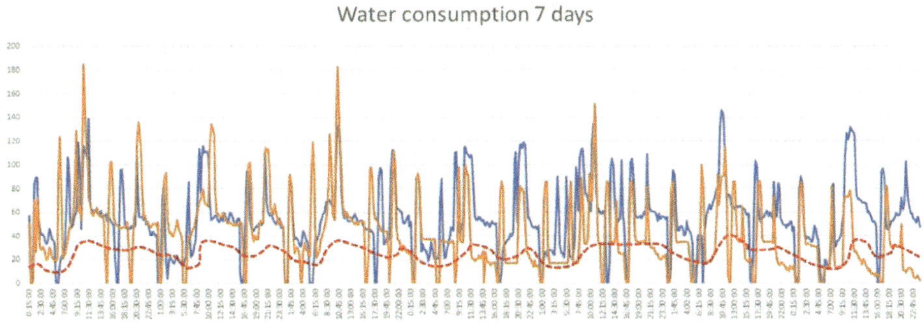

Fig. 2. Comparison of consumption two similar weeks in different years.

The dotted red line shows the general trend in water consumption, where it can be seen that it is very similar for both weeks. This same effect occurs in general throughout the year, except for extraordinary situations such as climatic phenomena, local holidays and other conditioning factors. This is why the consumption histories, where the past consumption is currently stored for each date-time, will be recoded to extract for each date-time a coding of the hour, day of the week and week of the year, as follows: the hour is coded as a number from 0 to 24; the day of the week as a number between 1 and 7; the week of the year as a number between 1 and 53. Figure 3 shows an example.

	Original data		Consumpt.	Encoded data		
Id	Date	Hour		HourNum	weekday	YearWeek
1	01/01/2021	0:00:00	81,4833333	0	5	1
2	01/01/2021	0:15:00	81,4166667	0,25	5	1
3	01/01/2021	0:30:00	75,8680002	0,5	5	1
4	01/01/2021	0:45:00	26,0640002	0,75	5	1
5	01/01/2021	1:00:00	36,516667	1	5	1
6	01/01/2021	1:15:00	81,2200012	1,25	5	1
7	01/01/2021	1:30:00	74,386001	1,5	5	1

Fig. 3. Example of date-time coding in hour number, day of the week and week of the year.

The prediction is defined as $P = Pred(now, long)$ where *now* is the date at which the prediction is to be calculated, and *long* is the size of the prediction. The result is a vector of prediction values $P = [p_1, p_2, \ldots, p_{long}]$, where each p_i indicates the predicted consumption for the instant from *now* to *now* + *long*. Each p_i is determined by means of a KNN algorithm that calculates a weighted average of values in instants of previous years, with $K = 3$ (*K* will be the number of years of the history, for simplicity we will use 3 years of history that will be the ones we have later in our real dataset), as follows:

- Given a date to calculate *now*, the year, week and day of the week are extracted from that date, and the coded time is calculated, obtaining: *year, week, dayweek, hour*. The prediction vector $P = []$ is initialised to empty.

1. From $i = 1$ to K, we obtain the historical consumptions HC of those same dates with similar (*week, daykeew, hour*), but from *year-i* to *year-K*. In our example of $K = 3$, and being the $year = 2024$, it would be $HC = [hc_{2023}, hc_{2022}, hc_{2021}]$.
2. The prediction p with KNN is calculated for this vector of consumptions using the equation:

$$p = KNN(k) = \frac{\sum_{i=1}^{k} \frac{hc_{year-i}}{i}}{\sum_{i=1}^{k} \frac{1}{i}} \quad (1)$$

3. The date *now* is shifted in a leap, until *long* elements are completed, repeating steps 1 and 2 at each shift.

The KNN calculation includes a weighting of each element from i to k of $1/i$. This is intended to give greater weight to values closer in time to the current year and less to those more distant, since experience indicates that the expected behaviour will be similar to the near past.

4 Control Module

The control is defined as a function that generates a control vector $C = Control\ (P, tl, ps, max, min, long)$ where: P is the vector of prediction values; tl is the level of water currently stored; ps is the state of the water impulsion groups (off = 0 / on = 1); *max* indicates the maximum permitted level of storage; *min* indicates the minimum level of storage; *long* is the length of the control to be generated, the maximum being that of the prediction. The output C is a vector of the type $[c_1, c_2,..., c_{long}]$, where each c_i is equal to the state of the pumps in that time slot i, which can be $on = 1$ or $off = 0$, and the cardinality of the vector C will be equal to $|C|=long$. In addition to these values, which will be variables for calculating the function, we know: *wcm* (water capacity per meter stored) or water capacity per meter in the storage, which indicates how many litres per meter of height the storage contains; *ppc*, water capacity produced by the booster pumps; and three cost constants A, B and C will be established. The cost constants will be used in the final cost calculation (step 7 of the algorithm) and will be used to establish the cost of each control proposal in order to obtain the minimum cost.

To calculate C, it will establish a set of principles, aligned with the objectives, which will allow it to create a search heuristic for the optimal control pattern. Each principle takes precedence over the lower ones.

- The amount of water stored plus the amount of water produced should cover the predicted consumption needs, ensuring supply.
- The amount of stored water shall never fall below the set *min*.
- The amount of stored water should never exceed the set *max*.
- The alternation between switching pumps on and off should be kept to a minimum, thus minimising stop/start cycles.

The control calculation algorithm will be based on the calculation of an optimisation matrix, M_O, through which we can obtain the optimal mode of operation. The algorithm is as follows:

1. We calculate the surplus water (that which we could spend from storage without exceeding the set minimum) as:

$$w_e = wcm * (tl - \min) \tag{2}$$

2. We calculate the amount of water needed to meet the consumption prediction as:

$$w_p = \sum_{i=1}^{long} p_i \tag{3}$$

3. We calculate the total water needed to cover the period to be monitored as follows:

$$w_n = w_p - w_e \tag{4}$$

If $w_n <= 0$, then $w_n = 0$.

4. We calculate the number of start-up cycles of the booster pumps as follows:

$$cicles = \frac{w_n}{ppc} \tag{5}$$

The value is rounded up to the next whole number. If the number of cycles is greater than or equal to *long*, the control vector is equal to $C = [1,\ldots, 1]$ and stops here. This means that the same or more water must be produced than will be consumed. Similarly, if the number of cycles is equal to 0, the control vector is equal to $C = [0,\ldots, 0]$ and also ends here. This would mean that no water must be produced because the stored water is sufficient.

5. We calculate the permutation matrix, M_P, of 0 s and 1 s, with a number of 1 s equal to *cycles*, and a number of 0s equal to *long - cycles*. This matrix contains $cols = long$ columns and the number of rows is calculated as:

$$\text{rows} = P_{|C|}^{cicles,|C|-cicles} \tag{6}$$

To refer to an element of the matrix M_P positioned in row i, column j, this is done by the expression $M_{P(i,j)}$.

6. We calculate the stored water level evolution matrix, M_L, of size *rows* and *cols* (the same as M_P), where each element of the matrix $M_{L(i,j)}$ is equal to:

$$M_{L(i,j)} = M_{L(i,j-1)} + \frac{\left(M_{P(i,j)} * ppc\right) - p_j}{wcm} \tag{7}$$

Where $M_{L(i,0)} = tl$ for any i.

7. We generate the optimisation matrix, M_O, of size *(4, rows)*, where the columns are *[emin, emax, changes, cost]*, where for $j = 1$ to *rows* we calculate:

$$emin_j = \sum_{i=1}^{long} max\left(\left(\min - M_{L(i,j)}\right), 0\right) \tag{8}$$

$$emax_j = \sum_{i=1}^{long} max\left(\left(M_{L(i,j)} - max\right), 0\right) \tag{9}$$

$$changes_j = \sum_{i=1}^{long} \left(M_{P(i,j-1)} - M_{P(i,j)}\right)^2 \tag{10}$$

$$cost_j = A * emin_j + B * emax_j + C * changes_j \tag{11}$$

5 Instance of the Model and Discussion of the Problem

We are going to apply the model on a real dataset of a city in the southeast of Spain, with approximately 5000 inhabitants. Our example takes place at 14:00 on 12/02/2024. This coded date corresponds to the day of week 1 (Monday), week of month 6, and the time 14:00. A prediction length equal to *long = 12* will be used. To calculate the KNN(3) for this date, data from the same dates for the years 2023, 2022 and 2021 are used. Figure 4 shows the data table and the result of KNN(3).

HourNum	DayWeek	WeekYear	Hour	2021	2022	2023	KNN(3)
14	1	7	14:00:00	23,70	17,39	27,88	24,26
14,25	1	7	14:15:00	75,23	64,32	9,55	36,43
14,5	1	7	14:30:00	141,70	122,55	7,17	63,10
14,75	1	7	14:45:00	68,99	57,52	90,05	77,35
15	1	7	15:00:00	13,96	13,18	24,73	19,62
15,25	1	7	15:15:00	50,72	44,05	83,97	67,04
15,5	1	7	15:30:00	13,56	12,22	24,08	18,93
15,75	1	7	15:45:00	56,02	47,99	57,34	54,55
16	1	7	16:00:00	7,63	6,65	4,43	5,62
16,25	1	7	16:15:00	61,53	57,34	41,02	49,20
16,5	1	7	16:30:00	94,95	76,00	62,70	72,19
16,75	1	7	16:45:00	0,00	98,89	26,61	41,49

Fig. 4. KNN(3) for Monday of week 7 at 14:00.

Column KNN(3) shows the result of applying Eq. (1) on the consumption columns for the years 2021, 2022 and 2023. Therefore *Pred(14:00 12/02/2023, 12)* is equal to: *P = [24.26,36.43, 63.10, 77.35, 19.62, 67.04, 18.93, 54.55, 5.62, 49.20, 72.19, 41.49].*

The following parameters will be used for the control calculation: *C = Control(P, 1, 0, 1, 1, 0, 5, 12)*, where the storage level is *tl = 1*, the booster pumps are *off = 0*, the *max = 1*, the *min = 0.5*, and also a control length of *long = 12* elements. These infrastructures where the calculations are made have a storage capacity of 1500 m^3, at a maximum height of 3.5 m, i.e. *wcm = 1500/3.5*. And they have a water impulsion group capable of generating *ppc = 100* m^3. According to this we perform the calculations in Eqs. 2 to 11:

- By (2): we = (1 - 0.5) * (1500/3.5) = 214.29
- By (3): wp = 24.26 + 36.43 + 63.10 + 77.35 + 19.62 + 67.04 + 18.93 + 54.55 + 5.62 + 49.20 + 72.19 + 41.49 = 529.77

- By (4): wn = 529.77 - 214.29 = 315.48
- By (5): cicles = 315.48 / 100 = 3.1548 $\approx$ 4
- By (6): rows = 12! / (4! * 8!) = 495

Here a permutation matrix M_P of 495 rows and 12 columns would be generated, with all possible permutations of 12 elements, where there are 4 on (number 1), and 8 off (number 0). With the permutation matrix M_P, the level matrix M_L is generated using Eq. (7) and the optimisation matrix M_O with Eqs. (8)(9)(19) and the cost calculation Eq. (11). For the cost calculation, A = 10^6, B = 10^3 and C = 1 have been used. Figure 5 shows a section of the matrices M_P, M_L and M_O with the first 10 rows and the jump to row 430 which is where the optimal sequence of pump operation occurs. In this figure the *index* column shows the row number, and in dark green the optimal sequence is highlighted.

		M_P = Permutation Matrix 1x4 0x8												
	ps	1	2	3	4	5	6	7	8	9	10	11	12	tl
1	0	1	1	1	1	0	0	0	0	0	0	0	0	1
2	0	1	1	1	0	1	0	0	0	0	0	0	0	1
3	0	1	1	1	0	0	1	0	0	0	0	0	0	1
4	0	1	1	1	0	0	0	1	0	0	0	0	0	1
5	0	1	1	1	0	0	0	0	1	0	0	0	0	1
6	0	1	1	1	0	0	0	0	0	1	0	0	0	1
7	0	1	1	1	0	0	0	0	0	0	1	0	0	1
8	0	1	1	1	0	0	0	0	0	0	0	1	0	1
9	0	1	1	1	0	0	0	0	0	0	0	0	1	1
10	0	1	1	0	1	1	0	0	0	0	0	0	0	1
428	0	0	0	0	0	1	1	1	0	0	1	0	0	1
429	0	0	0	0	0	1	1	1	0	0	0	1	0	1
430	0	0	0	0	0	1	1	1	0	0	0	0	1	1
431	0	0	0	0	0	1	1	0	1	1	0	0	0	1
432	0	0	0	0	0	1	1	0	1	0	1	0	0	1

	M_L = Water Level Matrix											
	1	2	3	4	5	6	7	8	9	10	11	12
1	1,18	1,33	1,41	1,46	1,42	1,26	1,22	1,09	1,08	0,96	0,79	0,7
2	1,18	1,33	1,41	1,23	1,42	1,26	1,22	1,09	1,08	0,96	0,79	0,7
3	1,18	1,33	1,41	1,23	1,18	1,26	1,22	1,09	1,08	0,96	0,79	0,7
4	1,18	1,33	1,41	1,23	1,18	1,03	1,22	1,09	1,08	0,96	0,79	0,7
5	1,18	1,33	1,41	1,23	1,18	1,03	0,98	1,09	1,08	0,96	0,79	0,7
6	1,18	1,33	1,41	1,23	1,18	1,03	0,98	0,86	1,08	0,96	0,79	0,7
7	1,18	1,33	1,41	1,23	1,18	1,03	0,98	0,86	0,84	0,96	0,79	0,7
8	1,18	1,33	1,41	1,23	1,18	1,03	0,98	0,86	0,84	0,73	0,79	0,7
9	1,18	1,33	1,41	1,23	1,18	1,03	0,98	0,86	0,84	0,73	0,56	0,7
10	1,18	1,33	1,18	1,23	1,42	1,26	1,22	1,09	1,08	0,96	0,79	0,7
428	0,94	0,86	0,71	0,53	0,72	0,8	0,98	0,86	0,84	0,96	0,79	0,7
429	0,94	0,86	0,71	0,53	0,72	0,8	0,98	0,86	0,84	0,73	0,79	0,7
430	0,94	0,86	0,71	0,53	0,72	0,8	0,98	0,86	0,84	0,73	0,56	0,7
431	0,94	0,86	0,71	0,53	0,72	0,8	0,75	0,86	1,08	0,96	0,79	0,7
432	0,94	0,86	0,71	0,53	0,72	0,8	0,75	0,86	0,84	0,96	0,79	0,7

	M_O = Optimization Matrix				
	emin	emax	changes	cost	index
1	0	2,442271	2	2444,27	1
2	0	2,208938	4	2212,94	2
3	0	1,975604	4	1979,60	3
4	0	1,742271	4	1746,27	4
5	0	1,524631	4	1528,63	5
6	0	1,43427	4	1438,27	6
7	0	1,35702	4	1361,02	7
8	0	1,35702	4	1361,02	8
9	0	1,35702	3	1360,02	9
10	0	1,975604	4	1979,60	10
428	0	0	4	4,00	428
429	0	0	4	4,00	429
430	0	0	3	3,00	430
431	0	0,07725	4	81,25	431
432	0	0	6	6,00	432

Fig. 5. Partial view of the matrices M_P, M_L and M_O and the optimal control vector.

The optimal sequence of operation is the one generated by this control C = [0, 0, 0, 0, 1, 1, 1, 0, 0, 0, 0, 1], i.e.: the pumps are off for the first 4 periods, then on for 3 periods, off for 4 more periods and finally on for the last period.

If we look at the result and plot the evolution of the expected consumption against the evolution of the level and the control sequence (Fig. 6) we can see the following:

- The control sequence coincides with the most pronounced areas of consumption.
- The water level in storage has been maintained at optimal levels.
- The water supply to the population is ensured as the reservoir is never empty.

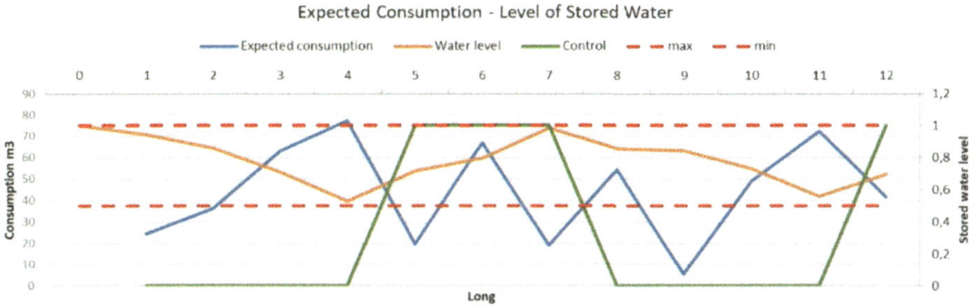

Fig. 6. Evolution of consumption, stored water and pump on/off using the optimal control.

By having the prediction as the basis for the calculations, the operation is adapted to current, but also future needs. And by finding the optimal solution, we ensure that the conditions will always be fulfilled, if possible, with existing resources.

6 Conclusions and Future Work

There are scenarios where it is necessary to find the optimal solution, for example, the calculation of control of operations in infrastructures that have a high energy consumption, but at the same time must ensure the supply. In this work we have presented the development of an algorithm for calculating the optimum based on artificial intelligence for predicting consumption. By means of the proposal it is possible to calculate, at each instant and based on the real state of the infrastructure (on/off state of the water pumps and total stored water) what is the optimal operation in the coming hours. The algorithm is configurable both in terms of the depth of the prediction, the size of *long*, and the impact on the prediction of the age of the data, the denominator in the KNN calculation, so that the algorithm can be adjusted. In times when there is a lot of variation in population or habits (e.g. holidays, days with extreme weather events, etc.), a small depth can be used and more relevance can be given to recent data. When the situation is stable (e.g. nights, working periods without holidays, etc.), a larger prediction depth can be used or older data can be given prominence to maintain similar behaviour.

Through this work, it is not only possible to calculate optimal control, but also to promote preventive maintenance integrated with the water production cycles of infrastructures, to observe anomalies in consumption and to help infrastructure operators in their decision-making by assisting them when there is uncertainty as to whether or not to stop a water pump.

As a line of future work, we are working on improving the algorithm in two aspects. When the depth of the prediction is greater, the search space expands exponentially, and it may be computationally unfeasible. To solve this problem, on the one hand, the parallelisation of the algorithm is proposed. To do this, it is necessary to establish a method that allows the permutation matrix to be divided into fragments that can be treated by independent nodes, generate local sub-optimums, and finally merge all the data into a single node that extracts the optimum. In order to do this fragmentation, it is necessary to systematise the generation of the matrix and its distribution. On the other hand, we are investigating the possibility of compressing the matrices, so that they can be searched in a smaller space. This would be done by compressing the length of the prediction, accumulating prediction ranges in a single value, for example, adding each 4 values by 1, and thus reducing a search space by 4.

Acknowledgments. This work has received funding from the UAIND22-01B project "Adaptive control of urban supply systems" of the Vice-Rectorate for Research of the University of Alicante.

Disclosure of Interests. The authors have no competing interests to declare that are relevant to the content of this article.

References

1. Ali, M., Shah, S.F., Khan, Y., Asim, S.M., Ullah, S.F.: A study on the consumption of safe drinking water and its determinants using the weighted multinomial logistic model. AQUA-Water Infrastruct. Ecosyst. Soc. **72**(12), 2422–2445 (2023)
2. Leveque, B., Burnet, J.B., Dorner, S., Bichai, F.: Impact of climate change on the vulnerability of drinking water intakes in a northern region. Sustain. Cities Soc. **66**, 102656 (2021)
3. Willaarts, B.A., De la Rúa, C., Cabal, H., Garrido, A., Lechon, Y.: The Water-Earth-Energy Nexus in Spain. Canal Isabel II Foundation. Spanish (2016)
4. Kali, S., et al.: Occurrence, influencing factors, toxicity, regulations, and abatement approaches for disinfection by-products in chlorinated drinking water: a comprehensive review. Environ. Pollut. **281**, 116950 (2021)
5. Calatayud Asensi, C., Berna Martinez, J.V., Arnau Munoz, L., Macia Perez, F.: Modelling and control of drinking water supply infrastructures through multi-agent systems for sustainability. Int. J. Softw. Eng. Knowl. Eng. (2024)
6. Calatayud Asensi, C., Berná Martinez, J.V., Arnau Muñoz, L., Macián Cervera, V.J., Maciá Pérez, F.: Simulation of drinking water infrastructures through artificial intelligence-based modelling for sustainability improvement. In International Conference on Simulation Tools and Techniques, pp. 131–146. Cham: Springer Nature Switzerland (2023)
7. Hossain, S., Chow, C.W., Cook, D., Sawade, E., Hewa, G.A.: Review of chloramine decay models in drinking water system. Environ. Sci. Water Res. Technol. **8**(5), 926–948 (2022)
8. Sanchez, M.O.: By-products of water disinfection using chlorine compounds. Health effects. Hig. Sanid. Ambient **342**, 335–342 (2008)
9. Gouws, R., Lukhwareni, T.: Factors influencing the performance and efficiency of solar water pumping systems: a review (2012).
10. Studziński, J., Ziółkowski, A.: Control of pumps of water supply network under hydraulic and energy optimisation using artificial intelligence. Entropy **22**(9), 1014 (2020)

Insights into Quantitative Evaluation Techniques for AI Explainability

Hugo Gomes, João Ferreira, and Manuel Rodrigues(✉)

LASI/ALGORITMI Centre, University of Minho, Braga, Portugal
{pg51242,pg50461}@alunos.uminho.pt,
manuel.rodrigues@algoritmi.uminho.pt

Abstract. The field of Artificial Intelligence (AI) has seen an exponential growth in recent years, with new models being created that achieve almost human-level accuracy in various tasks. However, this increase in accuracy sacrifices a key aspect for critical fields like medicine and finance, which is interpretability. This led to the rise of Explainability in Artificial Intelligence (XAI), which works towards giving more trustworthiness and transparency to these black-box models by generating explanations that translate the decision-making process of these models. One of the posing challenges in this field is to properly quantitatively evaluate these XAI methods due to limited research. This systematic review gives an overview of the current trends in the state-of-the-art for quantitative evaluation methods for evaluating XAI methods, giving more insight on what a good explanation should be and what properties it should meet. Then, we propose an evaluation benchmark based on the reviewed literature, comprising the properties we deemed most important to be met by an explanation. Finally, we present the limitations and challenges this field still poses and directions for future research.

Keywords: Explainability in Artificial Intelligence · XAI · Quantitative evaluation

1 Introduction

In recent years, the field of Artificial Intelligence (AI) has seen exponential growth, revolutionizing the way real-world decisions are made and tasks are accomplished. However, the inherently complex nature of these models frequently results in a lack of interpretability, causing concern about their trustworthiness, especially in fields that require transparency to hold the models accountable. Those challenges brought some opportunities to the field of Explainable Artificial Intelligence (XAI) which led to the emergence of this field as a critical research area.

While several qualitative processes have been proposed to evaluate AI models, the importance of quantitative methods cannot be overstated. Quantitative approaches offer rigorous methodologies for the evaluation of different XAI models, which allow the comparison of explainability between different artificial intelligence models. This systematic review synthesizes the existing literature, providing an overview of the current

R. Chinthaginjala et al. (Eds.): DCAI 2024, LNNS 1259, pp. 292–306, 2025.
https://doi.org/10.1007/978-3-031-82073-1_29

state-of-the-art methods and metrics, including their applications and the impact they had on the field.

The objectives of this systematic review are threefold. Firstly, in Sect. 2 we provide an overview of the current state-of-the-art quantitative evaluation methods for XAI algorithms, providing a detailed description of which properties the various authors try to evaluate and which metrics or methods they use. Then, in Sect. 3 we go over the properties we consider the most important and propose an evaluation benchmark based on the reviewed literature. Finally, in Sect. 4 we identify the limitations, gaps, and challenges still posed in this field, offering insights for future research directions.

As we explore the literature, we expect to uncover various quantitative evaluation methods, ranging from model particular to international transparency metrics. The analysis of this literature has the final objective of making contributions to the continued revolution in this field, fostering a deeper knowledge of the strengths and limitations of these evaluation methods.

2 Methodology

During the literature research for this work, we initially gathered 29 papers about quantitative evaluation of XAI methods, performed using the following platforms: ACM, Google Scholar, arxiv, Scopus, and IEEE Xplore. We performed our initial search by the following terms or similar: "quantitative evaluation of xai methods", "quantitative evaluation of xai methods for deep learning", "quantitative metrics for xai methods". Then, from those 29 papers we selected 12 which we considered to better match the scope of this work, being more focused on the properties and metrics rather than benchmarks, and that offered what we considered quality and interesting insights of these evaluation methods.

3 State of the Art

In 2020, Li et al. [9] made an experimental study of quantitative evaluations on saliency methods for XAI. They proposed to classify these metrics into 4 different categories: Faithfulness, Localization, Sensitivity Check, and Stability.

Faithfulness: Whether the relevance scores of the saliency maps reflect the true importance of the features.

- *AUC, Petsiuk et al.* [13]*:* The features are sorted from the least to the most important, and then they're gradually inserted into a reference input. A faithful explanation is expected to have greater AUC.

Localization: Whether the highly relevant region locates the actual object. Two metrics were studied for this category.

- *PG, Zhang, et al.* [22]*:* This metric measures if the most important features highlight the area defined as the most important in the ground-truth. This is done by measuring the ratio of the most important pixel in each image that is inside the ground-truth.

- *IntIoSR:* PG only focuses on the single highest most important point, and not the entire saliency map, and cannot well describe the unnecessarily captured information. This issue was resolved by revising the PG metric and taking this into account. Mathematically, is defined as the integral ratio of Intersection between the salient area and the ground truth mask over the Salient Region (IntIoSR).

Sensitivity Check: Whether the explanation methods have desired sensitivities to a given property (i.e.: classes).

- *Class Sensitivity (CS), Guillaumin et al.* [6]*:* Different classes usually have different discriminative regions. A good XAI method should have clear class sensitivity. The authors took the classes with the highest and lowest confidence and computed the similarity between their saliency maps. A good explanation method should have a CS score close to or below zero (but it does not necessarily work the other way around).

Stability: Whether the explanations are stable even with undesired variations. A good explanation is expected to be stable when the input is perturbed slightly but still has a similar model prediction and confidence distribution.

- *Max-Sensitivity, Yeh et al.* [21]*:* This metric is defined as the difference between 2 saliency maps of an original image and a perturbed image. The bigger the difference, the less stable the method is. Testing stability usually requires multiple variations, which makes this metric time-consuming to perform.

The authors tested this framework with 2 pre-trained models: ResNet50 [7] and VGG16 [18]; and 2 datasets: VOC, and COCO. The evaluated methods were Gradient, GBP, IntGrad [19], Grad-CAM [17], CEBP, Occlusion [20] and RISE [13]. They concluded that a fair evaluation of XAI methods could only be conducted under a fixed scenario and mode setting. Moreover, they consider a sound evaluation metric should reside deeply on cognitive rationales, and currently, there is a gap between the cognitive findings and their computational implementations, suggesting that continuous research attention could bring more insightful results.

In 2021, Budding *et al.* [3] proposed a framework to quantitatively evaluate saliency maps and applied this to evaluate datasets with different backgrounds. They used one with 2D brain MRI slices and another where the brain area was masked with Perlin noise. Each dataset has 2 different classes: class 1 without any lesions, and class 2 with two small circular lesions. They quantitatively evaluated the performance of the XAI methods only using the class with lesions since the metrics they used require binary ground truth maps. The metrics used were the following: ROC-AUC, mean average precision (mAP), and PREC99. The ROC-AUC is biased since it assigned mostly high scores to the explanations and did not match the qualitative evaluation previously made. The mAP and PREC99 seem more reliable, better matching the qualitative evaluation. The authors suggest further experimentation using this framework, which provided aid in a more objective evaluation of XAI methods to improve the correctness and validity of the obtained heatmaps that have not been sufficiently evaluated using ground-truth data.

In 2021, Mohseni *et al.* [11] introduced a benchmark for image and text XAI domains using multi-layer human attention masks aggregated from multiple human annotators. Unlike the commonly used saliency maps that score the important features of the data equally, the proposed multi-layer human attention mask attributes different scores to each feature based on their importance, allowing for a higher level of granularity in the evaluation of saliency maps. They evaluated the error in the model explanations by calculating the pixel-wise Mean Squared Error (MAE) between the model saliency score maps and the ground truth. They also studied the explanations **correctness** and **completeness**. The **correctness** was calculated by the MAE of the model saliency map scores outside the object's segmentation mask. The **completeness** was evaluated by the MAE of the model saliency map scores inside the object's segmentation mask of the ground-truth. The proposed benchmark was found to be more effective than single-layer object segmentation masks as the evaluation ground truth for saliency explanations, which shed some light on user biases in the subjective rating of explanations. These findings contribute to the understanding of the relationship between objective and subjective evaluations of saliency explanations.

In 2022, Rao *et al.* [14] proposed quantitative evaluation metrics for XAI methods applied to Graph Neural Networks for molecular property prediction. Rao et al. employed four popular GNN architectures (GraphSAGE, GAT, GraphNET, and CMPNN) that differ in their message-passing strategies. Then, they used six XAI methods, five attribution-based (CAM, SmoothGrad, GradInput, GradCAM, and IG), and one subgraph recognition method (MCTS). The evaluation involved an examination of the explanatory efficacy of XAI techniques through the quantification of the congruence between their interpretations and the ground-truth key substructures. The authors proposed two metrics:

Attribution-ACC: Metric that quantifies the accuracy of node-level predictions regarding important nodes/edges in explanation predictions ($\widehat{y_i}$) compared to the ground truths (y_i). It is calculated using the formula:

$$\text{Attribution - ACC} = \sum\nolimits_{i}^{N} (I(y_i = (\hat{y}_i))) \tag{1}$$

Attribution-AUROC: Metric that quantifies the performance of a model based on atom attribution scores, which consists of the contribution of each atom node for the output prediction. This can only be applied to the attribution-based XAI methods since the subgraph recognition method outputs are directly subgraphs rather than attribution scores. For p atoms in the substructure and n atoms not in the substruction, the formula for computing Attribution AUROC is as follows:

$$\text{Attribution} - \text{AUROC} = \frac{\sum_{i \in \text{substructures}} rank(i) - \frac{p(1+p)}{2}}{p \times n} \tag{2}$$

In 2023, Lu *et al.* [10] proposed a visual saliency explanation for Face Verification called CorrRISE, which is produce a visual saliency map that highlights the similar regions between two different photos, and a quantitative evaluation metric for this method. For the quantitative evaluation, they proposed new *Deletion* and *Insertion* metrics to better fit explanation methods for face verification, measuring the change in the verification accuracy after modifying the input image according to the importance map generated by the explanation method. In this new *Deletion* metric, the saliency maps' features are sorted by importance, then for n iterations, an increasing percentage of the most important features are masked from the inputs and the accuracy of the model is measured. The final score value is measured by the AUC score of the n registered accuracies. The *Insertion* metric is very similar to the *Deletion* metric, except in each iteration a percentage of the most important features are inserted in a plain image, instead of removed.

In 2023, Zhang et al. [23] introduced a comprehensive Visual Explanation pipeline, which enables researchers to conduct fair comparisons of various visual explanation techniques using quantitative metrics. The proposed benchamark assesses the **faithfulness** and **stability** as the fairness of visual explanation methods. These metrics can be categorized into 3 classes: 1) Performance-based metrics, 2) Perturbation-based metrics, and 3) Consistency-based metrics.

Performance-Based Metrics: These are quantitative indicators that primarily focus on the model's predictive accuracy, by evaluating its effectiveness in accomplishing the task it was trained for, regardless of the interpretability aspect. Since they are ground truth-centered evaluation metrics, they rely on subjective evaluations from human experts. These metrics are: *IoU, Precision, Recall, F1, Pointing Game.*

Perturbation-Based Metrics: These metrics evaluate how modifications made to the input data influence the result. They can aid in comprehending the significance of various features in the decision-making process of the model. These metrics are *Average % score, Deletion, Insertion, Adversarial Perturbation, Sufficiency and Comprehensiveness.*

Consistency-Based Metrics: These metrics measure how well the explanation aligns across similar instances. It measures the similarity of explanations of similar instances, given the assumption that model explanations should be similar for similar inputs.

In 2023, Nauta et al. [12] considered that instead of interpretability and explainability being presented as a subjectively validated binary property, it should be a multi-faceted concept. They identify 12 conceptual properties that should be evaluated for truly assessing the quality of an explanation in a comprehensive way. They argue that a quantitative way of measuring interpretability should result in a multidimensional view that could indicate and measure how certain properties are satisfied. The authors present the Co-12 Explanation Quality Properties, which are a group of 12 desired explanation properties identified by them that together present an aggregated view of what to evaluate.

Correctness: Describes how faithful the explanation is w.r.t the black box. This property can be evaluated by the metrics: *Model Parameter Randomization Check, Explanation Randomization Check, Controlled Synthetic Data Check, Single Deletion,* and *Incremental Deletion.*

Completeness: Describes how much of the black box behavior is described in the explanation. This property can be evaluated by the metrics: *Preservation Check, Deletion Check, Fidelity,* and *Predictive Performance.*

Consistency: Describes how deterministic and implementation-invariant the explanation method is. This property can be evaluated by the metric *Implementation Invariance.*

Continuity: Describes how continuous and generalizable the explanation function is. This property can be evaluated by the metrics: *Stability for Slight Variations, Fidelity for Slight Variations,* and *Correctedness.*

Contrastivity: Describes how discriminative the explanation is w.r.t. to other events or targets. This property can be evaluated by the metrics: *Target Sensitivity, Target Discriminativeness,* and *Data Randomization.*

Covariate Complexity: Describes how complex the (interaction of) features in the explanation are. This property can be evaluated by the metrics *Covariate Homogeneity* and *Covariate Regularity.*

Compactness: Describes the size of the explanation. This property can be evaluated by the metrics: *Size (absolute or relative), Redundancy,* and *Counterfactual Compactness.*

Composition: Describes the presentation format and organization of the explanation. This property can be evaluated by the metric *Perceptual Realism.*

Confidence: Describes the presence and accuracy of probability information in the explanation. This property can be evaluated by the metric *Confidence Accuracy.*

Context: Describes how relevant the explanation is to the user and their needs. This property can be evaluated by the metrics *Pragmatism* and *Simulated User Study.*

Coherence: Describes how accordant the explanation is with prior knowledge and beliefs. This property can be evaluated by the metrics *Alignment with Domain Knowledge* and *XAI Methods Agreement.*

Controllability: Describes how interactive or controllable an explanation is for a user. This property can be evaluated by the metric *Human Feedback Impact.*

With the proposal of the Co-12 properties, the authors identified a research opportunity to develop new evaluation methods for the properties that are currently insufficiently addressed and to develop new variants of the already existing ones. They also acknowledge that it might be unreasonable to expect an XAI method to score well in all 12 properties, but they suggest that in practice trade-offs between the desired ones should be made. As they mention, Coherence might contradict Correctness, and Completeness and Compactness might be considered diametrical opposites.

In 2021, Ding et al. [4] conducted a comprehensive and quantitative evaluation of saliency methods for NLP models, more specifically neural language models, which still largely remains an open problem. They evaluate the quality of prediction interpretations from two perspectives:

- **Plausibility**: measures how much the interpretations align with human intuitions about how a specific model makes decisions. A plausible interpretation would assign high importance to the features considered important for the decision of the model. For example, when a model predicts a verb with the singular morphological number, the singular nouns in the prefix should be considered important features.
- **Faithfulness**: measures whether the feature importance assigned is consistent with the internal decision mechanism of a model. An interpretation is considered faithful when the input features or the model decision mechanism is perturbed slightly, and

the features assigned with the highest importance remain consistent. Other authors use this definition for the Stability property.

The authors identified two limitations. The first limitation is that their plausibility tests only concern the words cue/attractor sets rather than other words in the input prefix. This can be mitigated by combining plausibility with faithfulness tests, which concern all the input prefix words. The second limitation is that the tests used in this benchmark need to be constructed in a case-to-case manner. Future work could explore automatic input consistency tests by utilizing adversarial input generation techniques in NLP.

In 2023, Batic et al. [2] presented a methodology for the explainability of regression-based Deep Learning NonIntrusive Load Monitoring (DL NILM), which refers to the process of estimating energy consumption of individual domestic appliances from aggregated household consumption. In this work, the authors implement a quantitative evaluation benchmark to compare different XAI methods. They define three desirable properties of NILM explanations: faithfulness, robustness, and effective complexity.

- **Faithfulness**: quantifies how well the explanations validate the predictive accuracy of a model. They measure faithfulness by observing the impact of a feature removal on the generated prediction. This property is measured by using the *Deletion* metric as seen by previous work.
- **Robustness**: measures whether a slight modification to the input causes the explanation to drastically differ from the original input. A robust explanation should give similar explanations to similar inputs. The authors modify the inputs by exposing the input to zero-mean Gaussian noise with controller standard deviation. Then, the robustness score (RS) is calculated by the average of the local Lipschitz constant over n steps. Methods with low Lipschitz value score display the characteristic of being more stable.
- **Complexity**: quantifies the entropy of the XAI output. If most of the input features are considered important, it does not provide the desired level of clarity and lowers the human interpretability of the explanation. The conciseness of the explanation is measured by the statistical dispersion of the output map, sorted in ascending order, using the Gini index computation. The authors consider that it's also important to consider the information about the complexity of the input variable, which is calculated by an adaptation of the noise aggregate measure (NAR). To quantify the complexity of explanation in the context of NILM, they formulate the"effective complexity" measure as:

$$EC^{(i)} = \frac{Gini}{1 - NAR^{(i)}} \tag{3}$$

Some limitations were acknowledged by the authors that need further exploration. First, the proposed evaluation framework displays an inability to present specific steps for enhancing the effectiveness of the explainability techniques, although it still facilitates the comparison between different XAI methods. Additionally, and suggested for future work, the authors suggest to extensively exploring the relationship and trade-offs between the properties of faithfulness, robustness, and complexity in XAI NILM approaches, which they believe a balance among these metrics is vital to ensure the utility, trust, and transparency in these methods.

Kohlbrenner et al. [8] published a study where they investigate and quantify the effect of the LRP method [16] on feedforward neural networks for visual object detection. They defined a good and representative XAI model as being able to reflect the model's decision in the explanation by highlighting the object being detected. The models used in this work were trained to categorize images rather than localize objects, but the authors assume that the models base their decisions on the target object instead of the image context. To test this, they compared the outputs when occluding the object areas with occluding the image backgrounds, and the results showed a steeper decrease in the output for a specific class, supporting their previous claim. Then, to evaluate the performance, they used a localization metric that measures how well the highlighted areas in the explanation reflect the object being classified:

$$\mu = \frac{R_{in}}{R_{tot}}, \mu_w = \mu \cdot \frac{S_{tot}}{S_{in}} \tag{4}$$

Here, R_{in} is the sum of positive relevance in the bounding box, R_{tot} is the sum of positive relevance in the entire image, S_{in} is the size of the bounding box, S_{tot} is the size of the image. In this metric, μ measures the ratio of positive relevance being attributed to the region of the object concerning the total image. It will be lower if there is positive relevance being attributed to the background, and higher if it is more concentrated in the object. μ_w also considers the size of the object, where a smaller object will be given a higher score than a bigger one. This is justified since smaller objects are harder to detect and classify. With this metric, the authors were able to properly measure how well the explanation is based on the object instead of the image context and background.

In 2022, Gevaert *et al.* [5] studied different existing metrics for feature attribution metrics evaluation to gain better insight into their behaviours, and then proposed a new metric based on the Sensitivity-n metric. The studied metrics are:

- *Deletion and Insertion*: Similar to the previously seen metrics, but removing or inserting the top k most important pixels
- *Minimal subset Deletion and Insertion:* The Deletion and Insertion metrics, while recording the smallest value for k that causes the prediction model to change.
- *Iterative Removal of Features (IROF):* An extension of the Deletion metrics, but removing the top k most important segments of a segmented image.
- *Sensitivity-n* [1]*:* The Sensitivity-n is satisfied when the sum of the attributions for a subset of features is equal to the variation of the output of a model caused by removing the features in that subset. In other words, it measures how relevant is the subset for the output confidence of the model.
- *Seg-Sensitivity-n*: This metric is an extension of the Sensitivity-n proposed by the authors, where it first segments the input image into multiple segments S, and then removes random subsets of segments instead of features.
- *Infidelity*: This metric compares the difference in output when applying a perturbation to the model input with the dot product of the perturbation and attribution map.
- *Max-Sensitivity* [4]: As previously seen in the work Li et al. [9].

- *Impact Coverage*: This is metric consists in calculating the Intersection-over-union (IoU) between the k most important features of the attribution maps of the original and perturbed images with patches, where k is the number of pixels covered by the patch. A score of 1 would suggest the most important pixels are perfectly covered by the adversarial patch.

Gevaert *et al.* also proposed a set of benchmarking guidelines to help developers select the best feature attribution method for their specific use case:

- *Baseline selection*: Select a baseline attribution method that must be defined, which can be a uniform random baseline or a more specific one depending on the use case.
- *Metric selection*: Select the metrics either manually or using a pilot study, which is done by selecting the metric implementations with high Krippendorff α and low inter-metric correlations on numerous metrics.
- *Rough statistical analysis*: A pairwise, nonparametric statistical test (like the Wilcoxon signed rank test) is made after the metric scores are computed in all the selected methods and baselines.
- *Detailed comparison*: Select a small number of the best performing attribution methods and compare them in more detail with new pairwise Wilcoxon tests, this time between two methods rather than a method and baseline. The Common Language Effect Size (CLES) can then be used to assess the method superior to the rest.

The authors concluded that the choice of the attribution metric should depend on the use case since the same metric can behave differently in different use cases. A statistical testing approach should also be made to draw global conclusions about these metrics. They also suggest a few different directions for future research. First, metrics do not necessarily measure the same underlying concepts of feature attribution maps, which leads to the question of what these concepts are exactly. Identifying these concepts could help better development of evaluation metrics and benchmarks. Secondly, using a combination of different attribution maps could be more informative than a single one. Finally, the application of benchmark procedures on new datasets could help identify the best attribution methods for a given use case.

In 2022, Rokade *et al.* [15] proposed an approach to quantitatively evaluate an explainable AI method. The main ideology behind this work is that the salient regions highlighted by the XAI methods (explanations) by themselves should be enough to obtain very similar accuracy. Initially, the Prediction Accuracy (pa) is measured by getting the predicted class label and accuracy. Then, the explanation is obtained, and the input image is masked by the highlighted salient regions and fed to the model. Finally, we obtain the Explanation Accuracy (ea) by measuring the predicted label and accuracy. This approach comprises four metrics:

- **Coherence** ($\bar{\alpha}$): This metric evaluates whether the explanation provided by the XAI algorithm is truly influential towards the prediction, by measuring the difference between accuracies averaged over N inputs.
- **Correctness** (β): This metric displays how well the explanation can capture the truly relevant features of the prediction with the highest accuracy.

- **Completeness** (γ): Evaluates how well the explanation represents the model's original prediction. This is done by comparing the Explanation Accuracy with the Prediction Accuracy.
- **Congruency** (σ): Calculates the standard deviation of the difference between the coherence for a specific input with the coherence averaged for N inputs.

The final quantification metric (Q) is obtained by aggregating the Correctness, Completeness, and Congruency. It incorporates weights (w_1 and w_2) to penalize the Correctness property more severely than the Completeness and Congruency. The final formula is translated as:

$$Q = w_1 \cdot \beta + w_2 \cdot \frac{\gamma}{\sigma} \tag{5}$$

As shown by the experimental results, this metric was able to provide sufficient information to quantitatively estimate the XAI algorithms LIME and Integrated Gradients and establish a comparative relationship between them.

4 Discussion

The quantitative evaluation of explainable AI methods is still a field without a strong consensus on which metrics are the best. One of the most important questions that should be answered is:" What is a good explanation?". Existing research suggests that a good explanation should satisfy different properties that evaluate its different aspects and characteristics.

Table 1. List of evaluated properties by each author

Source	Properties
Li *et al.* [9]	Faithfulness, Localization, Sensitivity Check, Stability
Budding *et al.* [3]	Localization
Zhang *et al.* [23]	Faithfulness, Stability, Fairness
Nauta *et al.* [12] (Co-12 properties)	Correctness (Faithfulness), Completeness, Consistency, Continuity, Contrastivity, Covariate complexity, Compactness, Composition, Confidence, Context, Coherence, Controllability
Ding *et al.* [4]	Faithfulness, Plausibility,
Batic *et al.* [2]	Faithfulness, Robustness, Complexity
Kohlbrenner *et al.* [8]	Faithfulness, Localization
Gevaert *et al.* [5]	Correctness (Faithfulness), Robustness
Rokade *et al.* [15]	Coherence, Correctness (Faithfulness), Completeness, Congruency

(*continued*)

Table 1. *(continued)*

Source	Properties
Mohseni *et al.* [11]	Correctness (Faithfulness), Completeness
Rao *et al.* [14]	Congruency
Lu *et al.* [10]	Faithfulness

The Table 1 displays the different properties measured by different authors. As stated by Nauta *et al.* [12], it is irrealistic to expect that a method scores well in all properties. Therefore we propose for future work a framework to for a quantitative evaluation of general XAI methods. For this framework, we selected the properties we considered the most imnportant to be evaluated: Faithfulness, Completeness, Coherence, Stability, Complexity, and for problems in the image domain, Localization.

We can see that the property the vast majority of studies deemed important to be measured is the **Faithfulness** (or Correctness) ([2, 4, 5, [8–12, 15, 23]), which measures how well the explanation captures the truly relevant features of the input data. We consider this as the most important property to be evaluated in an explanation since it measures whether the XAI algorithm gives a transparent explanation that the users can trust. The majority of metrics used in the review studies are based on the *Deletion* and *Insertion* metrics [2, 5, 9, 10, 12], but different metrics should also be considered, like the *Model Parameter Randomization Check, Explanation Randomization Check, White Box Check, Controlled Synthetic Data Check* [12], or the approaches by Ding *et al.* [4] and Mohseni *et al.* [11].

The **Completeness** ([11, 12, 15]), assesses how much of the model behavior is described in the explanation, helping evaluate the transparency of the explanation, which is one of the most desired properties in the XAI field. This property is measured by metrics that compare the decisions made by the model when given an original input and when given a modified input based on the explanation obtained for the original input, like the *Preservation Check* and *Deletion Check* [12], and by evaluating the performance of the model with resource from the explanations, like the *Fidelity, Predictive performance* [12], the approach of Mohseni *et al.* [11], and the *Completeness metrics* by Rokade *et al.* [15].

The **Coherence** (or Plausibility) was defined in slightly different ways, by Nauta *et al.* [12], Ding *et al.* [4] and Rokade *et al.* [15]. Nauta *et al.* and Ding *et al.* defined Coherence as an explanation being consistent with relevant background knowledge, beliefs, and general consensus, while Rokade *et al.* as capturing the extent to which the explanations are truly influential to the prediction of the model. We believe the definition of Nauta *et al.* and Ding *et al.* contributes to assuring more trust in the explanations since it measures how aligned they are with human intuitions. The Coherence metrics essentially aim to compare the obtained explanations with preexisting domain knowledge, either given by a ground-truth (*Alignment with Domain Knowledge* [12] and the *Plausibility metric* by Ding *et al.* [4]), or given by other XAI methods (*XAI Methods Agreement* [12]). This way, the metrics are able to evaluate how coherent the explanations are with prior knowledge.

The **Stability** (Continuity or Robustness) [2, 5, 9, 12, 23] is a very desirable property to ensure the XAI method is trustworthy in its explanations. This property measures how stable the explanation is when the input is slightly changed. In other words, it evaluates if the explanations are similar when given similar inputs. A stable XAI method should generate similar explanations for similar inputs, ensuring the trustworthiness of the method. The majority of the metrics used to evaluate this property are based on applying slight perturbations to the input and comparing both explanations, meaning that the bigger the difference, the less stable the XAI method is. Some of the metrics that are utilized are: *Max-sensitivity* [5, 9, 21], *Adversarial perturbation* [23], *Stability for Slight Variations, Fidelity for Slight Variations* [12], and *Robustness score* [2]. Additionally, Nauta *et al.* also used counterfactual explanations to measure the Stability with the *Correctedness* metric, by evaluating how connected they are to samples in the training data.

The **Complexity** property [2, 12] quantifies how complex the explanations are. A good explanation should be simple enough, so an easy and simple human interpretability is ensured. If an explanation is too complex, it might not provide the desired level of information and clarity the user would expect. This property can be evaluated by resourcing to the entropy of the explanation, like the *Effective Complexity* [2] and *Covariate Regularity* [12] metrics, and by evaluating the consistency in which a feature of the explanation represents an already predefined human interpretable concept, like the *Covariate Homogeneity* [12] metric.

When obtaining explanations from problems where the model relies on identifying a certain object in an image, the **Localization** property [3, 9] is also important to be met. This property measures how well the highlighted features locate the important object. This is useful to guarantee that the model bases its decisions on the actual object and not the rest of the image. This property is measured by essentially verifying if the regions assigned with high relevance correspond to the location of the relevant objects (or regions) for the decision. This property can be measured with the following metrics: *PG* [9, 22], *IntIoSR* [9], *ROC-AUC, mAP* and *PREC99* [3].

Different XAI methods could be evaluated by these standalone properties, to obtain more detailed information about these algorithm's characteristics to make an objective quantitative evaluation, but we considered we should be able to calculate one performance metric that agglomerates these properties to enable an easier comparison between methods. This can be achieved similarly to what Rokade *et al.* [15] proposed, by weighting these metrics together. This way, we can adjust the weights hyperparameters w_i for each property p_i in a way that reflects the ones we deem most important being met in the explanations. Given N_p as the number of properties being evaluated, the final score can be defined as follows:

$$Q = \sum_{i<N_p}^{i=0} w_i \cdot p_i \tag{5}$$

Additionally, Gevaert *et al.* [5] stated that metrics for the same property do not necessarily measure the same underlying concepts, which suggests that each of the studied properties would be better evaluated if multiple metrics were used. This would help improve the coverage of those underlying concepts, giving a more accurate measure for each property. The final score for a property could be obtained by averaging the metrics. Given m_i as a metric score and N_m as the total number of metrics measuring a property, we can obtain the property score p_{score} as follows:

$$p_{score} = \frac{\sum_{i<N_m}^{i=0} m_i}{N} \quad (6)$$

As studied, more properties can be utilized to evaluate XAI methods to obtain a more detailed evaluation depending on the use case and the needs of the evaluators. We consider that a good evaluation should be concise, and the suggested properties offer a good evaluation basis for what we believe can be a good consensus on quantitative evaluation for XAI methods. With Faithfulness, Coherence, and Stability, we can ensure the trustworthiness of the XAI algorithm, guaranteeing that the obtained explanations truly reflect the most relevant features and that the decisions made by the model were only affected by those features. The Completeness and Complexity assure the transparency of the model by evaluating how much of its behavior was described in the explanations, and by making sure the explanations are easy to interpret and understand by a user, respectively. For problems that rely on detecting objects, the Location property provides a way to ensure that the actual objects are properly identified by the explanation and therefore used in the model's decision. Finally, the quantifiable metric aims to give a way to objectively evaluate and compare different XAI algorithms.

Gevaert *et al.* also proposed a very interesting benchmarking pipeline, which gives a good direction for the different stages of selecting the most appropriate XAI methods for a given use case. This pipeline is composed of four stages: 1) Baseline selection, 2) Metric selection, 3) Rough statistical analysis, and 4) Detailed comparison. We can easily integrate the properties and metrics discussed in this paper. First, the second step (Metric selection) should be modified to"Properties and metrics selection", where in addition to the metrics being selected, we could also define the properties we want the explanation methods to meet and how they should weigh in the overall quantitative score. Then, in step 3 (Rough Statistical analysis) we could integrate the overall quantitative score in Eq. 1 and select the methods based on this score, simplifying this step. Then, in step 4 (Detailed comparison) the more detailed Wilcoxon test can be made to assess the superior method among the selected ones from step 3.

5 Limitations and Future Work

In this comprehensive review, several limitations have been identified that still pose a challenge to establishing a robust consensus for quantitative evaluation methods for explainable Artificial Intelligence (XAI) methods and suggestions for future research. First, we noticed that different authors use different property names for the same definitions, which can generate confusion in the field [24]. This issue be addressed to facilitate the cooperation and the consensus we seek. Second, new metrics should be developed, and variants of already existing ones should be further explored, especially for certain properties that still lack some coverage. This would make these evaluation methods more suitable for a more diverse type of data, problems, and explanations [12]. The third limitation, as stated by Gevaert *et al.* [5], is that different metrics measure different underlying concepts. These concepts should be better investigated and identified to help make a better and more informed decision on the choice of the metrics, making the evaluation method itself more transparent for the evaluators. Nauta *et al.* point out that it might be unreasonable to expect an XAI method to score well in every property, and it would be more practical to make trade-offs between these properties. These trade-offs should be further studied to help the users of these evaluation methods better decide the properties they consider more important for their use case. Finally, the evaluation benchmark proposed for future work in this paper should be formally studied and tested to assess its practical validity.

This work has been supported by FCT - Fundação para a Ciência e Tecnologia within the R&D Units Project Scope: UIDB/00319/2020.

References

1. Ancona, M., Ceolini, E., Öztireli, C., Gross, M.: Towards better understanding of gradient-based attribution methods for deep neural networks. In: International Conference on Learning Representations (2018)
2. Batic, D., Stankovic, V., Stankovic, L.: Towards transparent load disaggregation – a framework for quantitative evaluation of explainability using explainable AI. IEEE Trans. Consum. Electron. 1–1 (2023)
3. Budding, C., Eitel, F., Ritter, K., Haufe, S.: Evaluating saliency methods on artificial data with different background types (2021)
4. Ding, S., Koehn, P.: Evaluating saliency methods for neural language models (2021)
5. Gevaert, A., Rousseau, A.J., Becker, T., Valkenborg, D., De Bie, T., Saeys, Y.: Evaluating feature attribution methods in the image domain (2022)
6. Guillaumin, M., Ferrari, V.: Large-scale knowledge transfer for object localization in imagenet. In: 2012 IEEE Conference on Computer Vision and Pattern Recognition, pp. 3202–3209 (2012)
7. He, K., Zhang, X., Ren, S., Sun, J.: Deep residual learning for image recognition (2015)
8. Kohlbrenner, M., Bauer, A., Nakajima, S., Binder, A., Samek, W., Lapuschkin, S.: Towards best practice in explaining neural network decisions with LRP. In: 2020 International Joint Conference on Neural Networks (IJCNN), pp. 1–7 (2020)
9. Li, X.H., et al.: Quantitative evaluations on saliency methods: An experimental study (2020)
10. Lu, Y., Xu, Z., Ebrahimi, T.: Towards visual saliency explanations of face verification (2023)
11. Mohseni, S., Block, J.E., Ragan, E.: Quantitative evaluation of machine learning explanations: A human-grounded benchmark. In: 26th International Conference on Intelligent User Interfaces, IUI 2021, pp. 22–31. Association for Computing Machinery, New York, NY, USA (2021)
12. Nauta, M., et al.: From anecdotal evidence to quantitative evaluation methods: A systematic review on evaluating explainable AI. ACM Comput. Surv. **55**(13s) (2023)
13. Petsiuk, V., Das, A., Saenko, K.: Rise: Randomized input sampling for explanation of black-box models (2018)
14. Rao, J., Zheng, S., Lu, Y., Yang, Y.: Quantitative evaluation of explainable graph neural networks for molecular property prediction. Patterns (N. Y.) **3**(12), 100628 (2022)
15. Rokade, P., Alluri, K.R.: BKSP. Building quantifiable system for xai models. Available at SSRN 4038039
16. Samek, W., Montavon, G., Binder, A., Lapuschkin, S., Müller, K.R.: Interpreting the predictions of complex ml models by layer-wise relevance propagation. arXiv preprint arXiv:1611.08191 (2016)

17. Selvaraju, R.R., Cogswell, M., Das, A., Vedantam, R., Parikh, D., Batra, D.: Grad-cam: visual explanations from deep networks via gradient-based localization. Int. J. Comput. Vision **128**(2), 336–359 (2019)
18. Simonyan, K., Zisserman, A.: Very deep convolutional networks for large-scale image recognition (2015)
19. Sundararajan, M., Taly, A., Yan, Q.: Axiomatic attribution for deep networks (2017)
20. Tang, Z., et al.: Interpretable classification of alzheimer's disease pathologies with a convolutional neural network pipeline. Nat. Commun. **10**(1), 2173 (2019)
21. Yeh, C.K., Hsieh, C.Y., Suggala, A.S., Inouye, D., Ravikumar, P.: How sensitive are sensitivity-based explanations? CoRR, abs/1901.09392 (2019)
22. Zhang, J., Bargal, S.A., Lin, Z., Brandt, J., Shen, X., Sclaroff, S.: Top-down neural attention by excitation backprop (2016)
23. Zhang, Y., Gu, S., Song, J., Pan, B., Bai, G., Zhao, L.: Xai benchmark for visual explanation (2023)
24. Henriques, A., Parola, H., Gonçalves, R., Rodrigues, M.: Explainable AI: breakthroughs in medical diagnosis and surgery. In: Rocha, Á., Adeli, H., Dzemyda, G., Moreira, F., Poniszewska-Marańda, A. (eds.) Good Practices and New Perspectives in Information Systems and Technologies. WorldCIST 2024. Lecture Notes in Networks and Systems, vol. 986, pp. 254–272. Springer, Cham (2024). https://doi.org/10.1007/978-3-031-60218-4_23

Comparison of Machine Learning-Enhanced Dynamic Hybrid Models for a Nanobody Scorpion Antivenom Production with *Escherichia Coli*

Irene Martínez-Menéndez[1(✉)], Juan C. Acosta-Pavas[1], David Camilo Corrales[2], Susana María Alonso Villela[1], Balkiss Bouhaouala-Zahar[3,4], Georgios K. Georgakilas[5], Konstantinos Mexis[6], Stefanos Xenios[6], Theodore Dalamagas[7], Antonis Kokosis[6], Michael O'donohue[1], Luc Fillaudeau[1], Nadia Boukhelifa[8,9], Alberto Tonda[8,9], and César A. Aceves-Lara[1(✉)]

[1] TBI, Université de Toulouse, CNRS, INRAE, INSA, Toulouse, France
{martinez-men,aceves}@insa-toulouse.fr
[2] INRAE, UMS (1337) TWB, 135 Avenue de Rangueil, 31077 Toulouse, France
[3] Laboratoire des Biomolécules, Venins et Applications Théranostiques (LBVAT), Institut Pasteur de Tunis, 13 Place Pasteur, BP-74, 1002 le Belvédère, Tunis, Tunisia
[4] Faculté de Médecine de Tunis, Université Tunis el Manar, 15 rue Djebel Lakhdhar, 10007 Bab Saadoun Tunis, Tunisia
[5] Laboratory of Genetics, Section of Genetics, Cell Biology and Development, Department of Biology, University of Patras, 26504 Patras, Greece
[6] Department of Process Engineering, NTUA, Iroon Politechniou 6 Zografou, Athens, Greece
[7] Athena Research Center, Marousi, Greece
[8] INRAE, AgroParisTech, Université Paris-Saclay UMR 518 MIA-PS, 91120 Palaiseau, France
[9] Institut des Systèmes Complexes de Paris Île-de-France (ISC-PIF), UAR 3611 CNRS, Paris, France

Abstract. This study compares several hybrid dynamic models built with offline data to optimize nanobody production in *Escherichia coli*. These models combine the insights of mechanistic knowledge and Machine Learning algorithms (ML). Four ML algorithms were tested: Support Vector Regressor (SVR), Random Forest (RF), Decision Tree (DT) and K-Nearest Neighbors (KNN). The efficacy of each model was analyzed with three metrics: mean absolute error (MAE), root mean squared error (RMSE), normalized root mean squared error (NRMSE). Results demonstrate that the radial basis function (RBF) SVR model performed best simulations, with a NRMSE between 0.063 and 0.665. The insights gained are pivotal for advancing computational modeling techniques in biomanufacturing, with a specific emphasis on the production of recombinant therapeutic proteins, such as nanobodies, using bacterial hosts.

Keywords: Hybrid model · Machine learning · Bioprocess · *Escherichia coli* · Nanobody · Antivenom

R. Chinthaginjala et al. (Eds.): DCAI 2024, LNNS 1259, pp. 307–316, 2025.
https://doi.org/10.1007/978-3-031-82073-1_30

1 Introduction

Molecularly targeted biologics have become the new generation of therapies due to their precise specificity, affinity and their effective therapeutic performance [1]. Nanobodies are single domain antibody fragments significantly smaller than conventional antibodies, which allows for better tissue penetration and targeting of cryptic epitopes that are inaccessible to larger molecules [2]. They are more thermostable and production is achieved using microbial cell hosts making them a more cost-effective alternative to monoclonal antibodies [3]. The bacteria *Escherichia coli* has been engineered as the preferred host for the efficient production of antivenom proteins, leveraging its scalability and cost-effectiveness [4, 5].

The Food and Drug Administration (FDA) has promoted the digitalization of the biomanufacturing industry in recent years, through knowledge-aided methodologies [6, 7]. The tendency towards reducing costs and improving the production and quality of therapeutic molecularly targeted biologics goes beyond legislative measures. Enhancing cell culture productivity within existing bioreactors has been proven an effective strategy [8].

The use of Machine Learning to complement mechanistic models is a promising tool to enhance upstream bioprocess understanding and control [9–11]. By analyzing the data generated during protein production, ML can identify complex protein-related patterns and relationships that traditional methods might miss [12, 13]. This approach has been preliminarily explored in various studies, showcasing its potential in improving process efficiency and product quality [14–16].

One of the challenges associated with offline datasets in protein production lies in the small size and heterogeneous nature of the experimental data. This requires the testing of several ML-enhanced hybrid models that can perform when datasets are not big enough to deploy artificial neural networks [17].

This study aims to identify the best ML enhanced hybrid dynamic model for the described biopharmaceutical production among Decision Trees (DT), K-Nearest Neighbors (KNN), Random Forests (RF) and Support Vector Regressor (SVR).

The paper is organized as follows:

- Sect. 2 outlines the methodology, detailing the development of the hybrid models, the experimental setup, and the specific ML algorithms employed.
- Sect. 3 presents the results and discussion from the comparative analysis, focusing on the practical implications and the efficiency of each model in the specific context of nanobody production.
- Sect. 4 concludes with a summary of the findings and suggests future directions for the application of ML-enhanced hybrid models in biotechnological processes.

2 Materials and Methods

2.1 Experimental Data

The experimental data were obtained from a previous work [18] and are summarized in Table 1. Two recombinant strains were used: WK6 *E. coli* NbF12–10 and CH10–12 scorpion anti-toxin bispecific nanobody and its humanized form that possesses portent neutralizing activity against whole venom [19, 20]. A total of seven experiments were carried out in a working volume of 2L at a biomass production temperature (T) of 37 °C for the batch and fed-batch modes. Dissolved oxygen ($pO2$) was kept over 15%, and the pH was regulated at 7 for the entire experiment. Induction was carried out in each reactor at the temperature displayed in Table 1. Nanobody concentration was monitored over time for each experiment, with a number of data points, between 15 and 28. Following NiNTA downstream purification step, the specific recombinant protein concentration was determined using SDS-PAGE and densitometry [21]. The bioreactor was equipped with a BioPAT® MFCS (Sartorius) software, capable of measuring every 5 s different online variables such as pH, agitation, $pO2$ and temperature.

Table 1. Protein instances measured at each experiment [18].

Experiment	*E. coli* Strain	Induction temperature (°C)	Experimental offline measurements
1	CH10–12	28	17
2		30	21
3		33	28
4		37	27
5		29	28
6		32	21
7	NbF12–10	29	25

2.2 Mechanistic Model

The bioreactor is assumed to be an infinitely mixed culture without any transfer limitation. The mass balances for biomass, glucose, proteins and volume are given by Eq. (1) – (4).

$$\frac{dX}{dt} = r_X - X * \frac{F_{in}}{V} \tag{1}$$

$$\frac{dS}{dt} = \frac{F_{in}}{V}(S_{in}) - r_S - S * \frac{F_{in}}{V} \tag{2}$$

$$\frac{dP}{dt} = \boldsymbol{r_P} - P * \frac{F_{in}}{V} \tag{3}$$

$$\frac{dV}{dt} = F_{in} \tag{4}$$

where X and S are the concentrations of biomass and glucose, in g/L, and P is the concentration of nanobody, in mg/L. F_{in} is the glucose flowrate, in L/h, V is the volume of the liquid phase in the bioreactor, in L, S_{in} is the concentration of glucose in the feed, in g/L, and r_X is the production rate of biomass, in g/h, and defined by Eq. (5). r_S is the consumption rate of glucose, in g/h, as defined by Eq. (6).

$$r_X = \mu * X \tag{5}$$

$$r_S = -\frac{1}{Y_{SX}} * \mu * X \tag{6}$$

With μ as the specific growth rate, in 1/h, as defined by a Monod kinetics in Eq. (7); Y_{SX} as the yield coefficient of substrate, in g_{wcdw}/Gs and g_{wcdw} is the wet cell dry weight.

$$\mu = \frac{\mu_{max} * S}{K_S + S} \tag{7}$$

$$\mu_{max} = \mu'_{max} * T + b \tag{8}$$

μ_{max} is the maximum specific growth rate in 1/h and K_S is the substrate saturation coefficient, in g/L. μ'max is a temperature-dependent factor in 1/h, T is the temperature of the bioreactor, in °C, and b is a temperature-independent factor, in 1/h.

r_p is the nanobody production rate. The variable predicted with ML is the protein specific productivity q_P of the nanobody production $mg_{protein}/(g_{wcdw} \cdot h)$, that is connected to the mechanistic model previously detailed in Corrales *et al.*, (2024) as follows:

$$r_P = \begin{cases} 0 & \textit{if there is not induction} \\ q_P * X & \textit{if there is an induction} \end{cases} \tag{9}$$

where X is the biomass in g/L. Lastly, $mg_{protein}$ refers to the nanobody produced mass in milligrams; and h the production time in hours. Both r_p and q_P are continuous variables, also known as float variables.

2.3 Machine Learning Algorithms

Four different machine learning algorithms were explored to build the hybrid models: Decision Tree CART (DT), Random Forest (RF), K-Nearest Neighbors (KNN), and Support Vector Regressor (SVR). All the algorithms were trained using the Scikit-learn Python library 1.3.0. A fivefold cross-validation with three repetitions was performed and each model was trained with the whole dataset given the reduced quantity of experimental data. The simulation conditions of each algorithm are listened bellow:

- DT: The maximum number of branches in the tree was set to 5.
- RF: The maximum number of trees and branches was set to 5.
- KNN: The number of neighbors was set at 2.
- SVR: Hyperparameters were optimized using a Grid Search. The range of values for each search space is detailed in Table 2.

Table 2. Grid spaces for each hyperparameter to optimize SVR hyperparameters. Where rbf stands for radial basis function.

Hyperparameter	Grid space
C	[0.1, 1, 10, 100]
Gamma	[0.001, 0.01, 0.1, 1, 10, 100]
Epsilon	[0.001, 0.01, 0.1, 1, 10, 100]
Kernel	['rbf', 'linear', 'poly', 'sigmoid']

The training dataset is the experimental data provided by Alonso Villela *et al.* (2021) for both recombinant strains NF12–10 and CH10–12 and was described in Table 1. The ML algorithms estimate the specific productivity q_P based on two variables:

$$q_P = f(T, \mu) \tag{10}$$

where T is the temperature in °C, μ the growth rate in 1/h. The input variables were selected according to expert knowledge, concretely based on published evidence [18], to avoid spurious correlations. They are both continuous or float variables.

2.4 Statistical Analysis

The ability of the models to predict nanobody production in g/L was evaluated by comparing with experimental values measured at identical time intervals. The metrics used were MAE, RMSE and NRMSE:

$$MAE = \frac{1}{n}\sum_{i=1}^{n}\left|Y - \widehat{Y}\right| \tag{11}$$

$$RMSE = \sqrt{\frac{1}{n}\sum_{i=1}^{n}\left(Y - \widehat{Y}\right)^2} \tag{12}$$

$$NRMSE = \frac{\frac{1}{n}\sqrt{\sum_{i=0}^{n}\left(Y_i - \widehat{Y}_i\right)^2}}{Y_{max} - Y_{min}} \tag{13}$$

With n being the number of instances, Y the testing label set for the protein concentration, $\widehat{Y}$ the predicted value of the protein concentration using the soft sensors, and $\overline{Y}$ the mean value of Y.

3 Results

3.1 Validation of the Construction of the Model

Figure 1 shows the scorpion venom specific nanobody concentration at different temperatures for both strains. Four ML-enhanced hybrid models are displayed, in comparison with the experimental data points. In the three last graphs, the model Random Forest diverges from experimental data compared to the others. Also, the SVR model generates the best fit of the majority of experimental data. is the best fitted model for the majority of experiments. Finally, DT and KNN display similar behavior with poorer ability to fit data.

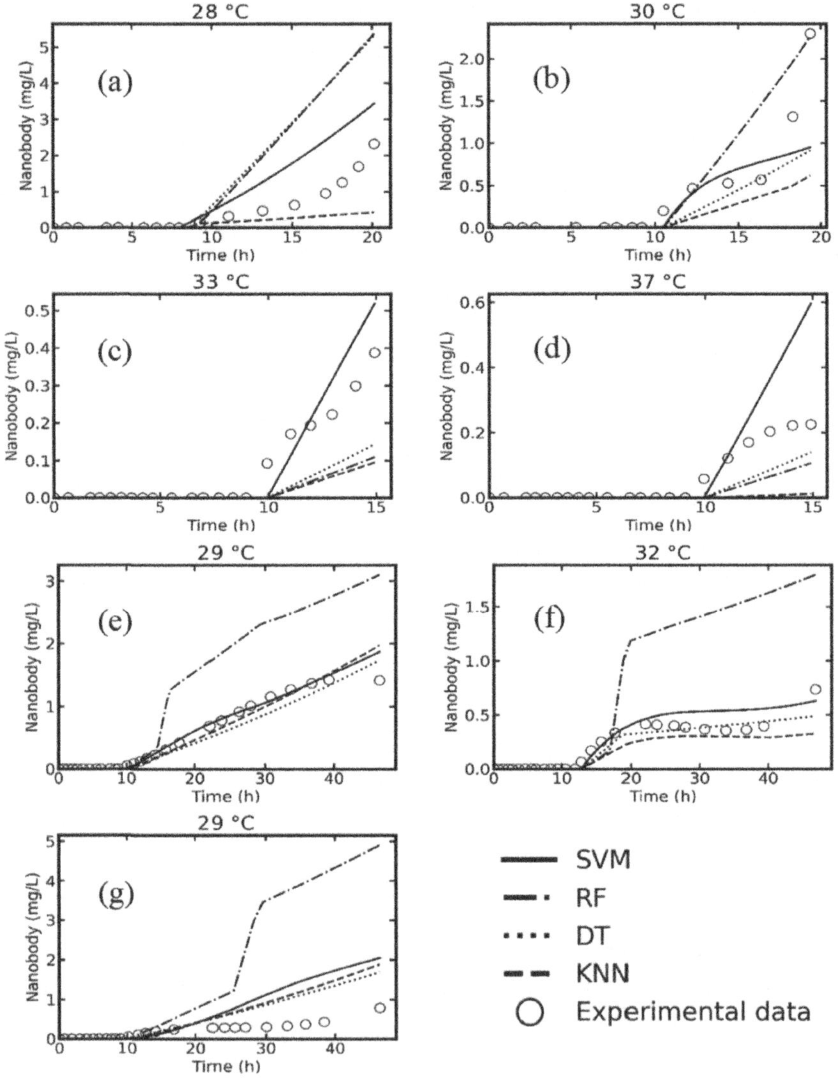

Fig. 1. Produced nanobody concentration in mg/L at different temperatures. Figure 1a is experiment number 1 at an induction temperature of 28 °C. 1b is the experiment number 2 with an induction temperature of 30 °C. 1c is 3rd experiment at 33 °C; 1d is the 4th at 37 °C; 1e is the 5th at 29 °C; 1f is the 6th at 32 °C; 1g is the 7th experiment at 29 °C with Nb1012.

Table 3 shows the different metrics to compare the performance of the four models. Concretely, MAE, RMSE and NRMSE for each model and in each experiment. The data reveals that the SVR model generally offers robust performance across a range of metrics, providing good fits of data generated in experiments 2, 3, and 5. This consistent low-error rate indicates that the SVR model has high predictive accuracy under certain conditions, validating its utility in scenarios similar to these experiments. However, the notable drop in the model's predictive performance of protein concentration in experiments 4 and 7 reveals potential areas for improvement, particularly in terms of overfitting and generalization, which are critical for the model's applicability to a wider range of bioprocess scenarios.

The RF model demonstrates variable performance, displaying good predictive fitness for data from experiment 2, but low suitability in the case of experiments 6 and 7. The insights gained from these observations are invaluable for refining the model-building process, particularly in understanding the parameters under which the RF model is likely to perform well or poorly.

The DT model shows a consistent and competitive performance in certain experiments, notably 5 and 6, suggesting that DT models can effectively capture and predict complex bioprocess dynamics under specific conditions. However, the increased occurrence of model simulation errors when fitting data from experiments 1 and 7 indicate limitations that must be addressed during the model construction phase, for example, using online data or other offline variables as estimators for the ML model.

The KNN model's performance in experiment 1, contrasted with more variable fitness in other experiments. This suggests that while it can be effective, its applicability may be limited by dataset specificity. This variability highlights the need for careful consideration of the KNN model's parameters and the characteristics of the dataset during the model construction process.

Table 3. *MAE*, *RMSE*, and *NRMSE* for each ML-enhanced hybrid model in each experiment compared to the experimental data.

Experiment	1	2	3	4	5	6	7
Induction T (°C)	28	30	33	37	33	32	29
SVR							
MAE	0.474	0.157	0.025	0.045	0.034	0.050	0.302
RMSE	0.762	0.375	0.051	0.108	0.090	0.083	0.527
NRMSE	0.329	0.163	0.132	0.478	0.090	0.083	0.665
RF							
MAE	0.988	0.148	0.049	0.032	0.462	0.360	0.924
RMSE	1.646	0.308	0.097	0.062	0.462	0.600	0.924
NRMSE	0.710	0.134	0.251	0.274	0.740	0.816	2.143

(continued)

Table 3. (*continued*)

Experiment	1	2	3	4	5	6	7
DT							
MAE	1.005	0.170	0.044	0.027	0.082	0.036	0.213
RMSE	1.646	0.394	0.087	0.052	0.127	0.064	0.369
NRMSE	0.710	0.171	0.224	0.231	0.089	0.087	0.466
KNN							
MAE	0.323	0.234	0.051	0.046	0.059	0.061	0.239
RMSE	0.627	0.501	0.102	0.091	0.121	0.107	0.420
NRMSE	0.270	0.218	0.102	0.404	0.085	0.146	0.530

4 Conclusions

Four machine learning enhanced hybrid models trained with temperature and growth rate data were compared to predict the specific production rate of two bispecific nanobodies expressed in recombinant WK6 *E.coli* strains NbF12-10 and CH10-12 [20]. SVR generates the best prediction of experimental data. This was revealed by low errors MAE between 0.025 and 0.474; RMSE between 0.051 and 0.762; and NRMSE between 0.063 and 0.665. The SVR-hybrid model showed good ability to handle non-linear and small-sized datasets, specifically for predicting variables such as the specific production rate of the nanobody concentration from the temperature and biomass growth rate.

Despite the constraints posed by the dataset size, the findings have advanced the application of ML-enhanced hybrid models for the production of nanobody-based antivenom with *E. coli*. The next steps to improve these hybrid models involves enriching and proving them with new data from experiments performed in a broader spectrum of biomanufacturing conditions, coupled with the integration of real-time data acquisition technologies. Using this approach, it will be possible to advance towards the creation of powerful digital twins for autonomous bioprocess control.

In summary, the shift towards hybrid computational models, supported by real-time protein production monitoring and digital twin technologies can be paradigm-changing if we are able to gather enough representative data. This approach promises not only to enhance the efficiency and yield of a specific recombinant protein bioprocess but also to pave the way for new innovations in the production of biologics and other complex biomolecules.

Acknowledgments. This work was funded (or co-funded) by the European Union under the Horizon Europe project Bioindustry 4.0, grant n. 101094287, the INRAE PhD scholarships and it has benefited from a State grant managed by the National Research Agency under the "Investissements d'Avenir" program with the reference ANR-18-EURE-0021.

References

1. Lee, Y.T., Tan, Y.J., Oon, C.E.: Molecular targeted therapy: treating cancer with specificity. Eur. J. Pharmacol. **834**, 188–196 (2018). https://doi.org/10.1016/j.ejphar.2018.07.034
2. Hamers-Casterman, C., et al.: Naturally occurring antibodies devoid of light chains. Nature **363**, 446–448 (1993). https://doi.org/10.1038/363446a0
3. Muyldermans, S.: Nanobodies: natural single-domain antibodies. Annu. Rev. Biochem. **82**, 775–797 (2013). https://doi.org/10.1146/annurev-biochem-063011-092449
4. Alonso Villela, S.M., Kraïem-Ghezal, H., Bouhaouala-Zahar, B., Bideaux, C., Aceves Lara, C.A., Fillaudeau, L.: Production of recombinant scorpion antivenoms in E. coli: current state and perspectives. Appl. Microbiol. Biotechnol. **107**, 4133–4152 (2023). https://doi.org/10.1007/s00253-023-12578-1
5. Bouhaouala-Zahar, B., Ben Abderrazek, R., Hmila, I., Abidi, N., Muyldermans, S., El Ayeb, M.: Immunological aspects of scorpion toxins: current status and perspectives. Inflamm. Allergy - Drug Targets. **10**, 358–368 (2011). https://doi.org/10.2174/187152811797200713
6. U.S. Food and Drug Administration: Pharmaceutical Quality for the 21st Century: A Risk-Based Approach Progress Report (2017). https://www.fda.gov/about-fda/center-drug-evaluation-and-research-cder/pharmaceutical-quality-21st-century-risk-based-approach-progress-report
7. Yu, L.X., Raw, A., Wu, L., Capacci-Daniel, C., Zhang, Y., Rosencrance, S.: FDA's new pharmaceutical quality initiative: knowledge-aided assessment & structured applications. Int. J. Pharm. X. **1**, 100010 (2019). https://doi.org/10.1016/j.ijpx.2019.100010
8. Rodrigues, M.E., Costa, A.R., Henriques, M., Azeredo, J., Oliveira, R.: Technological progresses in monoclonal antibody production systems. Biotechnol. Prog. **26**, 332–351 (2010). https://doi.org/10.1002/btpr.348
9. Nikita, S., Mishra, S., Gupta, K., Runkana, V., Gomes, J., Rathore, A.S.: Advances in bioreactor control for production of biotherapeutic products. Biotechnol. Bioeng. **120**, 1189–1214 (2023). https://doi.org/10.1002/bit.28346
10. Khanal, S.K., Tarafdar, A., You, S.: Artificial intelligence and machine learning for smart bioprocesses. Bioresour. Technol. **375**, 128826 (2023). https://doi.org/10.1016/j.biortech.2023.128826
11. Puranik, A., Dandekar, P., Jain, R.: Exploring the potential of machine learning for more efficient development and production of biopharmaceuticals. Biotechnol. Prog. **38**, e3291 (2022). https://doi.org/10.1002/btpr.3291
12. Gangadharan, N., et al.: Data intelligence for process performance prediction in biologics manufacturing. Comput. Chem. Eng. **146**, 107226 (2021). https://doi.org/10.1016/j.compchemeng.2021.107226
13. Corrales, D.C., et al.: Dinamic hybrid model for nanobody-based antivenom production (scorpion antivenom) with E. coli CH10–12 and E. coli NbF12–10. submitted. In: European Symposium on Computer Aided Process Engineering, ESCAPE 34, pp. 1–6 (2024)
14. Nikita, S., Thakur, G., Jesubalan, N.G., Kulkarni, A., Yezhuvath, V.B., Rathore, A.S.: AI-ML applications in bioprocessing: ML as an enabler of real time quality prediction in continuous manufacturing of mAbs. Comput. Chem. Eng. **164**, 107896 (2022). https://doi.org/10.1016/j.compchemeng.2022.107896
15. Pinto, J., Mestre, M., Ramos, J., Costa, R.S., Striedner, G., Oliveira, R.: A general deep hybrid model for bioreactor systems: combining first principles with deep neural networks. Comput. Chem. Eng. **165**, 107952 (2022). https://doi.org/10.1016/j.compchemeng.2022.107952
16. Pham, T.D., Manapragada, C., Sun, Y., Bassett, R., Aickelin, U.: A scoping review of supervised learning modelling and data-driven optimisation in monoclonal antibody process development. Digit. Chem. Eng. **7**, 100080 (2023). https://doi.org/10.1016/j.dche.2022.100080

17. Mondal, P.P., et al.: Review on machine learning-based bioprocess optimization, monitoring, and control systems. Bioresour. Technol. **370**, 128523 (2023). https://doi.org/10.1016/j.biortech.2022.128523
18. Alonso Villela, S.M., Ghezal-Kraïem, H., Bouhaouala-Zahar, B., Bideaux, C., Aceves Lara, C.A., Fillaudeau, L.: Effect of temperature on the production of a recombinant antivenom in fed-batch mode. Appl. Microbiol. Biotechnol. **105**, 1017–1030 (2021). https://doi.org/10.1007/s00253-021-11093-5
19. Hmila, I., et al.: Pre-clinical studies of toxin-specific nanobodies: evidence of in vivo efficacy to prevent fatal disturbances provoked by scorpion envenoming. Toxicol. Appl. Pharmacol. **264**, 222–231 (2012). https://doi.org/10.1016/j.taap.2012.07.033
20. Hmila, I., et al.: A bispecific nanobody to provide full protection against lethal scorpion envenoming. FASEB J. **24**, 3479–3489 (2010). https://doi.org/10.1096/fj.09-148213
21. Alonso Villela, S.M., Kraïem, H., Bouhaouala-Zahar, B., Bideaux, C., Aceves Lara, C.A., Fillaudeau, L.: A protocol for recombinant protein quantification by densitometry. MicrobiologyOpen **9**, 1175–1182 (2020). https://doi.org/10.1002/mbo3.1027

Computational Cost Reduction for MobileViT with Token Merging

Mitsuhiko Yasukura(✉), Michifumi Yoshioka, and Katsufumi Inoue

Graduate School of Informatics, Osaka Metropolitan University, Osaka, Japan
sb23982a@st.omu.ac.jp, {s21386n,inoue}@omu.ac.jp

Abstract. We focus on speeding up MobileViT, a model known for high accuracy in image classification tasks with fewer parameters. Our goal is to reduce the processing time of the Transformer part of MobileViT, which is the main slowdown factor, by using the Token Merging (ToMe) technique. However, because the number of tokens varies before and after merging, we cannot apply ToMe to MobileViT without making some modification. Therefore, we propose a simple method that facilitates the application of ToMe. By employing this method, we have successfully applied ToMe to MobileViT-s, leading to a slight drop in top-1 accuracy by 3.44% but reducing computational time by about 14.5%. While slightly inferior to the MobileViT-xs (smaller model), our model achieves similar speed and accuracy. Our method is simple and leaves the Token Merging algorithm unchanged. Thus, by improving the algorithm in addition to our proposed method, there is a potential to create models in the future that operate faster and with higher accuracy than MobileViT-xs. Additionally, since this modification can be adapted to other models facing similar issues, our proposal serves as a method that expands the adaptability of ToMe.

Keywords: Deep Learning · Transformer · Computational Cost Reduction · MobileViT

1 Introduction

High precision machine learning models, such as Swin Transformer V2 [1] and CoAtNet [2], often require substantial memory and computational resources. Consequently, the development of lightweight models is crucial for efficient inference in resource-limited environments.

Lightweight models for image processing tasks can be broadly categorized into three main types: convolutional neural network (CNN) based models, Transformer based models [3], and hybrid models that combine CNN and Transformer. Among these models, the hybrid models demonstrate higher accuracy in image processing tasks. This superior performance is attributed to the balanced ability of CNNs and Transformers, with CNNs specializing in capturing local features and Transformers in grasping global information. However, the hybrid models

R. Chinthaginjala et al. (Eds.): DCAI 2024, LNNS 1259, pp. 317–326, 2025.
https://doi.org/10.1007/978-3-031-82073-1_31

generally require longer computational times compared to CNN-based models. For instance, MobileViT [4], which is a hybrid model, demonstrates higher accuracy than MobileNetv2 [5] (CNN-based), and it has approximately eight times the computational time. To address this issue, we work on making the lightweight hybrid models faster.

The large number of tokens (image patches) in Transformer is one of the primary reasons for the slow computational speed of the hybrid models. For instance, the floating-point operations (FLOPs) of self-attention in Transformer are proportional to the square of the number of tokens. Therefore, reducing the number of tokens could effectively improve computational speed of MobileViT. Quantization [6], knowledge distillation [7], and pruning [8] are well-known techniques for accelerating machine learning models. Among them, there is a technique called Token Merging [9], which reduces tokens based on the idea of pruning. While pruning typically speeds up models by removing unimportant layers or weights, Token Merging speeds up models by removing redundant tokens. Since this technique reduces the computational time by half while maintaining accuracy in ViT-L [10], a model that contains many Transformers, it is considered an effective method for speeding up models with Transformers. Hence, to improve the computational speed of the lightweight hybrid models, we introduce Token Merging to them.

In our research on introducing Token Merging into lightweight hybrid models, we focus on MobileViT, which demonstrates very high performance. This is because introducing it into the high performance models will result in superior models. However, we cannot directly apply Token Merging to MobileViT. This is because the number of tokens changes before and after merging, which could cause mismatches in the flow of data between CNN and Transformer layers. In this paper, we propose a simple method that changes data shape to solve this issue. With this method, we can apply Token Merging to MobileViT, and make the model faster with a slight drop in accuracy. Furthermore, since this method is applicable to other hybrid models, we expect that the models will become faster with Token Merging.

2 Baseline Method

This section describes two methods that are crucial to this paper.

2.1 MobileViT

MobileViT [4] was proposed in 2021. The model introduces MobileViT blocks, a unique architecture for MobileViT, and MV2 blocks, which contain several convolutional layers. MobileViT structure is shown in Fig. 1, and MobileViT block is shown in Fig. 2. This model uses convolutional layers to extract local information and Transformers to extract global information. Combining these two architectures enables it to achieve higher accuracy than other lightweight models, such as MobileNetV3 [11] and Mobile-former [12]. However, it also faces the

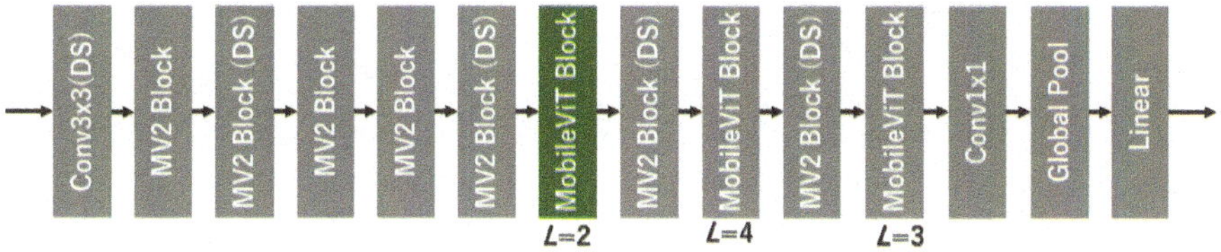

Fig. 1. MobileViT Structure. L is the number of layers of Transformers in the corresponding block. DS means down-sampling.

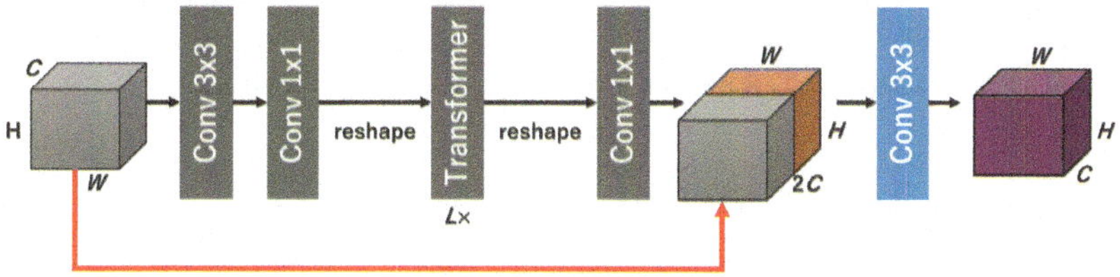

(a) Original Structure

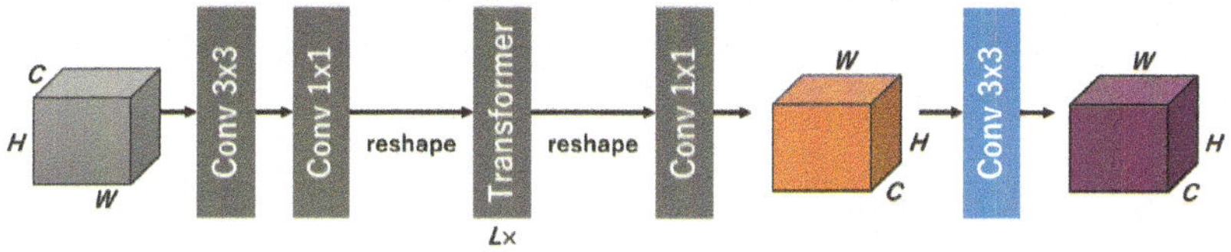

(b) Our Structure

Fig. 2. MobileViT Block Structure. A problem arises with skip-connections (red arrows in Figure (a)) in introducing Token Merging. Thus, we remove the skip-connection in our proposed method (Figure (b)).

issue of longer computational times compared to CNN-based lightweight models. This paper focuses on addressing this issue by introducing Token Merging to MobileViT.

2.2 Token Merging

Token Merging has been introduced to speed up Vision Transformer models. This approach merges tokens with similar feature representations, aiming to cut down computational redundancy. Using this method decreases the overall number of tokens, which in turn reduces the number of Floating Point Operations (FLOPs) required. This reduction contributes to improved processing speed. Token Merging is facilitated by introducing an architecture for it immediately after the Transformer [9]. The number of tokens to be merged can be adjusted according to the model or task, making it a highly versatile acceleration method.

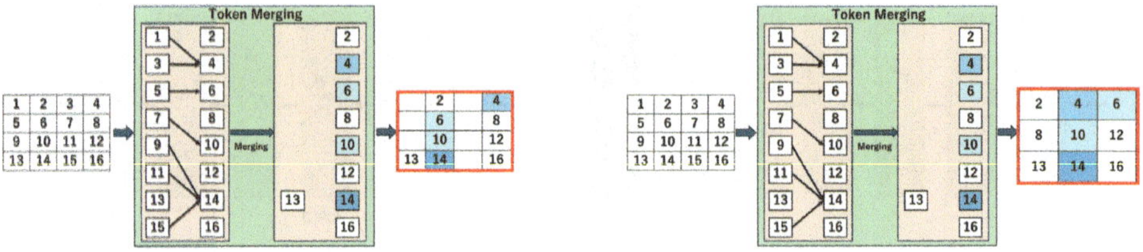

(a) Original Reshape (b) Our Reshape

Fig. 3. Reshape Features After Token Merging. In the reshaping process shown in Figure (a), a portion of the features located at the end (highlighted within the red frame) is missing. However, in the reshaping process we propose, as depicted in Figure (b), all features are retained, with no missing features at the end.

3 Adapting Token Merging to MobileViT

Although we aim to accelerate MobileViT by introducing Token Merging, it cannot be directly applied to the model. This section discuss the reasons for this limitation and outline the necessary modifications and processes required for implementation.

3.1 Reasons Why Token Merging Cannot Be Directly Applied to MobileViT

Convolutional layers are designed to capture information from images and typically require input feature shaped like squares or rectangles. Therefore, in MobileViT, the features produced by the Transformer are reshaped back to their original shapes to use in the subsequent convolutional layers (for example, the layers highlighted in blue in Fig. 2).

However, when Token Merging is applied, some features are lost, making it impossible to revert to the former image-like shape as shown in Fig. 3a. Therefore, simply introducing Token Merging, as is done in ViT base models, causes the model to fail to function.

3.2 About Modifications

To address the issues mentioned in Sect. 3.1, we propose simple modifications. These modifications can be applied to other models facing the same issues as MobileViT, such as PiT [13] and the successor models of MobileViT [14,15].

Feature Reshaping. The first modification is reshaping the features produced by the Transformer. As mentioned in Sect. 3.1, introducing Token Merging prevents the features produced by the Transformer from being reverted to the former shapes. Therefore, as illustrated in Fig. 3a, we eliminate the spaces and rearrange the features into a more compact square formation. This technique is simple and easy to implement.

Delete Skip-Connection. Another modification concerns the skip-connections within the MobileViT block. As indicated by the red arrows in Fig. 2a, the MobileViT block includes skip-connections. When reshaping tokens as described in Sect. 3.2, the size of the features changes before and after the Transformer. Therefore, the features from the skip-connection and those from the Transformer differ in size. This size difference makes it impossible to execute convolution in the convolutional layer indicated in blue in Fig. 2a.

To address this, we remove the skip-connections within the MobileViT block where Token Merging is introduced, as shown in Fig. 3b. Eliminating skip-connections not only solves the size mismatch issue but also reduces the computational time of the convolutional layers shown in blue. This elimination aligns with our aim to speed up MobileViT.

At first glance, these modifications seem to significantly decrease accuracy. Therefore, in Sect. 4, we discuss not only the time efficiency but also the accuracy in detail.

4 Experiments

We introduce Token Merging into MobileViT using modifications described in Sect. 3 to evaluate the performance of this model for the image classification task. MobileViT is available in three sizes: s, xs, and xxs. For our experiments, we have chosen MobileViT-s as the baseline model to introduce Token Merging.

In the first experiment, we examine how performance varies with the number of tokens merged. In the second experiment, we evaluate the impact of Token Merging by comparing the performance of the modified models incorporating this technique against the baseline model (MobileViT-s) and the smaller model (MobileViT-xs). Performance metrics include top-1 accuracy, inference time, FLOPs, and the number of parameters. As discussed in Sect. 1, computational time is important for this research. Therefore, we focus on evaluating performance in terms of top-1 accuracy and inference time.

4.1 Number of Tokens to Be Merged

Purpose. The number of tokens to be merged in one step is not fixed. Consequently, it is necessary to adjust the number of tokens to be merged according to the model and task. In fact, in the acceleration of the Stable Diffusion [16] with Token Merging, a large number of tokens are merged [17].

In this experiment, we aim to determine the optimal number of tokens to be merged for MobileViT-s by comparing various merging settings based on inference time and preliminary top-1 accuracy calculations.

Location and Amount to Merge. Regarding the placement of the Token Merging, we targeted only the first MobileViT block (green block in Fig. 1) of the three. The reason for focusing on this block is that by reducing the number

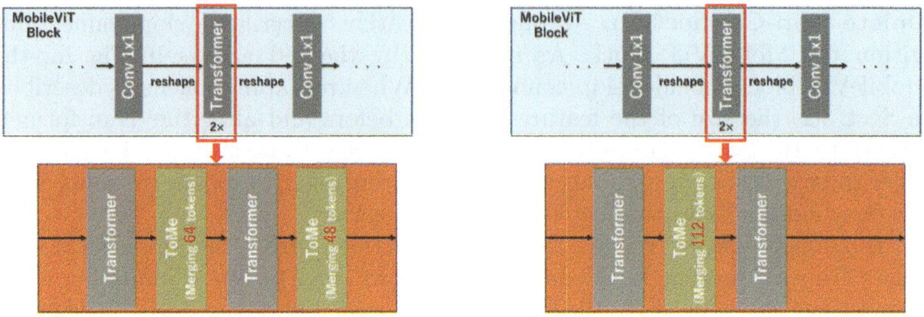

(a) $r = [64, 48]$ (b) $r = [112, 0]$

Fig. 4. MobileViT Block with Token Merging

of features through Token Merging in the preceding layer, we can decrease both the FLOPs and the computational time for all subsequent layers.

Since there are two Transformers within that MobileViT block, we experimented with varying the number of tokens to merge immediately following each Transformer. For the notation regarding the number of tokens to merge right after each Transformer, we use $r = [a, b]$. Here, 'a' denotes the number of tokens to merge right after the first Transformer, and 'b' denotes the number of tokens to merge right after the second Transformer. For instance, as shown in Fig. 4a, $r = [64, 48]$ means that 64 tokens are merged right after the first Transformer and 48 tokens are merged right after the second Transformer. Similarly, as in Fig. 4b, $r = [112, 0]$ means that 112 tokens are merged right after the first Transformer, and no Token Merging is performed right after the second Transformer. Thus, $r = [0, 0]$ means no introduction of Token Merging (i.e., the model remains unchanged).

Implementation Details. We varied the number of tokens to be merged and evaluated four evaluation metrics (top-1 accuracy, inference time, FLOPs, and the number of parameters). The baseline model for this study is MobileViT-s. Because training on the ImageNet-1k dataset takes a very long time, we utilized ImageNet-100, a subset of ImageNet-1k [18]. ImageNet-100 consists of 100 classes, with 130,000 training images and 5,000 test images.

The input images were resized to 256×256 pixels. We employed random resized cropping and horizontal flipping as data augmentation techniques. We used SGD as the optimization method with a learning rate of 0.001. The batch size was set to 30 to measure top-1 accuracy and inference time. We set the number of training epochs to 50.

These experimental conditions were set based on the conditions used for training the original MobileViT [4], with some modifications. Although the number of epochs is not sufficient for convergence, we think it is adequate to judge the relative performance of the models. This belief is supported by the fact

Table 1. Performance for Various Amounts of Merging. Base model is MobileViT-s.

amounts of merging	top1(50Epoch)	inference time[ms]	FLOPs[G]	params[M]
$r = [0, 0]$ (Base Model)	70.18%	1.878(0.0%)	1.86	5.00
$r = [60, 0]$	64.14%	1.842(−1.9%)	1.63	4.92
$r = [112, 0]$	64.20%	1.606(−14.5%)	1.41	4.92
$r = [30, 30]$	60.98%	1.894(+0.9%)	1.71	4.92
$r = [56, 56]$	61.66%	1.679(−10.6%)	1.49	4.92

that extending training by an additional 100 epochs for some models did not change their relative performance. Regarding the computing resources, inference time measurements were performed using an Intel Core i7-12700 CPU and an NVIDIA GeForce RTX 3070 GPU. The memory capacities for the CPU and GPU are 16GB and 8GB, respectively.

Result. The experimental results are shown in Table 1. Here, the inference time is reported as the average processing time per image.

Accuracy. Both configurations $r = [60, 0]$ and $r = [112, 0]$ achieve an accuracy of approximately 64%. Similarly, configurations $r = [30, 30]$ and $r = [56, 56]$ achieve an accuracy of around 61%. Generally, a smaller total reduction in data through merging tends to result in better accuracy. However, the configuration with a larger total reduction, $r = [112, 0]$, achieved comparable accuracy to that with the smaller total reduction, $r = [60, 0]$. Similarly, the accuracy of configuration $r = [56, 56]$ is equivalent to that of the smaller total reduction configuration $r = [30, 30]$. These results suggest that accuracy does not significantly change based on the total amount reduced.

Attention is then turned to the number of layers implementing the introduced Token Merging. Hereinafter, layers performing Token Merging will be referred to as ToMe layers. A configuration with no ToMe layers, $r = [0, 0]$ (i.e., the base model), achieves an accuracy of approximately 70%. Moreover, configurations with one ToMe layer, specifically $r = [60, 0]$ and $r = [112, 0]$, achieve an accuracy of around 64%. Configurations with two ToMe layers, such as $r = [30, 30]$ and $r = [56, 56]$, experience a further decrease in accuracy to 61%. These result indicates a gradual decline in accuracy with an increased number of ToMe layers.

Thus, we can infer that accuracy is more influenced by the number of ToMe layers introduced than by the number of tokens merged. This insight could be beneficial for future applications of Token Merging.

Inference Time. Configurations with a smaller total number of merged tokens, $r = [60, 0]$ and $r = [30, 30]$, show little change in inference time. However, configurations with a larger total number of merged tokens, such as $r = [112, 0]$ and $r = [56, 56]$, demonstrate a reduction in inference time. Moreover, when the total number of merged tokens is the same, configurations with fewer ToMe layers result in shorter inference times. Therefore, the configuration $r = [112, 0]$ is superior to $r = [56, 56]$ in terms of reducing inference time.

Inference time has increased with the configuration $r = [30, 30]$. This increase is because the time required for processes for Token Merging, such as reshaping data and calculating dot products, exceeds the reduction in time.

Better Merging Setting. As indicated in Table 1, a comparison of the performance across the four merging configurations, $r = [60, 0], [112, 0], [30, 30]$, and $[56, 56]$, reveals that the configuration $r = [112, 0]$ produces superior performance. These findings suggest that significant token reduction with minimal ToMe layers is crucial when applying Token Merging to lightweight models.

4.2 Comparison with MobileViT-s and -xs

Purpose. In Sect. 4.1, it was determined that employing the configuration $r = [112, 0]$ is advantageous when applying Token Merging. However, the efficacy of Token Merging for MobileViT remains unclear due to the lack of comparisons with already existing models lighter than MobileViT-s. Therefore, performance will be compared among three models: the base model (i.e., MobileViT-s), the base model with configuration $r = [112, 0]$, and MobileViT-xs, which is smaller than the base model.

Implementation Details. Measurements of four evaluation metrics (top-1 accuracy, inference time, FLOPs, and the number of parameters) were conducted under the same conditions as described in Sect. 4.1, except that the training was carried out twice, with each session consisting of 400 epochs. The reason for setting the number of training epochs to 400 is that the highest accuracy of MobileViT-s was not updated after 350 epochs. The models evaluated include MobileViT-s, MobileViT-s with the $r = [112, 0]$ configuration, and MobileViT-xs.

Table 2. Performance of The Three Models. The model with $r = [0, 0]$ represents the original model without Token Merging applied. The model with $r = [112, 0]$ is the model with Token Merging applied (ours).

Model	top1(400Epoch)	inference time[ms]	FLOPs[G]	params[M]
MobileViT-s, $r = [0, 0]$	(80.52±0.06)%	1.878(0.0%)	1.86	5.00
MobileViT-s, $r = [112, 0]$	(77.08±0.08)%	1.606(-14.5%)	1.41	4.92
MobileViT-xs, $r = [0, 0]$	(77.92±0.06)%	1.477(-21.4%)	0.93	1.97

Result. The results are shown in Table 2. The inference time is reported as the average processing time per image. Additionally, the top-1 accuracy is the highest accuracy measured on the test images every two epochs of training. This is because comparing the highest accuracies enables the evaluation of the performance by eliminating the effects of overfitting.

Comparison with MobileViT-s. From Table 2, MobileViT-s with Token Merging decreases by 3.44% in top-1 accuracy and reduces inference time by 14.5% compared to the original MobileViT-s. These results suggest that our method can reduce computational time with a small loss in accuracy for CNN and Transformer hybrid models that are not directly compatible with Token Merging.

Comparison with MobileViT-xs. Comparing MobileViT-s with Token Merging and MobileViT-xs, our model with Token Merging is inferior to MobileViT-xs. However, the differences in top-1 accuracy and inference time, which are important for this study, are slight. Therefore, from the perspective of inference time and accuracy, our method of applying Token Merging to MobileViT-s can achieve performance close to that of MobileViT-xs. This suggests that the simple modifications we propose and the adaptation of Token Merging to MobileViT can have impacts on computational time and accuracy, comparable to those of scaling down the model size.

5 Conclusion

By applying our method, we successfully implement Token Merging in MobileViT-s, reducing the computational time by 14.5% with a small loss in accuracy of 3.44%. However, the performance of the model with Token Merging was inferior to that of MobileViT-xs. Nevertheless, our method focused on modifications to the process after merging, employing the Token Merging algorithm without any changes. Hence, this approach leaves significant potential for further improvements. Indeed, there are potential improvements for the Token Merging algorithm depending on the purpose. For example, while the original algorithm is capable of merging only up to approximately half of all tokens in one step, a modified version aimed at improving Stable Diffusion allows for the merging of up to approximately three-quarters of all tokens [17]. Incorporating this change is expected to further reduce the inference time in the MobileViT model. Additionally, our method can be utilized for other hybrid models that are unable to introduce Token Merging for reasons similar to those of MobileViT. Consequently, our method may contribute to the development of various models capable of achieving high accuracy image recognition in environments with resource constraints in the future.

References

1. Liu, Z., et al.: Swin Transformer V2: scaling up capacity and resolution. In: Proceedings of the International Conference on Computer Vision and Pattern Recognition (CVPR) (2022)

2. Dai, Z., Liu, H., Le, Q.V., and Tan, M.: CoAtNet: marrying convolution and attention for all data sizes. In: Advances in Neural Information Processing Systems, vol. 34, pp. 3965–3977 (2021)
3. Vaswani, A., et al.: Attention is all you need. In: Advances in Neural Information Processing Systems, pp. 5998–6008 (2017)
4. Mehta, S., Rastegari, M.: MobileViT: light-weight, general-purpose, and mobile-friendly vision transformer. In: Proceedings of the International Conference on Learning Representations (2022)
5. Sandler, M., Howard, A., Zhu, M., Zhmoginov, A., Chen, L.-C.: MobilenetV2: inverted residuals and linear bottlenecks. In: Proceedings of the IEEE Conference on Computer Vision and Pattern Recognition, pp. 4510–4520 (2018)
6. Wu, J., Leng, C., Wang, Y., Hu, Q., Cheng, J.: Quantized convolutional neural networks for mobile devices. In: Proceedings of the IEEE Conference on Computer Vision and Pattern Recognition (CVPR) (2016)
7. Hinton, G., Vinyals, O., Dean, J.: Distilling the knowledge in a neural network. arXiv:1503.02531 (2015)
8. Han, S., Pool, J., Tran, J., Dally, W.: Learning both weights and connections for efficient neural network. In: Advances in Neural Information Processing Systems, vol. 28 (2015)
9. Bolya, D., Fu, C.-Y., Dai, X., Zhang, P., Feichtenhofer, C., Hoffman, J.: Token Merging: Your ViT but faster. In: Proceedings of the International Conference on Learning Representations (ICLR) (2023)
10. Dosovitskiy, A., et al.: An Image is Worth 16×16 words: transformers for image recognition at scale. In: Proceedings of the International Conference on Learning Representations (2021)
11. Howard, A., et al. Searching for MobileNetV3. In: Proceedings of the IEEE/CVF International Conference on Computer Vision, pp. 1314–1324 (2019)
12. Chen, Y., et al.: Mobile-Former: bridging MobileNet and transformer. arXiv:2108.05895 (2021)
13. Heo, B., Yun, S., Han, D., Chun, S., Choe, J., Oh, S.J.: Rethinking spatial dimensions of vision transformers. In: Proceedings of the IEEE/CVF International Conference on Computer Vision (ICCV) (2021)
14. Mehta, S., Rastegari, M.: Separable self-attention for mobile vision transformers. arXiv:2206.02680 (2022)
15. Wadekar, S.N., Chaurasia, A.: MobileViTv3: mobile-friendly vision transformer with simple and effective fusion of local, global, and input features. arXiv:2209.15159 (2022)
16. Rombach, R., Blattmann, A., Lorenz, D., Esser, P., Ommer, B.: High-resolution image synthesis with latent diffusion models. In: Proceedings of the IEEE Conference on Computer Vision and Pattern Recognition (CVPR) (2022)
17. Bolya, D., Hoffman, J.: Token merging for fast stable diffusion. In: CVPR Workshop on Efficient Deep Learning for Computer Vision (2023)
18. Russakovsky, O., et al.: ImageNet large scale visual recognition challenge. Int. J. Comput. Vis. **115**(3), 211–252 (2015). https://doi.org/10.1007/s11263-015-0816-y

Clues for Improving Database Systems Quality-of-Service – A Process Mining Approach

Bárbara Teixeira and Orlando Belo(✉)

ALGORITMI Research Centre/LASI, University of Minho, 4710-059 Braga, Portugal
a47038@alunos.uminho.pt, obelo@di.uminho.pt

Abstract. One of the tools most used by database administrators in the analysis of database systems quality of service problems are system event logs, which contain detailed information about the actions (SQL transactions) performed over time in a database system. Using these files, applying their experience and appealing to the expertize of data analysis tools, database administrators seek to establish event patterns that may identify the causes and reveal a path for establishing an appropriate solution to the problems. In this paper, we present and discuss a process mining system specifically designed and implemented to detect operational bottlenecks in relational database systems using transaction logs, providing expert information for their resolution, and contributing to improve system's quality of service.

Keywords: Database Systems · Database Systems Performance · Querying Analysis · Quality of Service · SQL Transaction Logs · Process Mining

1 Introduction

Current database systems transaction logs [1] are extremely rich data files, containing very pertinent information, which allow us to analyze over time what actually happened in the system in terms of actions to describe, manipulate or control its data elements. A careful analysis of this information may reveal situations of operational constraints, which may occur during the execution of one or more queries, whose characteristics – owners, permissions, types of data, volume of data, etc. – caused long waiting times, or even situations in which system's users are unable to carry out their activities on a regular basis. Throughout the operational life of a database system, query logs host very relevant data elements in terms of time, which allow us to observe all actions triggered in the system each day and hour of work, the resources that used, the duration of each of these actions and, obviously, the users who triggered them.

Sometimes operational bottlenecks happen in database systems do not have an obvious explanation, even when analyzed based on the performance indexes and functional characterization that traditional log analysis tools provide. In these cases, the experience of the system administrator makes a difference. A large part of these situations are caused by sequences of events, not by isolated events, which at some point in time combine and cause unwanted operational situations. In fact, although these situations are not frequent, they raise many obstacles, which are not always easy to overcome, even when the level

R. Chinthaginjala et al. (Eds.): DCAI 2024, LNNS 1259, pp. 327–336, 2025.
https://doi.org/10.1007/978-3-031-82073-1_32

of experience of the database system administrator is high. The adoption of process mining techniques [2, 3] within an organization aims to discover, monitor and improve unforeseen real processes through knowledge extraction practices performed on records of events occurring in a given system. In most of its applications, we use process mining for detecting problems in event logs – e.g. performance degradation, operational freezes, computational resource conflicts or inefficient tasks – and, consequently, to analyze how such problems affect the system. From the analysis of the data obtained in a process mining process, it is possible to improve the behavior of the system and consequently its performance [4]. In short, improve system's quality of service.

In this paper, we present the inception and implementation of a process mining system especially conceived for improving the quality of service of a relational database system using transaction logs contents. The goal is not to replace the expertise of a database administrator or his ability for tracking and troubleshooting system's poor performance or service dissatisfaction by their users. With this work, we provide some means to complement his services, analyzing in a more intensive and detailed way the transaction logs of a database system, seeking to find problematic operational patterns that may contribute to lower the quality of service of the system or even jeopardize its general operability. Applying process mining techniques and using a real-world transaction log, enriched it with other data elements considered pertinent, we identified some problematic situations in the event process map generated and made some suggestions for improving the performance of the database system under analysis. The remaining part of this paper is organized as follows. Section 2 exposes the domain of process mining, approaching its origins, evolution, and applications. Section 3 presents and discusses the process we conceive and implement for analyzing the quality of service of a relational database system. Finally, Sect. 4 presents some remarks and conclusions.

2 Applying Process Mining

A quick analysis of the work carried out over the last decade in the field of mining and process analysis reveals a very varied set of works with great relevance in diverse application domains. Most of them were the result of research processes promoted and carried out by Wil Van der Aalst [3, 5] in conjunction with a large community of researchers involved in his project network and initiatives. In fact, we could rely solely on his works to provide a very concrete and detailed idea of the field of process mining, and of the broad range of works and concrete applications that were developed using the techniques, methods and tools designed and developed. However, we think that it is important to cross the boundaries of Van der Aalst's works, to provide a broader, more relevant and fair insight about the domain of process mining.

Over the last 25 years, numerous researchers have investigated process mining in order to find new approaches for improving the performance of systems and consequently the processes that are integrated. To the best of our knowledge, Cook and Wolf [6] conducted the first noteworthy study in the domain. These authors attempted to discover software process models from data contained in event logs. In the same year, Van der Aalst [7] analyzes the use of Petri nets in workflow management processes to analyze process performance. Subsequently, other authors such as Medeiros et al. [8] analyzed the work carried out by Van der Aalst, studying and discussing what they considered the weaknesses and strengths of his proposal, proposing some improvements in relation to the use of information from noisy data in the process of analysis. Later, Weijters et al. [9] and Turner et al. [10] presented process mining as a technique that attempts to extract useful, non-trivial information from records of events, of actions, performed by some kind of system. According to another perspective of use, presented by Medeiros et al. [11], the objective of process mining is the discovery of process models from event records, which can be used to extract information about tasks that were performed at some point, as well as about their chance relationships. Lassen et al. [12], in our view, reinforced this perspective a bit more, claiming that process mining would allow dealing with event records with the objective of discovering control flows of one or more process models. Today, there is many concrete evidences of practical applications of process mining. Garcia et al. [13] analyzed about 1200 works in order to identify the main areas and domains of application of process mining. In this study, they concluded that the most relevant applications might be found in healthcare [14, 15] and information technology [16,17]. In the next section, we present a process mining application initiative inside the information technology domain, which was conceived especially oriented to help improving the quality of service of a relational database system.

3 Improving Database Querying Services

3.1 Mining SQL Log Events

The adoption and integration of process mining techniques by an organization allows for extracting effective models of the behavior of a system, through the analysis of the records of operational processes, tasks, actions or events that the system performs over time. With these models we can get a very concrete description of the processes carried out by the system, which allows to better understand its functioning and, if necessary, help make decisions about possible changes in its behavior, whether it is good or bad [18].

In database systems, it is common to record all the actions carried out over time, required either directly by their users or by software applications that use them as information sources. These actions can be of several types. However, transactions, which are sequences of data manipulation operations, are those that probably have the greatest importance and impact in monitoring and analyzing the overall operability of the database system and in evaluating its quality of service. Transactions are defined and executed in order to satisfy a given set of operational requirements, by performing a well-defined set of operations over databases. With a proper configuration, a database system records and stores all these operations in specific files called transaction logs. The information contained in these files may aid database system administrators to correct operational performance problems and improve the quality of database services. This will have a positive impact on their user community. A community of "happy" users is an excellent indicator of the usefulness and performance of the system. However, it is not always easy to find effective ways to analyze the information in the transaction logs, in a way that gives us important clues that can contribute for solving problematic situations and improving the overall operability of the system [19]. Current database system market offers a very diverse range of tools for monitoring and analyzing data contained in transaction logs [20]. Without wishing to disregard the value of such tools, in fact, we find them very valuable and useful tools, we decided to follow a different approach (Fig. 1) for analyzing and evaluating the quality of service of a database system: to identify non-trivial and non-evident operational patterns occurring in a database system. Thus, we decided to design a process mining based solution, believing that we would be able to establish such patterns of anomalous behavior through intensive processing and analysis of the transaction log of a relational database system.

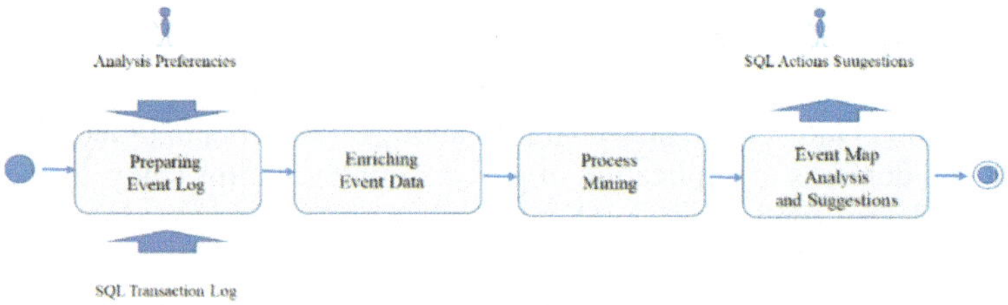

Fig. 1. Illustration of the mining and analysis process.

3.2 Log Mining and Analysis

Before starting and carrying out the first phase of the project, we had to arrange the "raw material" for the work: a transaction log. Initially, we thought to synthetize one, using the structure of a database manufacturer's transaction log or using a generative artificial intelligence tool to do so. However, knowing that this possibility allows us to have a valid transaction log file for using in the mining process, we consider that the problems we would encounter (and later present some solutions for them) would not have a much practical effect.

Preparing Event Log. We start working preparing a transaction log provided by an external organization – its name and activity we will not mention here due to confidentiality reasons. The transaction log contains a set of SQL operation records, quite extensive and diverse, in JSON format. We made a first analysis of its content. The transaction file had been generated by an Oracle system and contains 98 attributes (columns) – if necessary, the structure of Oracle's transaction files can be consulted online [21]. Then, we extracted from the original file only the information identified as useful and relevant (11 attributes) (Table 1) for mining and system analysis we want to carry out. Next, we performed a set of transformation operations over a few attributes, in order to separate specific data elements considered crucial to the analysis process.

For each transaction log record, we split the attribute SQL_TEXT into two new attributes, namely STATEMENT and OBJECT. In the first we store data manipulation operations – SELECT, INSERT, UPDATE or DELETE – included in the transaction, and in the second attribute the designation of the table (database object) on which the query

Table 1. The transaction log attributes.

Log Attribute	Description
SQL_TEXT	Executed query
SQL_ID - ID	Identifier generated by the database system for each SQL statement cached
SORTS	Number of ordered lines for the statement
FETCHES	Number of lines got by the statement
EXECUTIONS	Number of times the query was executed since it entered the cache
LOADS	Number of times the query was loaded or reloaded to the cache
ROWS_PROCESSED	Number of rows processed
SERVICE	Applied service
FIRST_LOAD_TIME	First time the system uploaded the query
LAST_LOAD_TIME	Last time the system uploaded the query
LAST_ACTIVE_TIME	Last time the system executed the query

was executed. Then we add these two new attributes to the previous set of attributes, previously selected in the original transaction log. During this process, we found that our event log did not contain information about some attributes included in log structure, not even showing the identification of the attributes themselves. This did not allow us to have an immediate and concrete perception of the meaning of the values that were contained in these attributes, even after we had analyzed their documentation. So, an event map would allow us to verify the quality of the log data, and if the information we have would be sufficient to move forward with the mining process. To do this, we selected the first 1000 records from the transaction log and processed them with the process mining tool Disco [22]. This tool was developed by Fluxicon in order to make process mining more accessible to users. It has been used in various application domains, by several organizations, for solving auditing, improvement and process optimization problems. The tool allows for easy and flexible event discovery due to its effective visualization and filtering capabilities. In [23] one can find an interesting list of other process mining tools.

After converting the logs to the data format recognized by Disk, we imported them into the tool environment and executed the event mining process, configuring the process according to our case identifier, the event, the temporal tags, the resources used, among other things. From the map of events obtained, we made the selection of some instances, according to our initial analysis criteria that we established to discover the content of the transaction log and the type of events that were recorded in it. In Fig. 2 we present one event map (instance) revealing the absolute frequency of the identified events. This first approach for analyzing the event maps, we found a few interesting things, namely the attribute(s): SQL_ID. Configured as the identifier of the cases, it restricted the type of analysis we intended to perform, since being a single value it did not allow the aggregation of the instructions of the cases; FIRST_LOAD_TIME and LAST_ACTIVE_TIME.

Timestamps cannot effectively represent the start and end date (and time) of each event. Consequently, the event average time is quite large, since the values of these attributes are mostly very far apart. Thus, it is necessary to prepare new time values to obtain more realistic transaction maps. Finally, STATEMENT, the operation carried out in each event, needed more information, in order to obtain a more detailed map of the transactions executed. Based on this preliminary analysis, we decided to enrich the events with some more information, in particular with data related to the execution periods of transactions. In the next section, we present the enrichment actions carried out.

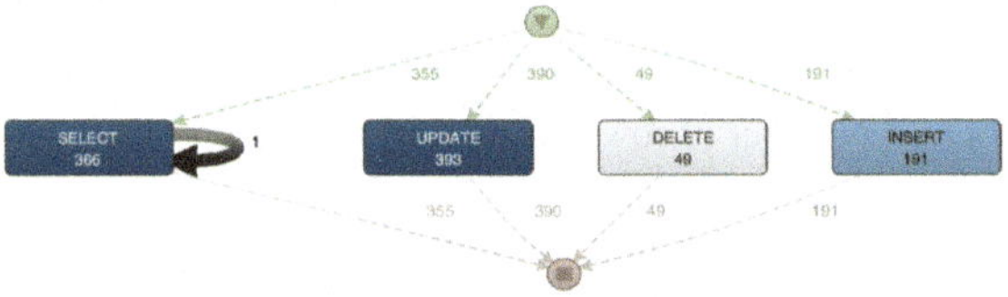

Fig. 2. Absolute frequency of SQL events.

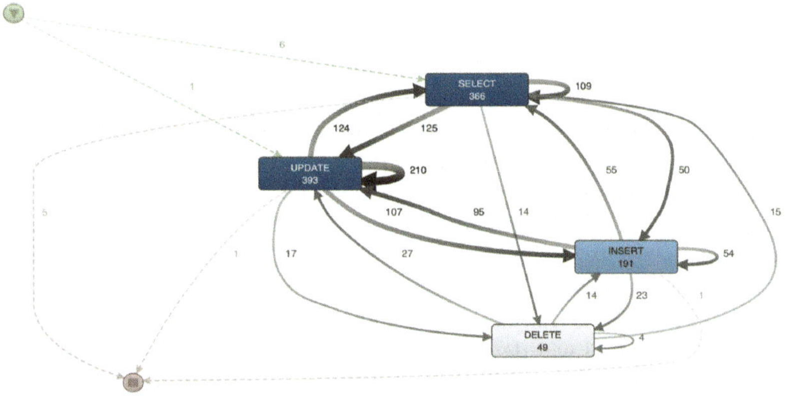

Fig. 3. Event map having as case identifier DAY_OF_THE_WEEK.

Enriching and Mining Event Data. In a traditional event log we can find a set of records about executed actions, which one usually contain a reference to a case identifier, a reference to an activity, and at least one timestamp (date and time) [24]. In the preliminary analysis of the event log, we found that it would be necessary to enrich it with other information, in order to get an event map having relevant information for allowing reaching the goals previously established. We started by adding an attribute that would serve as the identifier of the cases. This is essential since this attribute defines the different possible variants for each case. Next, we added a new attribute, DAY_OF_THE_WEEK, to characterize the execution period of a transaction in weekly terms, getting its values deriving the values of the attribute FIRST_LOAD_TIME. With these new elements, we got a new event map (Fig. 3), more valuable, obtained by choosing the latter attribute as case identifier.

In order to allow us to view the transactions carried out from different perspectives, we decided to add another attribute to our log, WEEK_OF_THE_MONTH, to describe how the transactions occurred in each week of the month. Although this identifier considers only four cases, which obviously correspond to the weeks of the month, we thought that it could reveal some patterns regarding the type of events that occur at certain times of the month that may be more overloaded, with a greater number of transactions carried out. Finally, we added two other attributes, quite generic, which do not distinguish many cases, but which could be of some use to the ongoing study. However, it was important to add a wider field of application that would allow the log to be separated into a greater number of cases. With this, we expected to get a more descriptive view of the data resulting from the mining process. Therefore, we added a CASE_ID attribute to represent each of the days mentioned in the log records using the values of the FIRST_LOAD_TIME attribute.

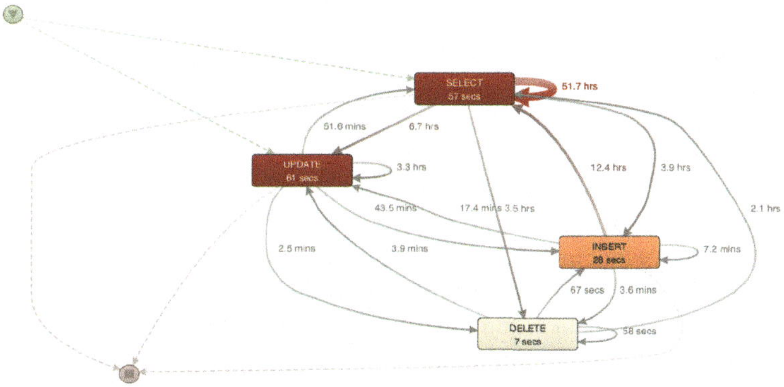

Fig. 4. System performance event map.

Although the changes made went towards our goals, we still had time values very similar to the initial values contained in the log. As we wanted to identify points of performance degradation, it was important to have time values that would allow us to carry out, in a more realistic way, the study of a transaction log, having concrete information about the execution time of each event. Our transaction log did not allow us to obtain such values. For this reason, we decided to synthesize the execution times for each of the recorded queries, assuming some correction values based on our experience in the field. From this process, a new attribute emerged, FINISHED_LOAD_TIME, which hosted the execution time generated for the query from the FIRST_LOAD_TIME attribute. In this way, we obtained an event map reflecting the performance of the system (Fig. 4). Finally, we merged the attributes STATEMENT and OBJECT into a single one, ACTIVITY, so that we could distinguish the various data manipulation operations executed and evaluate their influence on the overall system's performance. Due to its size, we cannot present the event map here or other instances.

Event Map Analysis and Suggestions. Although we have used a real transaction log in our work, for confidentiality reasons, we do not have access to the application domain of the relational database system that produced it. This knowledge could help us to understand better the contents of the transaction log, as well as to understand in a more focused way the event maps produced. However, this did not hinder our work. After getting a global event map, making a detailed analysis, we created a set of small application scenarios we assumed were likely to occur in the application environment of the system in analyses. In those scenarios, we selected some queries that had longer execution times, which may indicate that something could be done to lessen their impact on overall system performance. Let see two of these scenarios (Fig. 5).

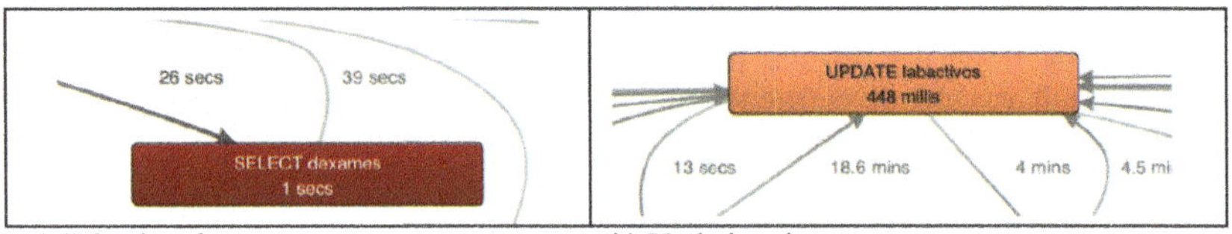

a) Selecting data b) Updating data

Fig. 5. Two analysis scenarios extracted from the global event map.

The first scenario involves a simple query used to select all the rows and columns of a table (Fig. 5a). The query in question does not have any projection or filtering criteria associated with it, which means that its execution will return the entire contents of the target table. If some of that information was not used, we could reduce the computational effort of the system by applying some kind of selection and projection criteria that would meet user or application data requirements that launched that query. This will allow to reduce the number of disk writes/reads, memory utilization, or CPU time, positively influencing the overall performance of the query. For a case like this, we suggest defining a concrete projection criterion, identifying the smallest set of attributes to include in the query result set, the most relevant attributes for the user, as well as applying filtering criteria to reduce the querying data space.

The second scenario considers a query to update data in a table (Fig. 5b), which is performed frequently, without involving any data index. Indexes are important, especially in database systems with large tables, helping to improve access to data in most operations that use them. However, they can have both positive and negative effects on updating operations. Usually, this kind of operations is performed in two steps. One to find the desired information and the other to modify that same information. This query includes a filter criterion. As such, it would be interesting to evaluate the definition of an index on the filtering attribute, so that the information could be found and modified faster. For this case, we would suggest identifying the attributes used in the filtering criterion, for selecting the attributes' values to be modified, and evaluating a possible definition of indexes over these same attributes, since it is a very regular execution query.

Many other cases could be analyzed. The map of events generated in this work revealed a very diverse range of situations that could be the target of a careful analysis and an eventual application of some kind of corrective action. However, we think that the cases we have selected provided a concrete idea of the potential and usefulness of the application of process mining techniques in the evaluation of the quality of service of a relational database system.

4 Conclusions and Future Work

In this paper, we presented the development of a process mining system especially oriented for discovering non-trivial and non-evident operational patterns that could help database systems administrators to identify anomalous operational situations, which may affect direct or indirectly the quality of service of the systems under their supervision. After improving and enriching the structure and contents of the transaction log we had access, we generated several event (SQL transaction) maps, which gives us a clear view of the functioning of the system during the period of analysis. With the map, it was possible to get some key performance indicators (e.g. absolute frequency, average time, and minimum and maximum time of SQL event) as well as identify SQL queries having some unusual characteristics (e.g. great execution frequency, large result sets, or high CPU times) according to the application scenario established. Based on two distinct cases extracted from the global event map, it was possible to evaluate potential performance problems and suggest some corrective actions for improving system operability. All the experience done and the results got were positive and are clear indicators that the research path followed can provide in a near future interesting clues for an effective and applicable solution for improving database systems quality-of-service in real world application scenarios.

Acknowledgements. This work has been supported by FCT – Fundação para a Ciência e Tecnologia within the R&D Units Project Scope: UIDB/00319/2020, and the PhD grant: 2022.12728.BD.

References

1. Li, Q., et al.: Transactions Make Debugging Easy (2022). https://doi.org/10.48550/arXiv.2212.14161
2. Van Der Aalst, W.: Process mining: overview and opportunities. ACM Trans. Manage. Inf. Syst. (TMIS) **3**(2), 1–17 (2012)
3. Van der Aalst, W.: Process Mining. Springer, Heidelberg, January 2016. https://doi.org/10.1007/978-3-662-49851-4
4. Belo, O., Dias, N., Ferreira, C., Pinto, F.: A Process Mining Approach for Discovering ETL Black Points. WorldCIST 2017. In: Advances in Intelligent Systems and Computing, vol. 570. Springer, Cham (2017). https://doi.org/10.1007/978-3-319-56538-5_43
5. Wan der Aalst, W., Carmona, J.: Process mining handbook. Springer Nature, 2022 (2022)
6. Cook, J., Wolf, A.: Discovering models of software processes from event-based data. ACM Trans. Softw. Eng. Methodol. (TOSEM) **7**(3), 215–249 (1998)
7. Van der Aalst, W.: The application of petri nets to workflow management. J. Circuits Syst. Comput. **8**(01), 21–66 (1998)
8. Medeiros, A., Van der Aalst, W., Weijters, A.: Workflow mining: Current status and future directions. In: OTM Confederated International Conferences "On the Move to Meaningful Internet Systems", pp. 389–406. Springer (2003)

9. Weijters, A., Van der Aalst, W., Van Dongen, B., Günther, C., Mans, R., Medeiros, A.: A Rozinat, MS Song, and HMW Verbeek. Process mining with prom. In: 19th Belgian-Dutch Conference on Artificial Intelligence (BNAIC 2007). Utrecht University (2007)
10. Turner, C., Tiwari, A., Olaiya, R., Xu, Y.: Process mining: from theory to practice. Business Process Management Journal (2012)
11. Medeiros, A., Van Dongen, B., Van der Aalst, W., Weijters, A.: Process mining: Extending the α-algorithm to mine short loops (2004)
12. Lassen, K., Dongen, B., Van der Aalst, W.: Translating message sequence charts to other process languages using process mining. In: Proceedings of the Workshop on Petri Nets and Software Engineering (PNSE'07), pp. 82–97 (2007)
13. Garcia, C., et al.: Process mining techniques and applications–a systematic mapping study. Expert Syst. Appl. **133**, 260–295 (2019)
14. Rebuge, A., Ferreira, D.: Business process analysis in healthcare environments: a methodology based on process mining. Inf. Syst. **37**(2), 99–116 (2012)
15. Rojas, E., Munoz-Gama, J., Sepúlveda, M., Capurro, D.: Process mining in healthcare: a literature review. J. Biomed. Inform. **61**,(2016). https://doi.org/10.1016/j.jbi.2016.04.007
16. Kapustova, J.: Process mining application in software process assessment. [Phd Thesis 1 (Research TU/e / Graduation TU/e), Industrial Engineering and Innovation Sciences]. Technische Universiteit Eindhoven (2012). https://doi.org/10.6100/IR739990
17. Samalikova, J., Kusters, R., Trienekens, J., Weijters, A.: Process mining support for capability maturity model integration-based software process assessment, in principle and in practice. J. Softw. Evol. Process **26**(7), 714–728 (2014)
18. Marin-Castro, H., Tello-Leal, E.: Event log preprocessing for process mining: a review. Appl. Sci. **11**(22), 10556 (2021). https://doi.org/10.3390/app112210556
19. Ghasemi, M.: What requirements engineering can learn from process mining. In: 2018 1st International Workshop on Learning from other Disciplines for Requirements Engineering (D4RE), pp. 8–11. IEEE (2018)
20. Kuc, R.: Top 10+ Best Log Monitoring Tools & Software: Free & Paid. Sematex. October 11. Homepage (2023). https://sematext.com/blog/log-monitoring-tools/. Accessed 14 Apr 2024
21. Oracle (2024). Database reference - v$sql. Homepage. https://docs.oracle.com/en/database/oracle/oracle-database/19/refrn/V-SQL.html#GUID-2B9340D7-4AA8-4894-94C0-D5990F67BE75. Accessed 14 Apr 2024
22. Fluxicon (2024). Process mining in practice. Homepage. https://fluxicon.com/book/read/. Accessed 14 Apr 2024
23. Viner, D.: Process Mining Tools: 16 Best Providers. July 11. Homepage (2022). https://www.botsandpeople.com/blog/process-mining-the-16-best-software-tools-in-the-world#the-most-important-process-mining-tools. Accessed 14 Apr 2024
24. Elkoumy, G., Pankova, A., Dumas, M.: Mine me but don't single me out: differentially private event logs for process mining. In: 2021 3rd International Conference on Process Mining (ICPM), pp. 80–87. IEEE (2021)

A Gamified Multi Agent Robot System for Physical Stimulation

Jesús García-Martínez(✉), Sara Carrasco-Martínez, Sofía Álvarez-Arias, Guillermo A. Arrojo, José C. Castillo, Fernando Alonso-Martín, and Álvaro Castro-González

Department of Systems Engineering and Automation, Carlos III University of Madrid, Calle de Butarque, 15. Leganés, 29811 Madrid, Spain
{jesusgar,sacarras,garrojo,jocastil,famartin,acgonzal}@ing.uc3m.es, sofalvar@pa.uc3m.es
http://roboticslab.uc3m.es

Abstract. Social robots have the potential to help people with mobility and coordination problems, especially in elderly people, in a fun and enjoyable way, as they can support physical stimulation exercises by enabling the gamification of games for this purpose. In this paper, we present a version of the computer game Pong as a serious game implemented as a multi-agent system, in which the social robot Mini guides the activity to promote the physical activity of the user. We track the user's hands to replace the classic paddle game controller, facilitating the assessment of physical stimulation and hand-eye coordination. To improve the game's engagement, the user can play against the robot, against another player, or against themselves. Incorporating gamification elements improves sustained user participation and skill development. The proposed system has been integrated into the robot's software architecture, and preliminary tests show potential for promoting physical activity among seniors.

Keywords: Social Robots · Serious Games · Multi-Agent System · Gamification · Physical Stimulation

1 Introduction

New technologies offer complementary alternatives to conventional physical stimulation therapies that can be more engaging and enjoyable for users. Among these technologies, the use of social robots can improve task adherence as they implement advanced interaction mechanisms capable of keeping the user engaged in the therapy for extended periods of time [11]. Gongora et al. [4] advocated for simple games to stimulate physical activity and enhance hand-eye coordination. Chien et al. [3] study how social robots can improve coordination and motor function in patients with brain damage.

Recently, different works introduced classic games on social robots for therapy in patients with different motor disorders. Burdea et al. [1] tested robots to

R. Chinthaginjala et al. (Eds.): DCAI 2024, LNNS 1259, pp. 337–345, 2025.
https://doi.org/10.1007/978-3-031-82073-1_33

improve patients' mobility with reduced upper limb mobility, finding improvements in patients who embraced the therapy. The authors proposed a series of therapies developed through different serious games. Tresser et al. [10] studied whether serious game-based therapies are effective in children with motor deficiencies, demonstrating that this type of rehabilitation improved patients' mobility by encouraging them through play. Chau et al. [2], however, presented a study on elderly people with upper limb disabilities. This work presented a series of games to the users, studying the effectiveness of these games as well as the acceptance of the users to perform these therapies. The study concluded that physical rehabilitation therapies were more effective in getting users to exercise more through the games.

This work proposes using the "Pong game" as a serious game for physical stimulation among elderly users implemented in a multi-agent robot system. Numerous studies use multi-agent systems to stimulate the elderly. Menezes and Rocha [6] study a system that uses different avatars and robots that interact to motivate users to perform physical exercises appropriate to their individual needs and capabilities. Oliveira et al. [7] combine activity recognition technology and multi-agent communication to assist visually impaired and elderly people, promoting greater independence and safety in their daily activities.

Pong is a classic arcade game that simulates a table tennis match where two players control paddles positioned at each end of the screen, moving them vertically to hit a virtual ball moving back and forth. The object of the game is to prevent the ball from going over your paddle while trying to get the ball past your opponent's paddle to score points. The game continues until one player fails to return the ball properly, scoring a point to the opponent, and a player wins if they reach a certain number of points. In our case, the classical controller has been replaced by a computer vision module able to track users' hands, which, together with the game's simplicity, allows enhanced physical activity and hand-eye coordination assessment. The multi-agent system involves two elements: a Game Agent for detection and gameplay and a social robot to assist and interact with the users.

The paper is structured as follows: Sect. 2 offers a description of the physical elements necessary to develop and test the game; Sect. 3 explains the multi-agent architecture and how it is integrated into the Mini social robot; Sect. 4 describes the game dynamics; And, finally, Sect. 5 discusses the main conclusions, limitations, and future directions derived from this work.

2 Materials

This section presents the hardware components of our work, a social robot, a Game Agent and a screen to display the game. The social robot Mini oversees the actions of the Game Agent, activating or stopping the game, and retrieving pertinent information associated with these games (e.g., user performance metrics). In turn, the Game Agent renders the game, and detects and tracks the user movements. is responsible for visualization, processing, and detection of the components involved during the game.

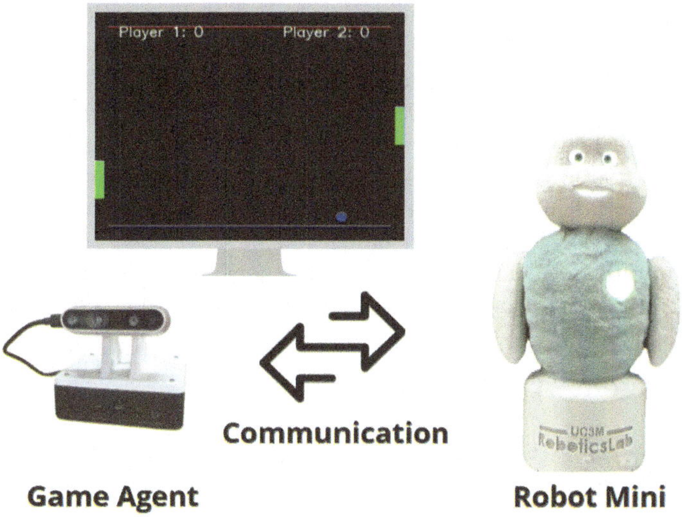

Fig. 1. Game Agent and social robot Mini.

2.1 Social Robot Mini

Mini [9] is a social robot developed at the Carlos III University of Madrid, shown in Fig. 1 right. Mini is a desktop robot originally designed with the main purpose of providing assistance and entertainment to elderly people with moderate cognitive impairment. Mini can convey its emotional state through its expressive eyes, heartbeat, cheeks, and movement of its arms, head, and base. It is capable of engaging with users through multimodal communication, utilizing verbal, visual, and tactile cues.

Mini's software architecture is implemented on ROS [8], which is a middleware composed of nodes that run independently and communicate by topics and is divided into four main modules, essential for monitoring and performing the robot's operations. The *Perception Manager* (PM) is responsible for acquiring and processing information from the environment and the user with whom it interacts. *The Decision-Making System* (DMS) executes and controls the robot's skills. The *Skills* are the abilities that the robot has, for example, the robot has entertainment, such as a multimedia platform capable of displaying photos or videos, and physical and cognitive stimulation through specific games. The *Human-Robot Interaction* (HRI) manages the information from the perception system and generates multimodal expressions to communicate with the user.

2.2 Game Agent

The Game Agent is the device responsible for managing the game logic and detecting and tracking user hand movements. As shown in Fig. 1 left, it comprises two main hardware components: an Intel NUC Mini PC Kit and an Intel RealSense Depth Camera D435[1], forming a portable system designed to connect to an everyday HDMI display, such as a TV. As with the robot, the software architecture is implemented in ROS and divided into two main systems, perception and action. The perception system is in charge of extracting the information from the camera, and is responsible for activating and deactivating the detector to extract the hand information required by the action system, which in turn is responsible for the logic and visualization of the game. In this multi-agent system, communication between the Game Agent and Mini is done via sockets and allows transmitting data between the systems with low delays.

3 Multi-agent Architecture

This section describes the game design and discusses the sequence diagram of the multi-agent interaction shown in Fig. 2. The sequence is divided into three phases: calibration, gameplay, and the game conclusion. On the left are the interactions performed by the user, in the centre are the actions of the Mini robot, and on the right are the parts corresponding to the Game Agent.

3.1 Game Design

In our game (see the screen in Fig. 1), two players can participate simultaneously, each controlling one paddle using their hand. Alternatively, players can use a player-versus-robot setup or a single-player mode where the user controls both paddles. Each paddle is represented by vertically opposing green rectangles at the sides of the screen. The ball is depicted as a blue circle that starts at the centre of the screen and moves in a random direction each round. The image boundaries are defined by two horizontal bars that act as upper (red bar) and lower (blue bar) edges, where the ball bounces upon collision, imitating the logic of the original game. In addition, the ball also bounces off the paddles. The opposing player scores a point when a player fails to hit the ball and passes into their region. The game continues until one player reaches the maximum score. Our game implementation incorporates several internal parameters of interest; for example, the ball speed gradually and steadily increases as the game progresses, making it more challenging for both players. The game's difficulty level, adjustable during the game, adds a calculation and trajectory prediction component, enabling the machine to anticipate the ball's future positions.

3.2 Configuration Phase

After Mini begins the interaction, the robot asks the user if they wish to review the rules, requesting confirmation through either voice or the tablet. If the user agrees, Mini explains the rules aloud. This initiates the Configuration phase,

[1] https://www.intelrealsense.com/depth-camera-d435i/.

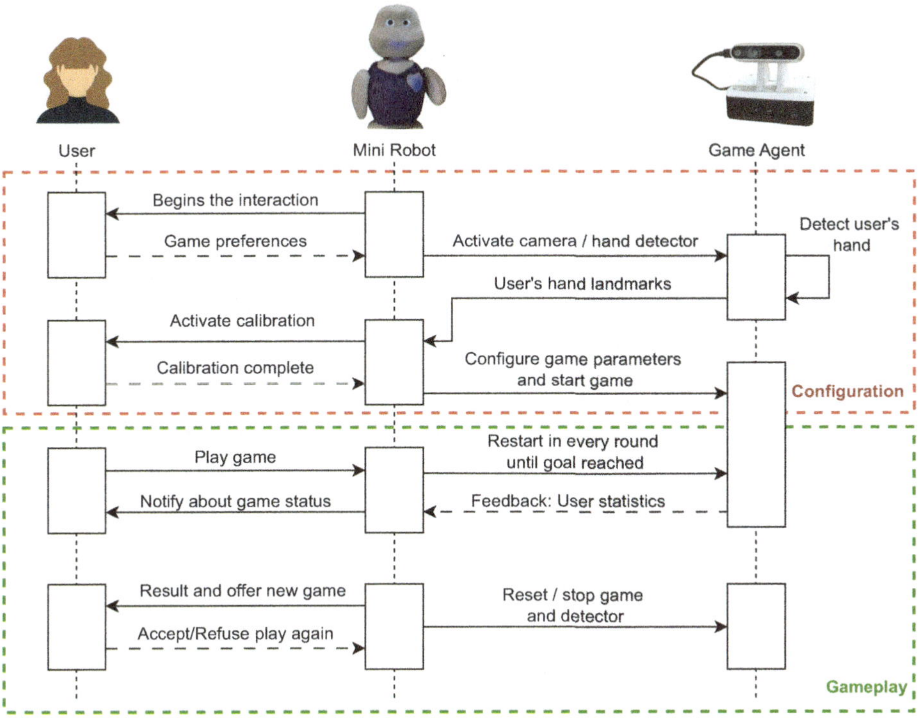

Fig. 2. Sequence diagram for the interactions between the actors in the Pong game: the user, the Game Agent, and the robot Mini running the Pong Game skill.

where the user selects their *game preferences*. For this purpose, the first thing that the robot does is ask different questions. The first question is the type of game; the user can play with the robot, with another user, or with oneself using both hands. The second question is which hand the user wants to play with.

Following the user's game preferences, Mini sends a command to the Game Agent to activate the camera and hand detector. Mini then instructs the user to perform hand position calibration. Calibration is done in two steps (activate calibration): First, the robot tells the user to raise their hands as high as possible and acquires this position. Secondly, the robot asks the user to lower their hands as low as possible to obtain the bottom limit. These upper and lower limits will be used during the game to match the top and bottom positions of the paddle. Utilizing the Mediapipe library [5], the Game Agent detects the user's hand landmarks for the Pong game, feeding this information back to Mini for calibration and gameplay adjustments. During calibration, the maximum and minimum hand placement positions are stored to interpolate with the maximum and minimum paddle movement values.

Upon completing the calibration process, Mini establishes the game parameters based on the calibration data and user preferences and sends the configuration to the Game Agent to start the game.

3.3 Gameplay Phase

If the user is playing against Mini and wins a round, the game's difficulty increases, making Mini harder to beat. If the user loses, the difficulty is maintained. In the case that the user plays with another user, the only thing that makes the game harder is the increment of the speed of the ball. Mini manages the game rounds, instructing the Game Agent to reset the speed and the position of the ball for each new round until a predetermined goal is achieved. The Game Agent is responsible for managing the hand detector, which is used to locate the position of the user's hands in the image. This information is utilized in the Pong game activity to control the sidebars' movement by positioning them according to the user's hand centroid positions. The Game Agent also manages the game execution and displays it on the screen.

During gameplay, the Game Agent sends real-time feedback on game statistics to Mini. This feedback includes paddle positions, paddle-ball contacts, current scores, game status (started or finished), and game difficulty level. With this information, Mini provides the current stores and the game status. The game pauses whenever a player scores a point until Mini sends a message to the Game Agent to continue. If the user plays with the robot, Mini moves an arm according to the movement of the paddle in the game.

The game resumes if the user misses the ball, Mini scores a point, and the robot celebrates the point with a happy gesture. If the user wins the point, the robot indicates that the user has leveled up, and a new round starts. If the user plays with another user or with themself, Mini encourages the user(s) along the game and announces who scores points.

After one of them scores two points first, the game concludes. Mini informs the user about their performance and game status: if the user wins, the robot congratulates them; if the user loses, the robot celebrates its victory and encourages the user to play again.

4 Use Case

To illustrate the operation of the proposed system, we present a scenario in which a user stands in front of a screen and the Game Agent device where the game takes place (see Fig. 3 left). Mini, as the user's opponent, is placed on top of a table nearby. Finally, to the user's front, we find the Game Agent facing user just below the screen.

At first, Mini initiates the interaction, asking the user if they want to review the rules of the game; in this case, the player knows them and chooses to continue without the rules explanation. The robot continues the configuration, asking the user to choose who they want to play against: robot, another person, or the user

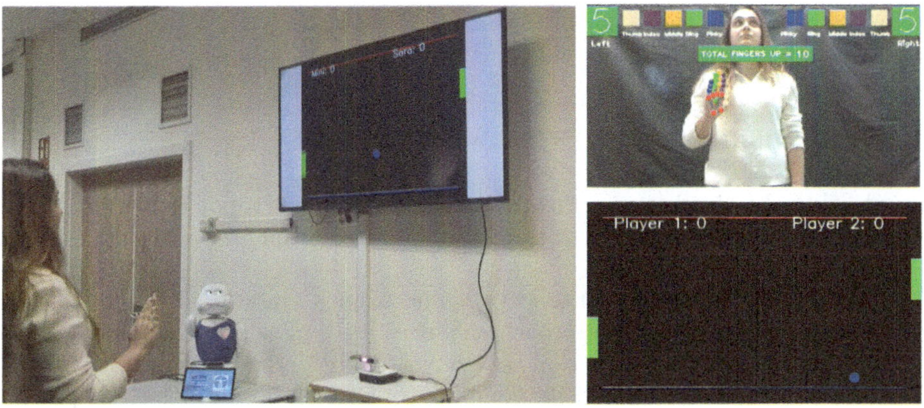

Fig. 3. User playing with the robot (left). Output of the hand detection module (upper right). Game interface (bottom right).

Table 1. Conceptual comparison between our approach game and other related studies.

Author	Description	Platform	Metrics employed
Burdea et al. [1]	Design and evaluation of the usability of a virtual rehabilitation system for the training of arm disabilities	Robotic table	Efficiency Duration Intensity Number of errors
Tresser et al. [10]	Exploring the efficacy of virtual gaming in aiding the rehabilitation process of children experiencing paralysis	Virtual reality	Efficiency User's behaviour Difficulty levels Participants' kinematics
Chau et al. [2]	Assessment of the reception, efficacy, and practicality of virtual reality training for seniors coping with upper limb impairments	Virtual reality	Health outcomes Duration Frequency Effects on upper extremity
J.Garcia et al. (Our proposal)	Design of a serious game for acceptance of the elderly and improvement of hand-eye coordination and physical activity	Social Robot Game Agent	Duration Frequency of movement Max-min marks Score Hand identification Longitudinal progression

themself; in this case, the user selects to play against the robot. Next, the robot asks the user to indicate the hand they want to play with, in this case, they choice their right hand.

Then the calibration phase begins, mini asks the user to raise their hands as high as they can, and then asks them to lower them. The game starts and the user moves the hand to bring the paddle towards the ball in order to prevent Mini from scoring. After several rebounds, the user manages to score. In the second round Mini scores. As both have managed to score a third round the user manages to beat Mini. The robot congratulates the user and proposes to play again, in this case, the user decides not to continue. This human-robot interaction is in the following video.

5 Conclusion

By utilizing the classic game Pong within a multi-agent robot system approach that integrates a social robot, we offer a practical solution for promoting physical activity in older adults.

Switching the control of the Pong game from a traditional controller to a hand-tracking vision-based approach enhances its applicability in physical stimulation and hand-eye coordination assessment. Moreover, our system allows users to compete against each other, the computer or oneself, promoting active participation and skill development. The game also includes gamification elements such as difficulty levels and progress tracking to increase user engagement towards the game. In future user-centred studies, the data collected during the game sessions will provide valuable insights into users' performance, such as their physical progress or hand-eye coordination. By analyzing parameters such as: the duration of each round, the frequency of hand movements, the maximum and minimum ranges achieved by the users, recorded game scores, hand identification, and the longitudinal progression.

With this parameters we can examine the level of physical performance and promote hand-eye coordination. Longer durations of rounds and increased frequency of hand movements suggest higher levels of physical activity, indicating active participation and potential improvements in motor skills. Variations in the maximum and minimum ratings attained by the users indicate their ability to control and manipulate their hand movements within the game environment. Higher game scores over time suggest improved skill acquisition and strategic decision-making by the users, reflecting enhanced coordination. By analyzing hand identification, we can provide valuable information about asymmetries or preferences in motor function, which can be further investigated to tailor gameplay experiences to individual user needs. Additionally, a consistent upward trend in these ratings over multiple sessions may signify enhanced hand-eye coordination and motor proficiency as users become more familiar with the game's mechanics.

In contrast to the approaches detailed in the referenced Table 1, our study introduces a multi-agent system that integrates the Mini social robot as an autonomous platform for engaging older adults in physical activity. Diverging from the hardware-centric solutions such as robotic tables and virtual reality. Also, in this paper, we propose the Pong Game as a serious game. This game allows us to acquire different metrics from the user to evaluate the user's progress furthers. This allows for a nuanced understanding of user engagement, skill development over time, and the potential for tailored interventions. By focusing on parameters like hand identification and longitudinal progression, we aim to offer insights into the personalized needs of users, thereby informing the ongoing refinement of game design and contributing to the development of targeted physical stimulation exercises that are both enjoyable and beneficial for the elderly population.

Our research contributes to the potential of integrating social robotics, serious gaming, and gamification to create engaging and effective interventions for promoting physical activity in elderly populations. Future work will focus on further refining our system, expanding the range of games available, and conducting longitudinal studies to evaluate its long-term efficacy and impact on user outcomes.

References

1. Burdea, G., Kim, N., Polistico, K., Kadaru, A., Roll, D., Grampurohit, N.: Novel integrative rehabilitation system for the upper extremity: design and usability evaluation. J. Rehabil. Assist. Technol. Eng. **8**, 20556683211012884 (2021)
2. Chau, P.H., et al.: Feasibility, acceptability, and efficacy of virtual reality training for older adults and people with disabilities: single-arm pre-post study. J. Med. Internet Res. **23**(5), e27640 (2021)
3. Chien, W.T., et al.: Robot-assisted therapy for upper-limb rehabilitation in subacute stroke patients: a systematic review and meta-analysis. Brain Behav. **10**(8), e01742 (2020)
4. Góngora Alonso, S., et al.: Social robots for people with aging and dementia: a systematic review of literature. Telemedicine e-Health **25**(7), 533–540 (2019)
5. Lugaresi, C., et al.: MediaPipe: a framework for building perception pipelines. arXiv preprint arXiv:1906.08172 (2019)
6. Menezes, P., Rocha, R.P.: Multi-agent approach to foster regular physical activity in elderly users. In: 2020 IEEE International Conference on Autonomous Robot Systems and Competitions (ICARSC), pp. 325–331 (2020)
7. Oliveira, J.D., Engelmann, D.C., Kniest, D., Vieira, R., Bordini, R.H.: Multi-agent interaction to assist visually-impaired and elderly people. Int. J. Environ. Res. Public Health **19** (8 2022)
8. Quigley, M., et al.: ROS: an open-source robot operating system. In: ICRA Workshop on Open Source Software, vol. 3, p. 5. Kobe, Japan (2009)
9. Salichs, M.A., et al.: Mini: a new social robot for the elderly. Int. J. Soc. Robot. **12**(6), 1231–1249 (2020)
10. Tresser, S., Kuflik, T., Levin, I., Weiss, P.L.: Personalized rehabilitation for children with cerebral palsy. User Model. User-Adap. Inter. **31**(4), 829–865 (2021). https://doi.org/10.1007/s11257-021-09296-6
11. Winkle, K., Caleb-Solly, P., Turton, A., Bremner, P.: Social robots for engagement in rehabilitative therapies: design implications from a study with therapists. In: Proceedings of the 2018 ACM/IEEE International Conference on Human-Robot Interaction, pp. 289–297 (2018)

Decision System in a Building to Coordinate the Energy Forecasting Tasks

Daniel Ramos, Pedro Faria(✉), and Zita Vale

GECAD - Research Group on Intelligent Engineering and Computing for Advanced Innovation and Development, LASI - Intelligent Systems Associate Laboratory, Polytechnic of Porto, Porto, Portugal

{dados,pnf,zav}@isep.ipp.pt

Abstract. Nowadays, energy buildings have a huge impact in society regarding the active role in the management of energy consumption. Hence, building owners are required to avoid energy losses and improve energy efficiency as high as possible. Therefore, it is required to plan an optimization strategy to buy and sell energy in the market ahead of time. To formulate this optimization plan, building owners require the work of specialists responsible for processing, training, forecasting, and evaluation tasks regarding the prediction of energy consumption data from a building for a specific target of time. Therefore, a multiagent-system is needed to allow the cooperation of various agents including the building owner, forecast provider, data structurer and error analysis. Moreover, forecasting algorithms such as artificial neural networks should be taken into consideration in order to process large quantities of energy consumption data during the training and forecasting phases.

Keywords: agents · artificial neural networks · building owner · energy building

1 Introduction

Demand response programs are usually considered by building owners responsible for the energy consumption management as means to match the demand for power with the supply thus reducing energy losses [1]. However, the uncertainties regarding the effect of these demand response programs in the supply side complicate the planning of energy management strategies [2]. Fortunately, nowadays there are various forecasting algorithms capable of predicting the energy that should be bought and sold previously on the energy market, thus ensuring that energy may be used more efficiently [3]. The application of artificial neural networks algorithm for the prediction of electricity consumption data is capable of processing large amounts of data during training and forecasting phases, thus resulting in more accurate forecasts than other applications [4]. Moreover, forecasting research regarding the prediction of energy consumption data from residential and commercial buildings shows that the use of recursive artificial neural networks results in more accurate performance than non-recursive models [5]. Feature selection is taken into consideration in energy consumption prediction researches such as [6] in

R. Chinthaginjala et al. (Eds.): DCAI 2024, LNNS 1259, pp. 346–355, 2025.
https://doi.org/10.1007/978-3-031-82073-1_34

order to improve the forecasting accuracy. The prediction of energy consumption for space heating in residential buildings has resulted in [7] with an accuracy of 99% for training and test data while using the ANN algorithm. Researches involving the use of the ANN algorithm in the context of the energy and building sectors usually results in low forecasting errors, thus ensuring higher energy savings to the building [8]. Data scientists and machine learning specialists take into consideration the evaluation of the forecasting error as seen with the support of various metrics as seen in [9] such as Mean Absolute Percentage Error (MAPE) and Root Mean Square Error (RMSE).

The innovation of this scientific paper consists of planning a multi-agent system capable of processing, training, forecasting and evaluation activities contextualized for the prediction of energy consumption for short periods of time from a target week considering an energy building monitoring consumption and sensors data.

The use of a multi-agent system in this paper has been inspired by a multi-agent ecosystem framework presented in [10] that simulates an energy community and evaluates the community's energy bill by comparing a scenario without battery energy storage systems with a scenario where storage systems are available for some members of the community. The selection of sensors to be added to the historic with energy consumptions during the training and forecasting phases is justified by previous research work developed by the authors of this paper in [11] which analyzes and selects the sensors displaying a higher correlation with the consumption variable. The configuration of Artificial Neural Networks (ANN) model used in this paper for training and forecasting activities is justified by previous research work developed by the authors of this paper in [12] which consists in the combination of parameters of ANN resulting in lower forecasting errors as concluded through various trial and error scenarios.

After the finalization of this introduction, the methodology is presented in Sect. 2, followed by the case study presented in Sect. 3. Afterwards, the agents' relations are explained in Sect. 4. The results of this scientific paper are presented in Sect. 5. Finally, the conclusions are presented in Sect. 6.

2 Methodology

The methodology illustrated in this paper is presented in Fig. 1. This includes different entities representing agents including a building owner, a forecast provider, an ANN agent, a data structurer agent, and an error analysis agent. Two components are also presented in the indicated diagram including the historic consumption database and the database.

The agents illustrated in Fig. 1 shows two different strong interactions as represented in dash lines. The first interaction considers the calculation of the forecasting consumptions with the support of the ANN agent that trains and forecasts data with ANN algorithm and a data structurer agent that prepares the data for training and forecasting tasks. The data structurer has access to an historic consumption database in order to prepare the data for training and forecasting tasks. Moreover, a forecast provider ends the first interaction by delivering the forecasted consumptions to the building owner (presented in the second interaction). The second interaction considers a building owner that makes the main request to the forecast provider (presented in the first interaction) to receive the forecasted consumption. Moreover, the second interaction also features an error analysis agent that evaluates the forecasting error to measure the forecasting accuracy. The building owner has access to a database to allow the other agents to perform their tasks concerning the processing, training, forecasting, and evaluation phases. The full process of the agents relation is explained with full detail in Sect. 4.

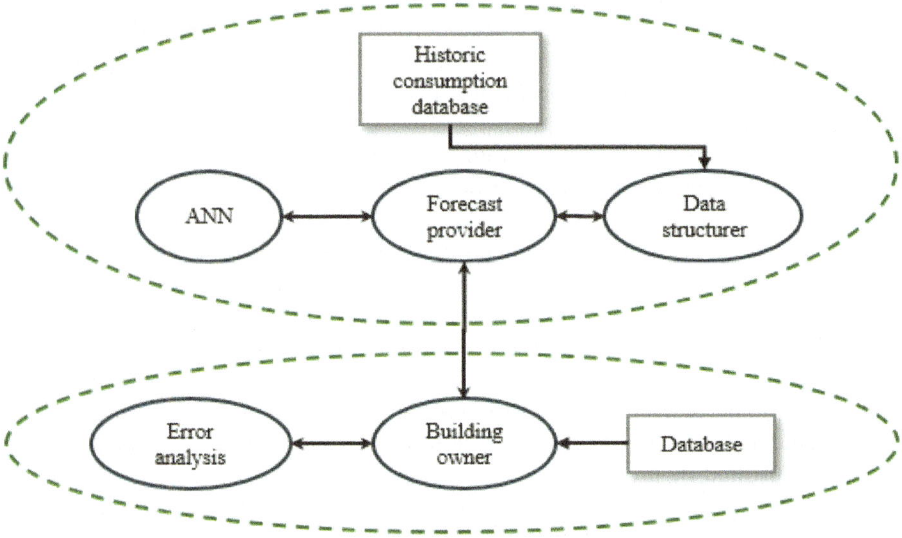

Fig. 1. Methodology diagram.

The configuration of the ANN model is composed of an input layer with twelve neurons, two hidden layers with sixty-four neurons and an output layer with one neuron.

Ten of the twelve neurons involved in the input layer are composed of ten consumption values contextualized for sequential periods of five minutes. The two remaining neurons involved in the input layer are associated to the CO2 and light sensor evidencing either the light presence (with the value 1) or absence (with the value 0). These two sensors involved in the input layer occur in the same period of five minutes as the last input consumption. The neuron presented in the output layer considers the consumption value occurring in the period of five minutes following the last consumption input.

The hidden layers use 64 neurons to process the information between the input and output layers with the support of a ReLU activation function which reassigns the received values to a positive value. The gradient descent algorithm is considered with a learning rate assigned to 0.001 to search accurately for local minimums and to minimize the forecasting error. Additionally, the training of the historic data is configured with a total of five hundred epochs and an early stopping procedure that stops the training automatically stops once no significant training improvements are detected within 20 sequential epochs.

The considered metrics used to evaluate the forecasting error in this research are the Mean Absolute Percentage Error (MAPE), Symmetric Mean Absolute Percentage Error

(SMAPE), Mean Absolute Error (MAE), and RMSE (Root Mean Squared Error). These are respectively represented in Eq. 1, Eq. 2, Eq. 3, and Eq. 4.

$$MAPE = \frac{1}{F} * \sum\nolimits_{t=n-F}^{n} \left| \frac{A(t) - F(t)}{A(t)} \right| * 100 \tag{1}$$

$$SMAPE = \frac{1}{F} * \sum\nolimits_{t=n-F}^{n} \frac{|F(t) - A(t)|}{((|A(t)| + |F(t)|) * 0.5)} * 100 \tag{2}$$

$$MAE = \frac{1}{F} * \sum\nolimits_{t=n-F}^{n} |A(t) - F(t)| \tag{3}$$

$$RMSE = \sqrt{\frac{1}{F} * \sum\nolimits_{t=n-F}^{n} (A(t) - F(t))^2} \tag{4}$$

- A – actual value
- F – forecasted value
- n – current moment
- F – frame used for calculation

The Mean Absolute Percentage Error (MAPE) represented in Eq. 1 calculates an average sum of absolute values, each one assigned to the difference between the actual and forecasted value divided by the actual value for the respective period of time. The Symmetric Mean Absolute Percentage Error (SMAPE) represented in Eq. 2 calculates an average sum of division calculation for different periods of time presenting in the numeration the absolute difference between the actual and forecasted value and in the denominator half of the sum between the actual and forecasted value. The Mean Absolute Error (MAE) presented in Eq. 3 calculates the average sum of the difference between the actual and forecasted values for different periods of time. The Root Mean Squared Error (RMSE) presented in Eq. 4 calculates the squared root of the average sum of the squared difference between the actual and forecasted values for the different periods of time.

3 Case Study

The building of this case study is composed by various agents that coordinate between themselves in order to process, train, forecast and evaluate the energy data for the intended target period using contexts of periods five minutes. The dataset is composed of consumptions and sensors data monitored in an energy building for periods of five minutes between 14 October and 15 November 2019. Moreover, only periods of five minutes from Monday to Friday are considered, meaning that periods of the weekend are excluded from the dataset. The consumption profile is illustrated in Fig. 2. The consumption profile illustrated in Fig. 2 displays the consumption daily behavior contextualized for periods of five minutes for a total of 25 five days. The consumption daily pattern features 288 periods of five minutes with activity between 500 and 2000 W. The maximum consumption activity of each day may between 1100 and 2000 W relying in the day at question.

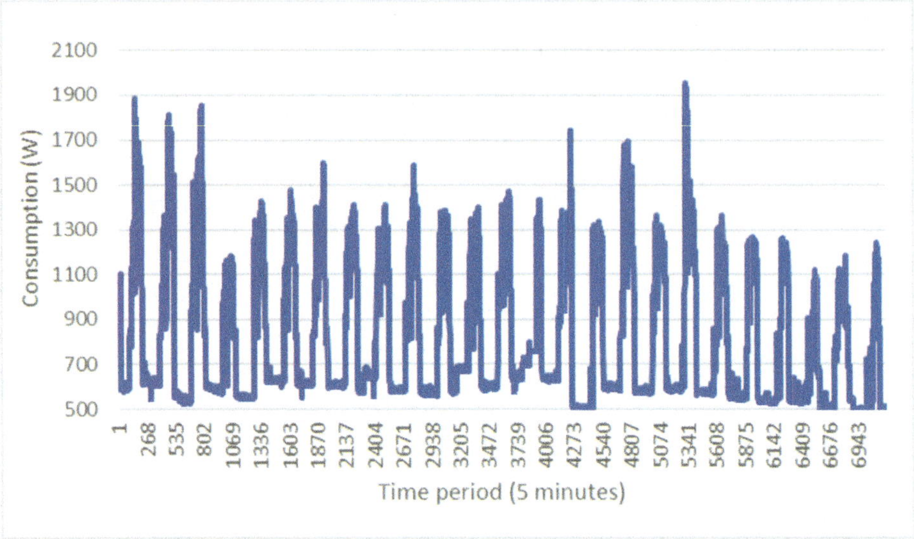

Fig. 2. Consumption of the building from zone 1 from 14 October to 15 November 2019.

Two additional sensors are considered in the historic of the dataset including a CO2 sensor and a light sensor that indicates either activity or no activity in the building.

The data structure considered for each input during the training and forecasting phases considers ten consumption values placed on sequential periods of five minutes and the two indicated sensors occurring the same period of five minutes as the last input consumption. The data structure for each output during the training and forecasting phases considers only a consumption value occurring in period of five minutes following the last consumption input. The building of the energy sector where we monitor energy and sensors data is split into three different zones, each one with three different rooms. Zone 1 of the building is illustrated in Fig. 3 as the train and forecasting of energy consumption are focused in this area.

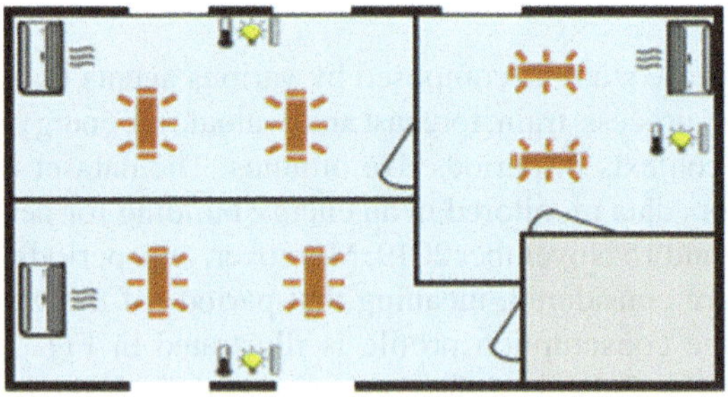

Fig. 3. Building plan of zone 1 of the building and respective sensors [11].

The presented sensors of this zone are the following: Four movement sensors; Three door status indicators; One air quality sensor; One temperature sensor; One humidity sensor; One CO2 sensor; Seven light power indicators.

4 Agents Relation

The various agents illustrated in Fig. 1 (building owner, forecast provider, data structurer, ann agent, and error analysis) are composed by the following interactions:

- The building owner gets in contact with the forecast provider to make a request with the forecasting of energy consumptions for various short periods of time occurring in a building;
- The forecast provider confirms the availability and requests the building owner to provide the dataset with the consumption and sensors data monitored in the building;
- The building owner sends the dataset requested to the forecast provider;
- The forecast provider requests the data structurer to rearrange the historic of energy consumptions and sensors data to be prepared for forecasting tasks;
- The data structurer confirms the availability and requests the forecast provider to provide the dataset with the consumption and sensors data monitored in the building;
- The forecast provider sends the dataset to the data structurer;
- The data structure agent rearranges the data structure on an array with 10 inputs presenting consumptions on sequential periods, 1 input with the value of the CO2 sensor, and 1 input with the value of the light sensor. The 10 inputs of consumptions correspond to consumption values separated in sequential periods of five minutes. The last input of consumption occurs in the same period of five minutes as the input with the light sensor value and the input with the CO2 value. The data structure rearrange also an output consumption with the consumption placed in the period of five minutes following the last input of consumption;
- The data structurer sends the prepared dataset for training and forecasting tasks to the forecast provider;
- The forecast provider requests to an ANN agent the training of an historic of energy consumptions and sensors and the forecasting for a target week for periods of five minutes with the Artificial Neural Networks;
- The ANN agent confirms the availability and requests the forecast provider to provide the prepared dataset for training and forecasting tasks;
- The forecast provider sends the prepared dataset to the ANN agent;
- The ANN agent uses the Artificial Neural Networks algorithm to train an historic with most of the consumption and sensors data presented in the prepared dataset and forecast the remaining consumption data for all periods of five minutes of an entire week;
- The ANN agent sends all the consumption forecasts of the target week to the forecast provider;
- The forecast provider sends the forecasted consumptions to the building owner;
- The building owner gets contact with an error analysis agent to request the evaluation of the forecasted energy consumptions;

- The error analysis agent confirms the availability and requests the building owner to provide the ANN forecasts and the real energy consumptions for the target week;
- The building owner sends the ANN forecasts and the real energy consumptions of the target week to the building owner;
- The error analysis agent evaluates the forecasting errors considering the final error of the target week and also the forecasting errors for each period of five minutes of the target week. Four metrics are considered to calculate the forecasting error: Mean Absolute Percentage Error (MAPE), Symmetric Mean Absolute Percentage Error (SMAPE), Mean Absolute Error (MAE) and Root Mean Square Error (RMSE);

The error analysis agent sends the forecasting errors to the building owner.

5 Results

The forecasts of energy consumption are obtained for all periods of five minutes from 11 to 15 November 2019 with the support of an historic of energy consumption and sensors considering only periods of five minutes from Monday to Friday from 14 October to 8 November 2019. The indicated forecasts are illustrated in Fig. 4.

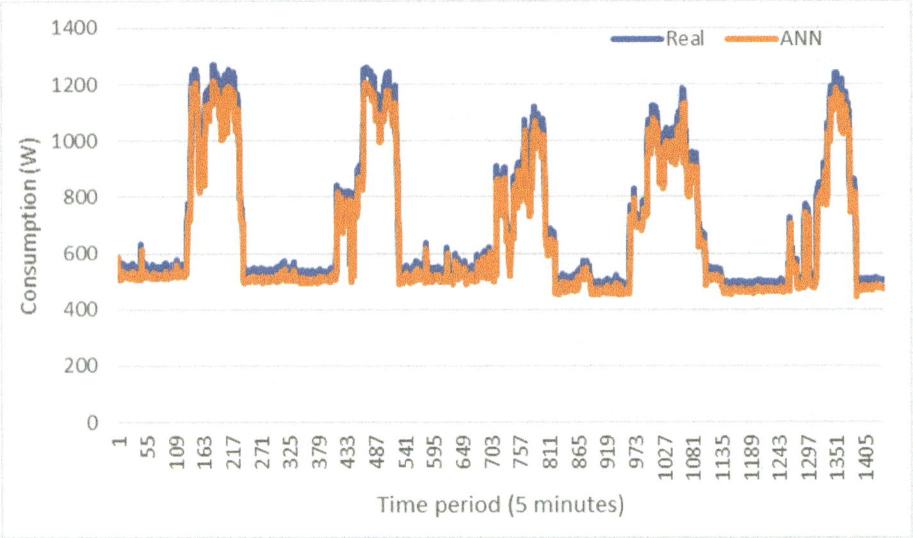

Fig. 4. Energy consumption forecasts from 11 to 15 November 2019 considering periods of five minutes.

The real energy consumptions and the respective forecasts obtained with the support of ANN algorithm show that the consumption behavior varies between 400 and 1300 W. The consumption forecasts are very accurate as the illustration in Fig. 4 shows that the deviation between the real and forecasted consumptions is too small for all the 1440 periods of five minutes from 11 to 15 November 2019. The obtained forecasting errors are displayed for the four forecasting metrics known as Mean Absolute Percentage Error (MAPE), Symmetric Mean Absolute Percentage Error (SMAPE), Mean Absolute Error (MAE), and Root Mean Squared Error (RMSE) as illustrated in Fig. 5.

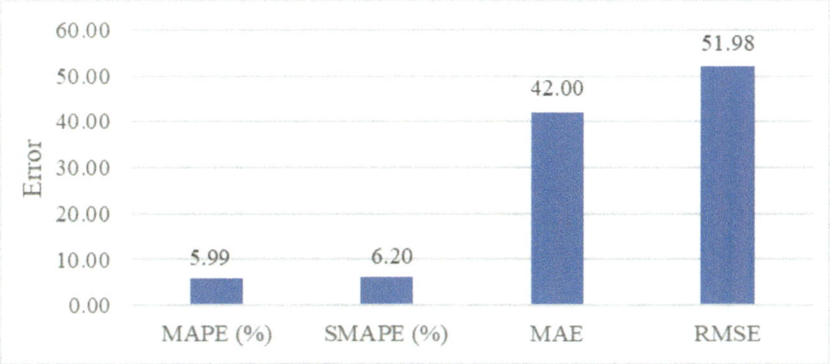

Fig. 5. Forecasting errors from 11 to 15 November 2019.

The percentage forecasting metrics known as MAPE and SMAPE result in low forecasting errors as illustrated in Fig. 5 between 5.99 and 6.20%. The forecasting errors MAE and RMSE show that the average deviation from the forecast to the real consumption is between 42 and 52.

The obtained forecasting errors are illustrated for periods of five minutes on Fig. 6 according to the metrics Absolute Percentage Error (APE) and Symmetric Absolute Percentage Error (SAPE). The obtained forecasting errors are illustrated for periods of five minutes on Fig. 7 according to the metric Absolute Error (AE).

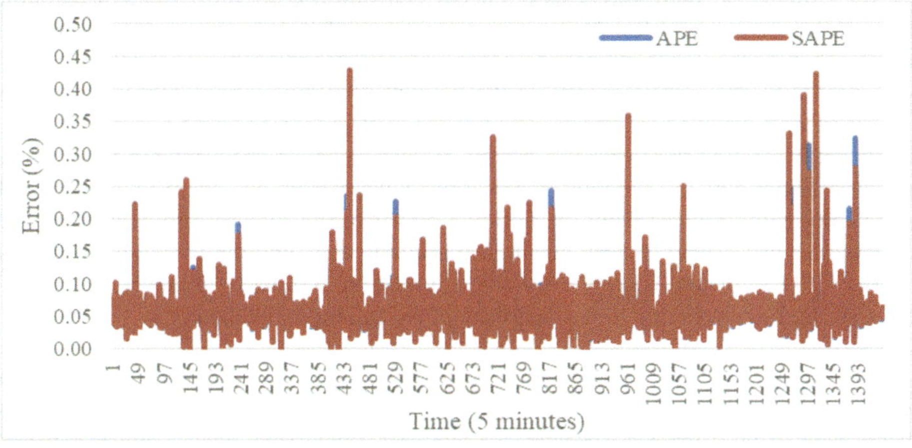

Fig. 6. Forecasting errors for periods of five minutes from 11 to 15 November 2019 (metrics APE/SAPE).

The obtained forecasting errors for all periods of five minutes illustrated in Fig. 6 demonstrate that the obtained error is usually between 0 and 0.2% with a maximum of 0.45%. The obtained forecasting errors for all periods of five minutes illustrated in Fig. 7 demonstrate that the forecasting error variates between 0 and 300 W.

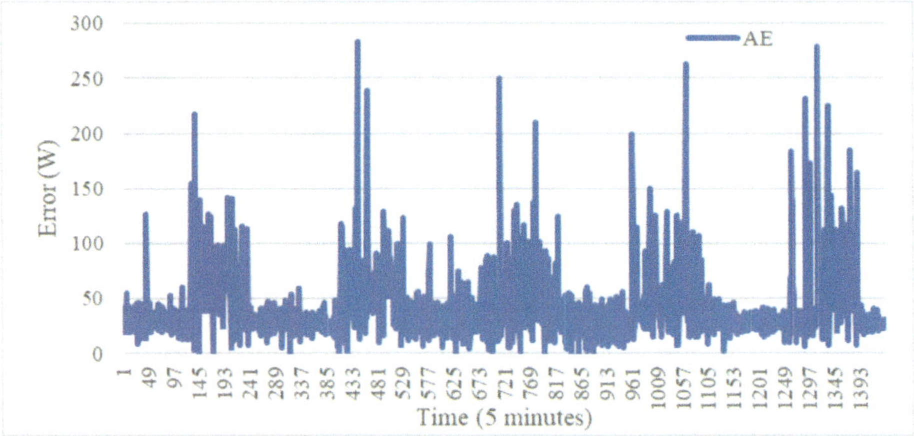

Fig. 7. Forecasting errors for periods of five minutes from 11 to 15 November 2019 (metric AE).

6 Conclusions

In conclusion, this scientific paper shows that the various agents of the building are successful on cooperating from the moment that the building owner sends the forecasting request until the moment the consumption forecasts and respective errors are obtained. The obtaining of accurate forecasts of energy consumption for all periods of five minutes and the obtaining of low forecasting errors show that the activities concerning the processing, training, forecasting and evaluation were successful. It is noted that the tasks performance by the various agents of the building (building owner, forecast provider, data structurer, ANN agent, error analysis) are well processed as demonstrated in the results section while the tasks processing of each agent is encapsulated to the other agents. Additionally, the send and receival of requests from one agent to another such as the forecasting request from the building owner to the forecast provider automates the activities processing of each agent. The accurate consumption forecasts imply also that the ANN algorithm is the most viable application in order to obtain forecasts with lower prediction errors due to its capability of processing large data.

Acknowledgments. This article was developed under the "NGS - New Generation Storage" Innovation Pact (C644936001-00000045), by the "NGS" Consortium, co-funded by the NextGenerationEU, through the investment "Agendas para a Inovação Empresarial" of the Portuguese Recovery and Resilience Plan (PRR)." The authors acknowledge the work facilities and equipment provided by GECAD research center (UIDB/00760/2020), DOI:https://doi.org/10.54499/UIDB/00760/2020 to the project team.

Disclosure of Interests. The authors have no competing interests to declare that are relevant to the content of this article.

References

1. Khorram, M., Zheiry, M., Faria, P., Vale, Z.: Energy consumption management in buildings in the context of voluntary and mandatory demand response programs in smart grids. In: 2020 IEEE PES Innovative Smart Grid Technologies Europe (ISGT-Europe), The Hague, Netherlands, 2020, pp. 275–279. https://doi.org/10.1109/ISGT-Europe47291.2020.9248750
2. Lei Wynn, S.L., Boonraksa, T., Boonraksa, P., Marungsri, B.: Optimal distribution level of energy management considering generation uncertainties and demand response. In: 2021 International Conference on Power, Energy and Innovations (ICPEI), Nakhon Ratchasima, Thailand, 2021, pp. 29–32. https://doi.org/10.1109/ICPEI52436.2021.9690652
3. Mariano-Hernández, D., Hernández-Callejo, L., García, F.S., Duque-Perez, O., Zorita-Lamadrid, A.L.: A review of energy consumption forecasting in smart buildings: methods, input variables. Forecasting Horizon Metrics. Appl. Sci. **10**, 8323 (2020). https://doi.org/10.3390/app10238323
4. Iruela, J.R.S., Ruiz, L.G.B., Capel, M.I., Pegalajar, M.C.: A TensorFlow approach to data analysis for time series forecasting in the energy-efficiency realm. Energies **14**, 4038 (2021). https://doi.org/10.3390/en14134038
5. Almalaq, A., Edwards, G.: Comparison of recursive and non-recursive ANNs in energy consumption forecasting in buildings. In: 2019 IEEE Green Technologies Conference (GreenTech), Lafayette, LA, USA, pp. 1–5 (2019). https://doi.org/10.1109/GreenTech.2019.8767130
6. Qiao, Q., Yunusa-Kaltungo, A., Edwards, R.: Feature selection strategy for machine learning methods in building energy consumption prediction. Energy Reports **8**, 13621–13654 (2022). ISSN 2352-4847. https://doi.org/10.1016/j.egyr.2022.10.125
7. Naim, M.M., Younes, R., Nahim, H.M.: Neural network to predict energy efficiency for space heating in residential buildings. In: 2023 IEEE 4th International Multidisciplinary Conference on Engineering Technology (IMCET), Beirut, Lebanon, 2023, pp. 85–90 (2023). https://doi.org/10.1109/IMCET59736.2023.10368226
8. Hamilludin, U.N.A., Abdullah, H.: Power usage modelling and prediction using artificial neural networks. In: 2022 4th International Conference on Smart Sensors and Application (ICSSA), Kuala Lumpur, Malaysia, pp. 126–129 (2022). https://doi.org/10.1109/ICSSA54161.2022.9870959
9. Putz, D. Gumhalter, M., Auer, H.: The true value of a forecast: Assessing the impact of accuracy on local energy communities, Sustainable Energy, Grids and Networks, vol. 33, 2023, 100983, ISSN 2352-4677, https://doi.org/10.1016/j.segan.2022.100983
10. Ribeiro, B., Pereira, H., Gomes, L., Vale, Z.: Python-based ecosystem for agent communities simulation. In: 17th International Conference on Soft Computing Models in Industrial and Environmental Applications (SOCO 2022), Notes in Networks and Systems, vol. 531. https://doi.org/10.1007/978-3-031-18050-7_7
11. Ramos, D., Teixeira, B., Faria, P., Gomes, L., Abrishambaf, O., Vale, Z.: Use of sensors and analyzers data for load forecasting: a two stage approach. Sensors **20**, 3524 (2020). https://doi.org/10.3390/s20123524
12. Ramos, D., Khorram, M., Faria, P., Vale, Z.: Load forecasting in an office building with different data structure and learning parameters. Forecasting **3**, 242–255 (2021). https://doi.org/10.3390/forecast3010015

Embedded AI and TinyML: A Practical Analysis of Workflows and Libraries

Ander Garcia[1](✉), Javier Tardos[1], Juan Luis Ferrando[1], Daniel Aguinaga[2], David Perez[2], and Leire Estanga[2]

[1] Vicomtech Foundation, Basque Research and Technology Alliance (BRTA), Mikeletegi 57, 20009 Donostia-San Sebastián, Spain
agarcia@vicomtech.org

[2] Ikor Technology Centre, Francisco Grandmontagne St 4th, 20018 San Sebastian, Spain

Abstract. Embedded AI has become a relevant research line with real applications involving the deployment of AI algorithms in units with limited computing resources, such as microcontrollers (MCUs). Embedded AI has led to TinyML, a recently emerged paradigm that proposes to embed optimized ML models in MCUs. Traditionally, this deployment has required a deep knowledge of low-level programming, but currently some higher-level software libraries ease this deployment. However, analysis and practical examples of these libraries are still scarce. The main objective of this paper is to present the practical deployment of some of these libraries on a Proof of Concept and a test bench to analyze their effects on the inference latency and accuracy of the AI models. Results show that the development and deployment of these embedded models is already a feasible task requiring only basic AI and programming experience. Furthermore, the inference latency and accuracy of these models meets the requirements of several real-world applications.

Keywords: Embedded AI · TinyML · microcontroller · ML

1 Introduction

The development of Artificial Intelligence (AI) algorithms, the increase of the computational power of affordable hardware, and privacy and latency concerns of users about Cloud Computing has led to Edge AI, where data is analyzed close to its source. While Cloud AI is suitable for processing-intensive applications that can tolerate latency, Edge AI meets requirements of time-sensitive scenarios where real-time processing is essential. Edge AI applications presents several advantages over Cloud AI applications [1]:

- Reduced Latency
- Real-Time Performance
- Improved Privacy Control
- Power-efficiency
- Cost-Effectiveness

R. Chinthaginjala et al. (Eds.): DCAI 2024, LNNS 1259, pp. 356–365, 2025.
https://doi.org/10.1007/978-3-031-82073-1_35

- On-Device Learning

Edge AI systems can be deployed in various configurations and across many different types of devices and data sources, such as local clusters of computers, gateways, IoT devices and fog nodes. Furthermore, some Edge AI applications deploy machine learning models within embedded systems, leading to Embedded ML (Machine Learning). TinyML is a recently emerged paradigm that proposes to embed optimized ML models in units with limited computing resources, such as those powered by micro-controllers [2]. TinyML deploys AI models on microchips and microcontrollers (MCUs) to enable extremely low-power on-device data analytics. MCUs are small single-board computers built around one or more processors with limited memory and computing capacity.

This paper analyses workflows, libraries and approaches to develop and deploy TinyML applications. The analysis is based on results obtained from a Proof of Concept (PoC) and a testbed. The paper presents an updated view on TinyML to foster future TinyML developments.

The paper is organized as follows. Section 2 reviews the state of the art. Section 3 recommends a TinyML deployment workflow. While Sect. 4 describes the TinyML deployment at the PoC, Sect. 5 is focused on the test bench. Finally, Sect. 6 highlights main conclusions, recommendations, and future work.

2 State of the Art

TinyML is already a relevant AI alternative focused on MCUs. A recent TinyML literature survey [3] describes main applications and related research efforts. Researchers [4] have previously analyzed several aspects to be considered to deploy artificial intelligence in embedded devices: (1) choosing a suitable deployment platform, (2) using hardware accelerators, (3) applying model compression techniques, and (4) enhancing hardware support for AI algorithms and models.

The current research of embedded AI is focused on two main areas: the optimization of AI models and algorithms to ease deploying them on MCUs, and hardware accelerators in MCUs to increase the performance of AI models and algorithms.

The construction of a TinyML system involves certain steps and methodologies [5]. As in traditional AI, collection and preprocessing of data is a crucial task for TinyML system, significantly impacting the accuracy and efficiency of the models. Several preprocessing techniques, such as normalization, noise removal, sampling, and feature selection, can be applied to increase data quality and ensure compatibility with the model. Again, selecting an appropriate AI/ML model and its training are crucial to obtain valid accuracy and efficiency measures in a TinyML system.

TinyML models deployed at MCUs must be lightweight, computationally efficient, and capable of delivering satisfactory performance. Decision trees, support vector machines (SVM), k-nearest neighbors (KNN), and Neural Networks (NN) are popular models well-suited for TinyML applications. After training the model to meet the desired performance criteria, it is deployed onto the target TinyML device. This deployment process involves transforming the model into an executable format, such as TensorFlow Lite [6] for MCUs. Libraries targeting this transformation consider both the constraints of the target device, including memory, processing power, and energy efficiency; and also, model optimization techniques, such as quantization, pruning, and compression.

This paper targets these steps, recommending a TinyML workflow, libraries and approaches and deploying them in two use cases and two different MCUs. The objective is to present an updated view on TinyML to foster future TinyML developments.

3 TinyML Recommended Workflow

General AI workflows must be adapted for TinyML scenarios. This adaptation would change according to the background of the AI team. This paper proposes the following workflow maximizing traditional AI knowledge and experience, where software tools ease the transformation of AI models to be compatible with and deployed to MCUs.

1) Data analysis. The initial step involves identifying the problem and gathering relevant data for model training. If data originates from a sensor connected directly to the microcontroller, developing a program for reading and labelling this data is required. This program is implemented on the board using platforms like Arduino [7] and CircuitPython [8], making use of appropriate libraries for sensor data acquisition.
2) AI model generation. The collected data must be studied and analyzed to perform the most suitable preprocessing for model training. This process takes place on a PC or Cloud, typically using Python along with libraries such as numpy or pandas. Model training can be executed using tools like TensorFlow, PyTorch, or Scikit-learn, preserving a traditional AI development workflow. Once the dataset is labeled and preprocessed, it is split into two parts: one for training the model and the other for validating its performance.
3) Model adaptation. The model generated needs to be adapted for use on a microcontroller. Tools such as M2Cgen [9] or EverywhereML [10] can be employed for this purpose, transforming the model into a series of operations and logic implemented in plain code. Alternatively, environments like TensorFlow Lite for MCUs can be utilized. It adjusts TensorFlow Lite implementation to enable execution of a simplified model on a microcontroller. When developing the model, it is crucial to consider that each tool is compatible only with certain types of models, thus requiring workflow planning tailored to the specific needs of the problem addressed.
4) Deployment. The final stage of the process involves deploying the model onto the microcontroller. Like the initial step, the chosen platform is utilized for deployment, supplemented with data acquisition and model processing libraries, if necessary.
5) Validation. The validation program must gather new real-time data from sensors, subject them to required preprocessing, and execute inference using the TinyML model. Based on the results obtained, the behavior of the model is validated under real operational conditions.

4 Proof of Concept (PoC): Movement Axis Detection

4.1 Description

The previous workflow has been first validated on a PoC where the axis of the movement of a prototype board has been detected, lighting a different LED for each axis. The objective of this PoC has been two-fold: to validate different software approaches and hardware combinations to execute the previous workflow, and to measure their inference latency.

Two widely used IoT devices have been selected to implement the different approaches to TinyML of this PoC, each with distinctive features. First one is the XIAO nRF52840 Sense [11], a compact development board produced by Seeed Studio (Fig. 1a). It utilizes the nRF52840 SoC from Nordic Semiconductor, tailored for IoT and low-power embedded applications. Featuring a 32-bit single-core processor clocked at 64 MHz, it includes 1 MB of flash memory, 2 MB of integrated QSPI flash memory, and 256 kB of RAM. Equipped with a digital microphone for real-time audio data and a 6-axis IMU detecting three-dimensional linear acceleration and angular velocity, the XIAO nRF52840 Sense stands out as a versatile device for TinyML projects.

The second device is the ESP32-C3-DevKitM-1 [12], a development board powered by the ESP32-C3 microcontroller by Espressif Systems (Fig. 1b). With a 32-bit single-core processor clocked at 160 MHz, it offers 4 MB of flash memory and 400 kB of RAM. Though lacking built-in sensors, it provides multiple communication interfaces including GPIO, SPI, UART, I2C, and I2S, facilitating external sensor connections for seamless integration into TinyML projects. To match functionalities, the same LSM6DS3TR IMU found in the XIAO nRF52840 Sense is utilized, communicating via I2C.

Fig. 1. Hardware devices (a) XIAO nRF52840 Sense and (b) ESP32-C3-DevKitM-1

The PoC has deployed the following libraries and preprocessing variants on these two MCUs. Both devices are compatible with popular development environments like Arduino and CircuitPython, enabling flexible implementation of diverse solutions.

Edge Impulse + Arduino. Edge Impulse [13] is a commercial platform facilitating AI deployments with a friendly user interface and user-guided workflow. The 6-axis IMU data of the XIAO nRF52840 is collected through the Data Forwarder tool [14] with a 104 Hz frequency, and split into 128-samples windows. Edge Impulse suggests a preprocessing method that converts time series to frequency spectrum through FFT. A Neural Network Classifier is trained, composed of 3 dense layers with 64, 32 and 4 nodes each. Both original and quantized (int8) models are exported as Arduino libraries. Edge Impulse signals that the original model uses 528 KB of flash memory, while the quantized one uses 144 KB.

TensorFlow + EverywhereML + Arduino. This approach emulates Edge Impulse's methodology while bypassing its associated costs. This time, data is acquired with the ESP32-C3-DevKitM-1 to check the compatibility of the IMUs in the opposite direction. Preprocessing of the data is applied with the Arduino FFT library [15], and a Neural

Network with 3 dense layers composed of 16, 16 and 4 nodes is trained with TensorFlow and Keras [16]. The model is transformed with EverywhereML into a series of weights, using 159 KB of flash memory, and the corresponding Eloquent Arduino library produces the inference on both devices.

TensorFlow Lite Micro + Arduino. Since TensorFlow Lite quantized models are not compatible with EverywhereML, this PoC directly implements TF Lite for Microcontrollers onto the devices. The previously trained Neural Network is converted into a TF Lite model, with optional model quantization. Using the 'xxd' Linux command, the TF Lite model is transformed into a manageable C array. This array utilizes 168 KB of flash memory for the original model, and 54 KB for the quantized one. Device processor architecture determines the Arduino inference library; XIAO nRF52840 employs tflm_cortexm library [17], while ESP32-C3 uses tflm_esp32 library [18].

Scikit-learn + M2Cgen/EverywhereML + Arduino/CircuitPython. The final comparison evaluates the performance of a lightweight AI model implemented with Arduino and CircuitPython. Random Forest from sklearn [19] is chosen for its accuracy and lack of FFT preprocessing, reducing inference time. It is compatible with M2Cgen and EverywhereML, requiring no inference library on devices. Data collected with ESP32-C3 is used for model training, resulting in four models: two with M2Cgen for Arduino and CircuitPython, and another two with EverywhereML for the same platforms.

4.2 Results

The PoC has prioritized the feasibility of the deployment of the TinyML workflow over the accuracy of the models. Thus, the analysis of the results is focused on the execution time of the inference for each of the previous variants instead of metrics like accuracy or recall. Table 1 displays the execution time for each scenario, including preprocessing of the live captured data (128 samples of the 6-axis IMU) and the model inference process.

Results clearly show the impact of the preprocessing on the execution time. These MCUs have low computation capabilities, so performing a FFT over a set of 128 samples from 6 axis is a high time-consuming operation. The relatively low difference between the quantized models and the original ones also remarks the fact that the model inference is taking less time to compute than the preprocessing.

Considering the Neural Networks, the models trained with Edge Impulse have an extraordinarily lower execution time with the nRF52840, while the ones trained with TensorFlow present lower execution times on the ESP32-C3. Although the motivation of this behavior should be further researched, it may be caused by a better optimization for the nRF52840 processor by the Edge Impulse conversion tool.

On the other hand, looking at the Random Forest models, two conclusions can be drawn. The models converted with EverywhereML are better optimized for these MCUs than the ones converted with M2Cgen. Also, the models generated for Arduino are faster than the models generated for CircuitPython. This behavior is due to the static typing and compiled nature of C++ (Arduino), that offers better memory management and reduced overhead for resource-constrained environments.

Table 1. Execution time (μs) for every variant of the PoC

Model	Data preprocessing	Training	Platform	Conversion Tool	Quantized	Device	
						nRF52840	ESP32-C3
Neural Network	FFT	Edge Impulse	Arduino	-	No	10743	117594
					Yes (int8)	9766	107132
		Tensor Flow	Arduino	Everywhere ML	-	100546	80669
				TensorFlow Lite	No	99883	80922
					Yes (int8)	99492	72039
Random Forest	None	Scikit-Learn	Arduino	Everywhere ML	-	1953	4133
				M2Cgen	-	2539	4300
			Circuit Python	Everywhere ML	-	6476	5451
				M2Cgen	-	16277	6638

Focusing on the axis detection accuracy, all models consistently infer the correct class when movements are replicated similarly to the training data. This level of precision is notable given the limited training data of only 40 sets of 128 samples.

As a conclusion of this section, the PoC has validated that the proposed workflow can be deployed at the selected MCUs based on different libraries and variants. Results also show that these TinyML deployments consistently fulfil requirements of inference latency higher than 100 ms.

5 Test Bench: Fan Working Mode Detection

5.1 Description

After the successful validation of the PoC, Ikor Technology Center has developed a test bench to simulate a more realistic setup. The objective of the test bench is to predict the speed of a fan based on the vibrations it generates. The system is composed of a small fan secured on a platform with screws, offering some vertical axis freedom via springs (Fig. 2). A plate intensifies vibration effects and houses the chosen components: the ESP32-C3-DevKitM-1 with the LSM6DS3 inertial measurement unit. These components are powered via the micro-USB connector of the ESP32-C3, while the fan is powered by an adjustable external voltage source. Implemented models aim to distinguish three working modes of the fan, related to three voltage levels: 0 V (stopped), 12 V (low speed), and 24 V (high speed).

For model training and data analysis purposes, 600 windows of 128 samples of 6-axis IMU data are collected with 104 Hz rate for each fan speed. Knowing the reading ranges of the IMU, the data is normalized between 0 and 1.

First, different properties of the captured data have been analyzed to find a preprocessing method that enhances the possibilities of a ML model to classify the different fan speeds: normalized acceleration, waveform of the acceleration, normalized angular speed, variance on each axis, and FFT. Then, multiple AI models have been trained with the previous properties: models trained with time series, models trained with time

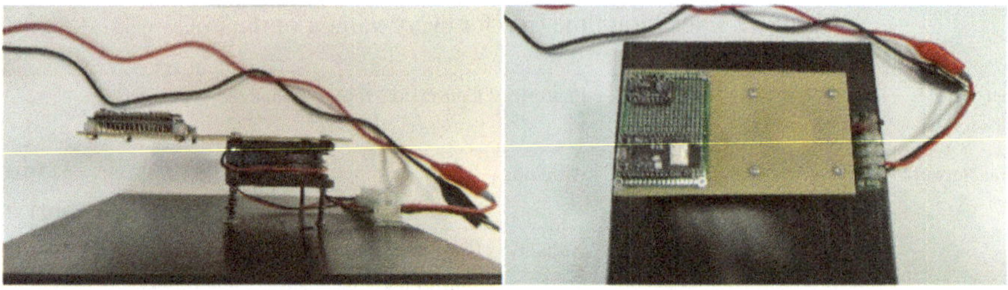

Fig. 2. Fan use case board (a) lateral and (b) top views

series variances, and models trained with FFT. The dataset is split taking 30% of the 600 windows as validation data. For the Random Forest models, multiple variants are trained varying the number of estimators and the maximum depth of the tree. The model with the higher accuracy is selected. On the other hand, a Neural Network with 3 dense layers is trained, with 16, 16 and 3 nodes for each layer. One hundred epochs are run with RMSprop optimizer and loss is calculated with mean square error (MSE).

5.2 Results

Table 2 shows the precision, recall and F-1 score of the models studied for every fan speed. Testing is carried out live on the device, running 100 measurements per fan speed, with 128 samples each. In the FFT preprocessing case, a study on the effect of modifying the sampling frequency was made. Given the sensor capabilities and the limited reading speed of the device, only 104, 208 and 416 Hz were considered. The best results are achieved with the 208 Hz sampling frequency.

Table 2. Performance metrics of the models trained for the fan test bench

Data preprocessing	Model	Conversion Tool	Platform	Fan speed	Metric		
					Precision	Recall	F-1 score
None	Random Forest	Everywhere ML	Arduino	Stop	0.730	1	0.844
				Low	1	0.630	0.773
				High	1	1	1
			Circuit Python	Stop	0.725	1	0.840
				Low	1	0.620	0.765
				High	1	1	1
		M2Cgen	Arduino	Stop	1	1	1

(continued)

Table 2. (*continued*)

Data preprocessing	Model	Conversion Tool	Platform	Fan speed	Metric		
					Precision	Recall	F-1 score
				Low	1	1	1
				High	1	1	1
			Circuit Python	Stop	1	1	1
				Low	1	1	1
				High	1	1	1
Variances	Random Forest	Everywhere ML and M2Cgen	Arduino and Circuit Python	Stop	1	1	1
				Low	1	1	1
				High	1	1	1
FFT	Neural Network	TensorFlow Lite	Arduino	Stop	1	1	1
				Low	1	1	1
				High	1	1	1
	Random Forest	Everywhere ML	Arduino	Stop	1	1	1
				Low	1	1	1
				High	1	1	1
		M2Cgen	Arduino	Stop	0.893	1	0.943
				Low	1	0.88	0.936
				High	1	1	1

The solution with the lowest inference time is the one with Random Forest model with no preprocessing method. Inference times are consistent with the results from the PoC: around 100 ms with preprocessing and 10 without it.

Regarding the accuracy of the variants, the model that best fits for this solution is that one trained with variances of the IMU data. Random Forest models obtained have an accuracy of 100% on the predictions made with any conversion tool and any deployment platform. The model trained with raw IMU data has the lowest inference time but is extremely sensitive to light inclination changes in the test bench.

6 Conclusions

Embedded AI and TinyML are active research areas of great relevance for real applications where AI and ML algorithms are deployed at low computing capabilities microcontrollers (MCUs). Thanks to latest advances both on hardware and on software tools and libraries, currently, TinyML deployment workflows compatible with traditional AI workflows are available. Furthermore, several Open Source tools and libraries are available for this deployment, which have been validated with the PoC and the test bench presented on this paper. Existing commercial solutions, such as Edge Impulse, have managed to greatly ease the TinyML deployment process and serve as a reference for Open Source initiatives.

The paper has deployed a PoC and a test bench, obtaining consistent latencies below 100 ms and accuracy metrics valid for several real applications. Although accuracy results are setup and use case dependant, the validation shows that embedded AI and TinyML are ready for inference at real applications and do not require very specialized electronic or low-level programming knowledge for their deployment.

However, there are still some working areas that should evolve to further ease TinyML deployments. Firstly, although there are libraries transforming models to be compatible with several MCU, hardware restrictions make that some AI models are not compatible with certain MCUs. Secondly, these libraries are improving to automate the transformation of AI models generated at PCs/Cloud into simpler models deployable at MCUs. This model transformation task is a very relevant research area, key to simplify development and deployment workflows.

Future work includes further validation with MCUs from different platforms and more algorithms with varying complexity tackling different use cases. Finally, authors have already begun research about the applicability of new AI paradigms such as Continuous, Federated or Transfer Learning in TinyML scenarios, which could be applied in new use cases.

Acknowledgments. This project has been partially funded by the Hazitek program of the Basque Government (ZE-2022/00042).

Disclosure of Interests. The authors have no competing interests to declare that are relevant to the content of this article.

References

1. Soro, S.: TinyML for Ubiquitous Edge AI (2020)
2. Warden, P., Situnayake, D.: Tinyml: Machine learning with tensorflow lite on arduino and ultra-low-power microcontrollers. O'Reilly Media (2019)
3. Abadade, Y., Temouden, A., Bamoumen, H., Benamar, N., Chtouki, Y., Hafid, A.S.: A comprehensive survey on TinyML. IEEE Access **11**, 96892–96922 (2023). https://doi.org/10.1109/ACCESS.2023.3294111
4. Zhang, Z., Li, J.: A Review of Artificial Intelligence in Embedded Systems. Micromachines, vol. 14, no. 5. MDPI, May 01, 2023. https://doi.org/10.3390/mi14050897
5. Hidalgo, C.H., Gonzalez-Vidal, A., Skarmeta, A.F.: Edge computing in Smart Agriculture scenario based on TinyML for irrigation control. In: IEEE 9th World Forum on Internet of Things, Universidad de Chile, pp. 183–209 (2023). https://doi.org/10.13039/501100011033
6. TensorFlow Lite. https://www.tensorflow.org/lite?hl=es. Accessed 12 Apr 2024.
7. Arduino. https://www.arduino.cc/en/software. Accessed 12 Apr 2024.
8. CircuitPython. https://circuitpython.org/. Accessed 12 Apr 2024.
9. M2Cgen. https://github.com/BayesWitnesses/m2cgen. Accessed 12 Apr 2024.
10. EverywhereML. https://github.com/eloquentarduino/everywhereml. Accessed 12 Apr 2024.
11. Getting Started with XIAO nRF52840. https://wiki.seeedstudio.com/XIAO_BLE/. Accessed 12 Apr 2024.

12. ESP32-C3-DevKitM-1 - ESP32-C3. https://docs.espressif.com/projects/esp-idf/en/stable/esp32c3/hw-reference/esp32c3/user-guide-devkitm-1.html. Accessed 12 Apr 2024.
13. Hymel, S., et al.: Edge impulse: an MLOps platform for tiny machine learning. In: Song, D., Carbin, M., Chen, T. (eds.) Proceedings of Machine Learning and Systems 5 (2023). http://www.edgeimpulse.com/
14. Data forwarder | Edge Impulse https://docs.edgeimpulse.com/docs/tools/edge-impulse-cli/cli-data-forwarder. Accessed 12 Apr 2024
15. ArduinoFFT. https://github.com/kosme/arduinoFFT. Accessed 12 Apr 2024
16. Keras: Deep Learning for humans. https://keras.io/. Accessed 12 Apr 2024
17. TensorFlow runtime for Cortex-M chips. https://github.com/eloquentarduino/tflm_cortexm. Accessed 12 Apr 2024
18. tflm_esp32. https://github.com/eloquentarduino/tflm_esp32. Accessed 12 Apr 2024
19. scikit-learn: machine learning in Python. https://scikit-learn.org/stable/. Accessed 12 Apr 2024

Author Index

R. Chinthaginjala et al. (Eds.): DCAI 2024, LNNS 1259, pp. 367–368, 2025.
https://doi.org/10.1007/978-3-031-82073-1

The manufacturer's authorised representative in the EU is Springer Nature Customer Service Centre GmbH, Europaplatz 3, 69115 Heidelberg, Germany. If you have any concerns regarding our products, please contact ProductSafety@springernature.com

Printed and bound by CPI Group (UK) Ltd, Croydon, CR0 4YY
30/09/2026
02225561-0011